Nickel in Soils and Plants

ADVANCES IN TRACE ELEMENTS IN THE ENVIRONMENT

Series Editor: H. Magdi Selim
Louisiana State University, Baton Rouge, USA

Nickel in Soils and Plants
edited by Christos D. Tsadilas, Jörg Rinklebe, and H. Magdi Selim

Trace Elements in Waterlogged Soils and Sediments
edited by Jörg Rinklebe, Anna Sophia Knox, and Michael Paller

Phosphate in Soils: Interaction with Micronutrients, Radionuclides, and Heavy Metals
edited by H. Magdi Selim

Permeable Reactive Barrier: Sustainable Groundwater Remediation
edited by Ravi Naidu and Volker Birke

Nickel in Soils and Plants

Edited by
Christos D. Tsadilas
Jörg Rinklebe
H. Magdi Selim

CRC Press
Taylor & Francis Group
Boca Raton London New York

CRC Press is an imprint of the
Taylor & Francis Group, an **informa** business

CRC Press
Taylor & Francis Group
6000 Broken Sound Parkway NW, Suite 300
Boca Raton, FL 33487-2742

First issued in paperback 2020

ISBN 13: 978-0-367-57120-7 (pbk)
ISBN 13: 978-1-4987-7460-4 (hbk)

Visit the Taylor & Francis Web site at
http://www.taylorandfrancis.com

and the CRC Press Web site at
http://www.crcpress.com

This book is dedicated

to

Christos Tsadilas' family

Jörg Rinklebe's family

&

Elliott, Emily, and Richard Selim

Contents

Preface

Nickel (Ni) has a widespread distribution in the environment; it is the fifth most common element on Earth. Recently, it has been proven that Ni is essential for the considerable growth of many species of vertebrates, plants, and microorganisms, and it may even be beneficial at low levels. At higher levels, Ni may become toxic to organisms, and accordingly, the concerns about toxicity are paramount. In several parts of the world, especially those with ultramafic or serpentine soils, the Ni content is high, causing serious environmental impacts. Plants grown in soils with high Ni content reveal various physiological alterations resulting in a reduction of crop yields.

The idea of publishing this book grew from the knowledge that elevated Ni content may cause certain problems for human and environmental health in many parts of the world. Those tasks have not yet been fully investigated, and thus, future efforts should be focused on achieving such goals. Also, powerful and sustainable remediation techniques of Ni-polluted soils should be developed. This is of special importance for areas with ultramafic and serpentine soils such as in Mediterranean ecosystems. Current research findings on the essentiality or usefulness of Ni to plants, impact of excess Ni on the physiology of plants, soil pollution with Ni, and remediation methods of soils contaminated with Ni such as phytomining or agromining need to be widely disseminated.

In this book, a broad range of topics dealing with Ni in the environment is presented. The understanding of the complex interactions of Ni in the environment is a prerequisite in the effort to predict their behavior in the plant system and in the vadose zone. Nickel essentiality to plants (Chapter 1); Ni biogeochemistry in soil, plants, and their interfaces (Chapters 2 and 3); Ni sources and forms of Ni in soils and sediments (Chapters 4 through 7); Ni in serpentine soils (Chapter 8); methods of Ni determination (Chapter 9); and Ni surface chemistry, mobility, and its kinetics, and advanced methods of Ni speciation (Chapters 10 through 12) are presented. Furthermore, Ni phytoavailability in soils (Chapter 12) and of its immobilization by using biochar as a green and recent remediation technique (Chapter 13) are discussed. In addition, Ni removal from wastes (Chapter 14) and methods of Ni mining by using appropriate plant species (Chapter 15) are presented.

We wish to thank the authors for their contributions to this book. We also would like to thank Dr. Thomai Nikoli of the Hellenic Agricultural Organization, General Directorship of Agricultural Research, Institute of Industrial and Forage Crops, Larissa, Greece, for her invaluable assistance in this project and to the staff of CRC Press/Taylor & Francis for their help and patience in getting this book published. Special thanks are also due to Ms. Irma Shagla Britton, senior editor, for her guidance and attention in getting this work published. Her help was indeed valuable and made the publication of this book possible.

Christos D. Tsadilas
Jörg Rinklebe
H. Magdi Selim

Editors

Christos Tsadilas, PhD, is a senior researcher of the Institute of Industrial and Forage Crops of the General Directorship of Agricultural Research of the Hellenic Agricultural Organization DEMETER, a public entity supervised by the Ministry of Rural Development and Foods of Greece. He graduated from the Agricultural University of Athens in 1975 and earned his master's degree and PhD from the same university. From 1977 to 1984, he worked at the Institute of Soil Mapping and Classification of the Ministry of Rural Development and Foods as a soil surveyor, and has been a senior scientist of the National Project of Soil Survey Mapping in Greece. Later, he worked as a researcher at the National Agricultural Research Foundation in Greece, which later merged with the current Hellenic Agricultural Organization DEMETER. He has taken post doc studies at the University of Reading (United Kingdom), University of Kentucky, and University of Lincoln (Nebraska), sponsored by the Royal Society of England, OECD, and the Fulbright Foundation, respectively. His main research interests include soil survey and classification, soil chemistry and fertility, soil pollution and remediation, and waste management. He also has work experience on land reclamation of disturbed lands. In the past few years, he has started to deal with the new technologies used in precision farming systems. Dr. Tsadilas is a well-known researcher in the field of trace elements in soils, having published more than 250 papers in peer-reviewed journals, chapters in books, and papers in international conference proceedings. A special mention is made on the book *The Soils of Greece* (Yassoglou, Tsadilas, and Cosmas, 2017, Springer). He has served as associate editor for several journals including *Environmental Pollution, Communications in Soil Science and Plant Analysis, Advances in Agriculture,* and the *African Journal of Agricultural Research.* He also served as president of the Hellenic Soil Science Society for several years.

Jörg Rinklebe, PhD, is a professor in soil and groundwater management at the University of Wuppertal, Germany. From 1997 to 2006, he worked as a scientist, research associate, and project leader in the Department of Soil Sciences at the UFZ (Centre for Environmental Research), Leipzig-Halle, Germany. Dr. Rinklebe studied ecology for one year at the University of Edinburgh, Scotland (1992–1993). He studied agriculture, specializing in soil science and plant nutrition, at Martin Luther University Halle-Wittenberg, Germany, where he earned a PhD in soil science. His main research is on wetland soils, sediments, waters, plants, and their pollution (trace elements and nutrients) and linked biogeochemical issues, with a special focus in redox chemistry. He also has expertise in remediation of toxic element–contaminated soils and soil microbiology. Dr. Rinklebe is internationally recognized for his research in the areas of biogeochemistry of trace elements in wetland soils. He has published numerous scientific papers in leading international and national journals, a book, *Trace Elements in Waterlogged Soils and Sediments* (2016), as well as numerous book chapters. He serves as associate editor of the international journal *Environmental Pollution* and as a guest editor of the international journals: *Journal of Hazardous Materials, Environment International, Science of the Total Environment, Chemosphere, Journal of Environmental Management, Applied Geochemistry,* and *Environmental Geochemistry and Health.* In addition, he is a member of several editorial boards including *Ecotoxicology, Geoderma, Water, Air, & Soil Pollution,* and *Archive of Agronomy and Soil Science,* and is a reviewer for many international journals. He has organized several special symposia at various international conferences such as "Biogeochemistry of Trace Elements" (10th, 11th, 12th, 13th, and 14th ICOBTE) and the "International Conference on Heavy Metals in the Environment" (15th, 16th, 17th, 18th, and 19th ICHMET). He has been an invited speaker (plenary and keynote) at many international conferences. In October 2016, he was appointed as honorable ambassador for Gangwon Province, South Korea. Since March 2017, he has been a visiting professor at the Department of Environment and Energy at Sejong University, Seoul, South Korea, and a guest professor at the Department of Environmental Engineering, China Jiliang University, Hangzhou, Zhejiang, China.

H. Magdi Selim, PhD, is professor of Soil Physics and George and Mildred Caldwell Endowed Professor of Soil Science, School of Plant, Environmental and Soil Sciences at Louisiana State University at Baton Rouge. He received his BS degree in soil science from Alexandria University, Alexandria, Egypt, and his MS and PhD in soil physics from Iowa State University, (Ames, Iowa). He is internationally recognized for his research in the areas of kinetics of reactive chemicals in heterogeneous porous media and transport modeling of dissolved chemicals in water-saturated and unsaturated soils. He is the original developer of several models for describing the retention processes of chemicals in soils. Pioneering work also includes multistep/multireaction and nonlinear kinetic models for trace elements, heavy metals, radionuclides, explosive contaminants, phosphorus, and pesticides in soils. Other research areas include water quality, soil health, and the fate of applied agricultural chemicals in soils at different scales, as well as best management practices that minimize nutrient and soil loss under different agricultural management.

Dr. Selim is the developer of *Chem_Transport*, a software package of models that describe the fate and transport of reactive chemicals and tracers in soils and geological media. The models represent recent advances made toward the understanding of transport characteristics of chemicals in soils with an emphasis on physical and chemical nonequilibrium.

Dr. Selim is the author or coauthor of numerous scientific publications in several journals. He is also editor and coauthor of several books. Dr. Selim is the recipient of several professional awards and honors. He is a member of the American Society of Agronomy, the Soil Science Society of America, the International Society of Soil Science, the International Society of Trace Element Biogeochemistry, the Louisiana Association of Agronomy, and the American Society of Sugarcane Technology. Dr. Selim was elected chair of the Soil Physics Division of the Soil Science Society of America. He has served on numerous committees of the Soil Science Society of America, the American Society of Agronomy, and the International Society of Trace Element Biogeochemistry. He also served as associate editor of *Water Resources Research* and the *Soil Science Society of America Journal* and as technical editor of the *Journal of Environmental Quality*.

Dr. Selim is a Fellow of the American Society of Agronomy and the Soil Science Society of America. Some of the awards he has received include the Phi Kappa Phi Research Award, the Gamma Sigma Delta Award for Research, the Joe Sedberry Graduate Teaching Award, the First Mississippi Research Award, the Doyle Chambers Achievements Award, and the EPA Environmental Excellence Award. More recent awards include the Soil Science Society of America Soil Science Research Award and the International Union of Soil Science von Liebig Award.

Contributors

Svetlana Antić-Mladenović
Faculty of Agriculture
University of Belgrade
Belgrade, Serbia

Ahamed Ashiq
Post Graduate Institute of Science
University of Peradeniya
Peradeniya, Sri Lanka
and
Ecosphere Resilience Research
    Center
Faculty of Applied Sciences
University of Sri Jayewardenepura
Nugegoda, Sri Lanka

Nikolaos Barbayiannis
Soil Science Laboratory
Faculty of Agriculture
Aristotle University of Thessaloniki
Thessaloniki, Greece

Rufus L. Chaney
USDA–Agricultural Research
    Service (Retired)
Crop Systems and Global Change
    Laboratory
Beltsville, Maryland

Zueng-Sang Chen
Department of Agricultural
    Chemistry
National Taiwan University
Taipei, Taiwan

David A. Dalton
Department of Biology
Reed College
Portland, Oregon

Ali El-Naggar
Korea Biochar Research Center
O-Jeong Eco-Resilience Institute
    (OJERI)
and
Division of Environmental Science
    and Ecological Engineering
    Korea University
Seoul, Republic of Korea
and
School of Natural Resources and
    Environmental Science
Kangwon National University,
Chuncheon, Republic of Korea
and
Department of Soil Sciences
Faculty of Agriculture
Ain Shams University
Cairo, Egypt

Dionisios Gasparatos
Soil Science Laboratory
Faculty of Agriculture
Aristotle University of Thessaloniki
Thessaloniki, Greece

Viraj Gunarathne
Post Graduate Institute of Science
University of Peradeniya
Peradeniya, Sri Lanka
and
Ecosphere Resilience Research
    Center
Faculty of Applied Sciences
University of Sri Jayewardenepura
Nugegoda, Sri Lanka

Paweł Harasim
Department of Agricultural and
    Environmental Chemistry
University of Life Sciences in Lublin
Lublin, Poland

Yohey Hashimoto
Tokyo University of Agriculture and
    Technology
Tokyo, Japan

Zeng-Yei Hseu
Department of Agricultural
    Chemistry
National Taiwan University
Taipei, Taiwan

Petra S. Kidd
Instituto de Investigaciones
    Agrobiológicas de Galicia (IIAG)
Consejo Superior de Investigaciones
    Científicas (CSIC)
Santiago de Compostela, Spain

Anna Sophia Knox
Savannah River National
    Laboratory
Aiken, South Carolina

Dien Li
Savannah River National
    Laboratory
Aiken, South Carolina

Lixia Liao
Louisiana State University
Baton Rouge, Louisiana

Theodora Matsi
Soil Science Laboratory
Faculty of Agriculture
Aristotle University of Thessaloniki
Thessaloniki, Greece

David H. McNear
Integrated Plant and Soil Science
    Department
University of Kentucky
Lexington, Kentucky

James W. Morris
Integrated Plant and Soil Science
    Department
University of Kentucky
Lexington, Kentucky

Nabeel Khan Niazi
Institute of Soil and Environmental
    Sciences
University of Agriculture Faisalabad
Faisalabad, Pakistan
and
Southern Cross GeoScience
Southern Cross University
Lismore, NSW, Australia

Thomai Nikoli
Hellenic Agricultural Organization
    "DEMETER"
Institute of Industrial and Forage
    Crops
Larissa, Greece

Yong Sik Ok
Korea Biochar Research Center
O-Jeong Eco-Resilience Institute
and
Division of Environmental Science
    and Ecological Engineering
Korea University
Seoul, Republic of Korea

Michael H. Paller
Savannah River National
    Laboratory
Aiken, South Carolina

Markus Puschenreiter
Department of Forest and Soil
  Sciences
Institute of Soil Research
University of Natural Resources
  and Life Sciences
Tulln, Austria

Anushka Upamali Rajapaksha
Faculty of Applied Sciences
University of Sri Jayewardenepura
Nugegoda, Sri Lanka

Jörg Rinklebe
School of Architecture and Civil
  Engineering
Institute of Foundation Engineering,
  Water and Waste Management
Laboratory of Soil- and
  Groundwater-Management
University of Wuppertal
Wuppertal, Germany
and
Department of Environment,
  Energy and Geoinformatics
Sejong University
Seoul, Republic of Korea

Theresa Rosenkranz
Department of Forest and Soil
  Sciences
Institute of Soil Research
University of Natural Resources
  and Life Sciences
Tulln, Austria

Kirk G. Scheckel
United States Environmental
  Protection Agency
National Risk Management
  Research Laboratory
Cincinnati, Ohio

Sabry M. Shaheen
Faculty of Agriculture
Department of Soil and Water
  Sciences
University of Kafrelsheikh
Kafr El-Sheikh, Egypt
and
School of Architecture and Civil
  Engineering
Institute of Foundation Engineering,
  Water and Waste Management
Laboratory of Soil- and
  Groundwater-Management
University of Wuppertal
Wuppertal, Germany

H. Magdi Selim
Louisiana State University
Baton Rouge, Louisiana

Christos D. Tsadilas
Hellenic Agricultural Organization
  "DEMETER"
Directorship of Agricultural
  Research
Institute of Industrial and Forage
  Crops
Larissa, Greece

Meththika Vithanage
Ecosphere Resilience Research
  Center
Faculty of Applied Sciences
University of Sri Jayewardenepura
Nugegoda, Sri Lanka

Shan-Li Wang
Department of Agricultural
    Chemistry
National Taiwan University
Taipei, Taiwan

Walter W. Wenzel
Department of Forest and Soil
    Sciences
Institute of Soil Research
University of Natural Resources and
    Life Sciences
Tulln, Austria

1

Essentiality of Nickel for Plants

David A. Dalton

CONTENTS

1.1 Introduction

The story of our understanding of the essential role of nickel for plants is an odd mix of old and new. The "old" part relates to the fact that urease (a Ni-containing enzyme) was the first protein to be crystallized (Sumner 1926). This was the first proof that a pure protein can function as an enzyme and led to Sumner receiving the Nobel Prize in chemistry in 1946. Despite years of intensive study of urease, it was not discovered that this enzyme contained Ni until 1975 (Dixon et al. 1975). This initial observation of Ni in urease used jack bean (*Canavalia ensiformis*) as the source of the protein; however, it now appears that Ni is universally present in both plant and bacterial ureases. Even before the discovery of Ni in jack bean urease, evidence that Ni might be essential for plants in general began to surface. Early work showed that low amounts of Ni led to slight increases in growth for a number of plant species including wheat, grape vines, cotton, paprika, tomato, Chinese hemp, potatoes, and soybean (Mishra and Kar 1974). These observed increases in growth were interpreted as being due at least in part to the fungicidal properties of Ni salts. The amounts necessary for stimulation were exceedingly small and the toxic effects of even moderate amounts were evident. Furthermore, other positive plant responses to low amounts of Ni include the increased germination rates for seeds of many plant species and the ability of Ni to substitute for other essential metals in a number of important metalloenzymes (e.g., yeast phosphatase, arginase, tyrosinase, polyphenol oxidase, and oxaloacetate decarboxylase); however, there is a long list of enzymes that Ni inhibits even at low concentrations (Mishra and Kar 1974). Other favorable plant responses to low amounts of Ni include increased synthesis of catalase, activation of ascorbic acid oxidase, enhanced ribonuclease activity, activation of amylase, increased activation of phosphodiesterase, increased activity of DNase, increased content of chlorophyll and carotenoids, increased rates of photosynthesis and respiration, increased sugar content, and increased total protein and N content (references cited in Mishra and Kar 1974). In general, these early studies showed marginal effects, the mechanisms for the putative responses were unknown, and often times other studies reported conflicting results. These early studies are reviewed in detail by Mishra and Kar (1974).

1.2 Establishment of Ni as Essential for Plants

Three criteria must be met in order for an element to be considered as essential: the element must be required for completion of the life cycle of the plant, a physiological role/mechanism must be established, and no other element can substitute for it (Arnon and Stout 1939). Some authors have also suggested a fourth criterion: that the requirement be for all species of plants, not just a few. The first two of these essentiality criteria were finally met in 1983 when Eskew et al. showed that soybeans deprived of Ni accumulated toxic concentrations of urea in necrotic lesions on the tips of leaves regardless of the source of N (Figure 1.1). Urease from soybean has been shown to contain Ni, just as in the case for urease from jack bean and presumably all plants (Polacco and Havir 1979).

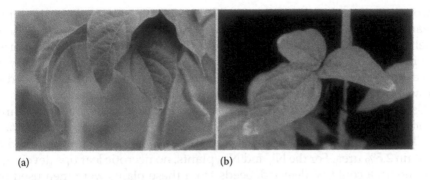

(a)   (b)

FIGURE 1.1
Leaves of soybean grown under Ni-deficient conditions and showing necrotic leaf tips due to accumulation of urea. (a) Leaves grown with NO_3^- and NH_4^+. (b) Leaves dependent on N^2 fixation showing more pronounced lesions. (From Eskew, D. L., R. M. Welch, and E. E. Cary. 1983. Nickel: an essential micronutrient for legumes and possibly all higher plants. *Science* 222:621–623.)

1.2.1 Ni Deficiency of Plants Grown with NH_4^+ or NO_3^- or N Fixation

Studies on the Ni requirement of plants can be understood best by separating those studies that used mineral N as the sole N source as opposed to studies where urea is the sole N source. There are much fewer studies in the former case. One of the first was a field study with soybeans in which Ni amendment of 40 g of Ni per hectare resulted in substantial increases in nodulation and seed yield (Bertrand and de Wolff 1973). Most other studies used carefully formulated nutrient solutions to control the Ni content. This was the case with the Eskew et al. (1983) study mentioned above that used soybeans with a nutrient solution of NH_4^+ and NO_3^- or plants grown symbiotically with *Bradyrhizobium* to allow for N fixation.

The demonstration of deficiency symptoms requires fastidious procedures to eliminate Ni from the growth media. It is straightforward to grow soybeans hydroponically, but simply using stock reagents is not sufficient to demonstrate a requirement for Ni because even the highest purity grade of reagents is likely to contain trace amounts of Ni. Ultrahigh-grade reagents are necessary as is chromatographic passage of nutrient solutions through a column of 8-hydroxyquinoline and controlled-pore glass beads, which removes Ni^{2+} ions from the solution. Eskew et al. (1983) reported that this column removed 99.99% of radiolabeled $^{63}Ni^{2+}$ from test solutions. They estimated their purified nutrient solution to have a Ni concentration of not greater than 0.06 μg L^{-1}. Careful attention to other sources of contamination must also be considered. For instance, water should be double distilled and then passed through a mixed-bed ion-exchange column. Plastic growing pots should be cleaned with 2M HCl. A further source of possible contamination is the seed itself. Ni is required in such low amounts that there may be enough Ni present in a seed for a plant to complete one generation. Consequently, it may be necessary to raise a plant under low Ni conditions and then harvest seed to grow a second generation with inconsequential Ni contamination.

Using these rigorous conditions, Eskew et al. grew soybeans under three conditions: no Ni (Ni_0), 1 μg L^{-1} (Ni_1), or 10 μg L^{-1} (Ni_{10}). After 56 days, 27.5% of the leaf tips of the Ni_0 plants developed necrotic leaf tips that were found to contain 2.5% urea. For the Ni_1 and Ni_{10} plants, no necrotic leaf tips developed and no urea could be detected. Seeds from these plants were then used for further study. Careful analysis of these Ni_0 seeds could not detect any residual Ni, indicating a Ni content of <10 ng g^{-1}, whereas the Ni_1 seeds contained 53 ng g^{-1} and the Ni_{10} seeds contained 637 ng g^{-1}. Comparison of the urease activity in these seeds gave values of 0.007 μmol mg^{-1} h^{-1}, 2.3 μmol mg^{-1} h^{-1}, and 36.9 μmol mg^{-1} h^{-1}, respectively. Seeds from the Ni_0 plants were used to grow a second generation.

The second generation of soybean grown from the Ni_0 seeds were grown under three conditions: (i) no Ni added, (ii) 1 μg of Ni per liter added, or (iii) a treatment in which analytical grade reagents were used for the nutrient solution but without further purification. Some plants were grown symbiotically with *Bradyrhizobium japonicum* used to form nitrogen-fixing root nodules. For the nitrogen-fixing plants receiving no Ni, 70.7% of leaves developed tip necrosis whereas no necrotic tips were observed in nitrogen-fixing plants receiving Ni. Some second-generation plants were grown with a supply of fixed nitrogen (NO_3^- and NH_4^+). Of these plants, 48.8% of the no-Ni plants developed leaflet tip necrosis. No necrosis was detected on plants receiving the treatment of 1 μg Ni per liter. Interestingly, plants grown with the unpurified nutrient solution also did not develop tip necrosis, indicating that standard nutrient solutions are sufficiently contaminated with Ni to meet the Ni requirements of plants.

At about the same time that work of Eskew et al. was published, there was a similar study by Klucas et al. (1983) that confirmed the role of Ni in soybean urease. Soybeans plants were grown hydroponically under Ni-deficient conditions with either N_2 or NO_3^- as the nitrogen source. Just as with Eskew et al., there was no effect on yield of soybean plants following the addition of Ni; however, there was a much higher activity of urease in leaves. In Eskew's case, this Ni-induced increase in leaf urease activity was 1.8-fold for plants grown with N_2 and 1.7-fold for plants grown with NO_3^- and NH_4^+. In the Klucas et al. study, the increases were 12-fold and 6.5-fold, respectively. The higher urease activity in plants grown with N_2 is consistent with expectations that nitrogen-fixing plants transport the products of fixation in the form of ureides that might require urease to be metabolized, a theory that has been largely discarded by more recent evidence (see Polacco et al. 2013 for a discussion on this point). Klucas et al. also tested other metals (Cr, Pb,

Sn, and V) and found no ability to increase urease. This addresses the last of the three criteria to establish essentiality, that is, that other metals cannot substitute for Ni, thus establishing Ni firmly as an essential element. This criterion was further met by Eskew et al. (1984) who demonstrated that neither Al, Cd, Sn, nor V was able to substitute for Ni in reducing leaf tip necrosis in Ni-deficient plants. More recently, Gerendás et al. (1998a) reported that Co does not replace Ni in urease. Thus, 1983 was the benchmark year that Ni was confirmed to be essential for plants, making it the first element to be so elevated since chlorine in 1954 (Broyer et al. 1954).

Ni deficiency has also been studied in other plants besides soybean. In cowpea (*Vigna unguiculata*), plants grown in Ni-deficient nutrient solutions developed necrotic leaf tip necrosis similar to those in soybean (Eskew et al. 1984; Walker et al. 1985). Tomato plants grown in the absence of Ni develop chlorosis of lower leaves and necrosis of the meristem (Checkai et al. 1986). A similar hydroponic study with barley (*Hordeum vulgare*) in the absence of Ni resulted in a 50% reduction in the seed germination rates (Brown et al. 1987a).

Of the seeds that did germinate, the seedling vigor was greatly reduced. The addition of Ni to the growth medium completely restored normal seed production. In this case, the deficiency symptoms were even more extreme than those in soybean because the plants were unable to complete their life cycle, another key criterion for establishing Ni as essential. A comprehensive table showing the range of plant species with evidence of Ni deficiency from these early studies can be found in Dalton et al. (1988).

1.2.2 Ni Deficiency of Plants Grown with Urea

Polacco (1977a) showed that growth of soybean in tissue culture with urea as the sole source of nitrogen could be strongly stimulated by low concentrations (10^{-4} mM) of Ni. Similarly, cultured cells of soybean, rice, and tobacco grew poorly on media deficient in Ni with urea as the sole source of N (Polacco 1977b). Gordon et al. (1978) observed that growth of *Lemna paucicostata* (duckweed) could be approximately doubled by providing low concentration of Ni when plants were grown with urea as the sole nitrogen source. Shimada and Ando (1980) and Shimada et al. (1980) worked with soybeans and tomatoes that were grown with urea as the sole nitrogen source. Under Ni-deficient conditions, the plants accumulated toxic levels of urea in their leaves that led to necrotic lesions on their leaf tips. Welch (1981) reported growth inhibition of *Lemna*, *Spirodela*, and *Wolfia* when grown with urea when Ni-limited. Spring rape (*Brassica napus*) grown under Ni-limited conditions with urea as the sole N source showed growth inhibition, lower amino acid content, and accumulation of urea in leaves (Gerendás and Sattelmacher 1999). Significantly, growth of rape with N supplied as NH_4NO_3 did not display any deficiency symptoms. Several other studies have reported significant growth reduction when urea-grown plants were Ni-limited. These include Shimada and Matsuo (1985) (cucumber and barley), Krogmeier et al. (1991)

(soybean), Gerendás and Sattelmacher (1997a) (rye, wheat, soybean, rape, zucchini, and sunflower), Gerendás and Sattelmacher (1997b) (zucchini), Gerendás et al. (1998b) (rice), Alibakhshi and Khoshgoftarmanesh (2016) (onion bulb), Arkoun et al. (2013) (oilseed rape), Kutman et al. (2013) (soybean), Khosgoftarmanesh and Bahmaniziari (2012) (cucumber), Khosgoftarmanesh et al. (2011) (lettuce), Gheibi et al. (2011) (maize), Moraes et al. (2009) (rice), Tan et al. (2000) (tomato), and Gheibi et al. (2009) (wheat).

1.3 Establishment of Critical Levels of Ni

The amounts of Ni required by plants are exceedingly slight. The best estimates are those of Eskew et al. (1984) who used seeds from plants grown under various Ni regimes to grow a second generation of plants from seeds with reduced Ni content. In the case of seeds from plants grown with high Ni (10 µg Ni L^{-1}), the amount of Ni in the seed was found to be 160 ng, per seed, an amount that was sufficient to prevent deficiency symptoms in plants even when grown in a second generation with nutrient solution with no Ni. The authors estimated that roughly 200 ng per seed would normally be required to carry plants through the next generation. Extrapolating these data to mature plants, they calculated that the critical Ni concentration in soybean tissue can be calculated to be between 2 and 4 ng of Ni g^{-1} dry weight (or 0.03 to 0.07 nmol g^{-1}). For plants grown with urea as the sole N source, the Ni content necessary to avoid deficiency symptoms is higher—around 100 ng g^{-1} (Gerendás et al. 1999). A similar estimate of 89–149 ng of Ni g^{-1} dry weight was reported for urea-grown rice (Gerendás et al. 1998b). An interesting comparison can be made with molybdenum, the micronutrient previously considered to be required in the least amount by plants. For Mo, the critical level seems to be 10 to 500 ng Mo g^{-1} dry weight. Thus, the requirement for Ni is much lower even than that for Mo.

1.4 The Role of Urease in Plants

The evidence that Ni is required for urease is firmly established, but this raises the question of whether urease itself is essential for plants. Soybean plants that are completely urease-negative mutants develop normally and produce a good yield of seeds that germinate well and produce a healthy second generation (Polacco and Holland 1993), apparently showing no ill effects of urease deficiency (or Ni deficiency) beyond necrotic leaf tips. Urease appears to be universal in the leaves of all plants (Hogan et al. 1983; Witte 2011) and especially abundant in the seeds of plants in the Fabaceae, Cucurbitaceae,

Asteraceae, and Pinaceae families. Further evidence for the ubiquity of urease comes from the presence of the structural genes for urease and Ni-insertion accessory proteins in all plant genomes sequenced to date (Polacco et al. 2013). Urease is absent in most animals, thus explaining the excretion of urea as a waste product. Urease is however present in many bacteria, including the intestinal microflora of animals and bacteria that are commensal on plants. The reaction catalyzed by urease is shown in Figure 1.2.

The presence of urea in plants can come directly from uptake in the soil. As explained in Section 1.7, urea is widely used as a crop fertilizer. Some of this urea is taken up directly by the plant, although most of it is broken down into ammonium and CO_2 by urease from soil microbes. An alternative source of urea is the production from the metabolism of arginine in the reaction catalyzed by arginase and shown in Figure 1.3. This reaction accounts for all endogenous urea in most plants (Polacco et al. 2013). The only other source of endogenous urea comes from the conversion of canavanine, an amino acid similar to arginine and found in only a few leguminous plants, into canaline and urea, a reaction that is also catalyzed by arginase (Witte 2011). Arginine and canavine are catabolized in germinating seedlings, thus producing urea that is processed by urease. Arginine is an important storage amino acid in many seeds and is the most abundant metabolite for nitrogen storage in many species, accounting for up to 50% of stored amino acid N (Polacco and Holland 1993; Vanetten et al. 1967). Canavine serves a dual purpose of N storage in seeds and protection from herbivory due to its toxic properties. Developing embryos have low arginase activity, and it appears that most arginase activity is located in mitochondria of cotyledons (Goldraij and Polacco 2000). This is also consistent with the findings that embryos do not produce urea and that urease activity in young seedlings takes place primarily in the cytoplasm of young cotyledons (Polacco and Holland 1993). Previous theories that urea can come from the catabolic breakdown of ureides such as allantoin and allantoate have been recently called into question (Polacco et al. 2013, or see Witte 2011 for a discussion).

FIGURE 1.2
The reaction catalyzed by urease.

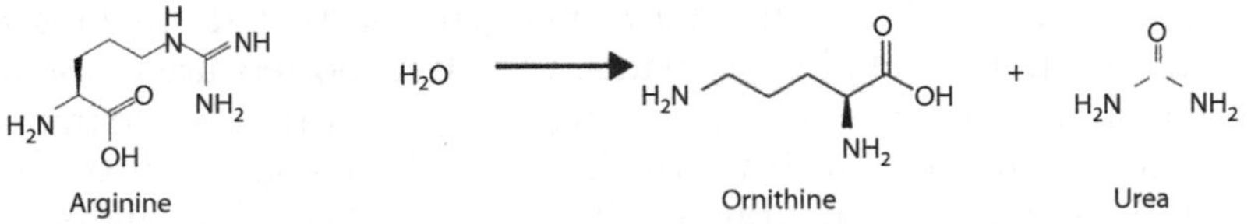

FIGURE 1.3
The reaction catalyzed by arginase.

Numerous mutants deficient in urease can give further evidence regarding the role of urease. In soybean, there are three classes of mutations in urease isozymes (Holland and Polacco 1992). Class I mutants result in the loss of an abundant embryo-specific urease. Class II mutations are pleotropic and cause complete loss of all ureases, including the "ubiquitous" enzyme present in tissue of all plants. Class III mutations affect only the ubiquitous isozyme. All of these mutations result in poor growth on urea in tissue culture and some toxicity due to urea accumulation, but the situation is not straightforward as some mutations are leaky and phyloplane bacteria living on the leaves may produce urease if sufficient Ni is present (Holland and Polacco 1992). The most revealing mutant is probably the one described by Stebbins et al. (1991) in which all classes of urease are lacking. Seeds of this urease mutant accumulated up to 51 µmol urea g^{-1} dry weight (a level approximately 250 times that of wild type) and developed necrotic leaf tips similar to those seen in Ni-deficient plants. The urea produced by urease mutants presumably represents a "dead-end in N metabolism" as plants are unable to recover the N stored as arginine in the seeds (Polacco and Holland 1993). Curiously, the embryo proper does not produce urea, so the role of the embryo-specific urease is puzzling (Stebbins et al. 1991). Polacco and Holland (1993) have suggested that the embryo-specific urease functions in plant defense rather than urea assimilation. Would-be herbivores could suffer from either "hepatic coma by subversion of the urea cycle or peptic ulceration by localized increases in NH_4^+ and OH- ions." Collectively, these observations indicate that urease is an essential part of nitrogen metabolism in all plants and thus supports the conclusion that the role of Ni is indispensible.

1.5 Urea Transport in Plants

Further evidence for the importance of urea comes from the presence of urea membrane transport proteins (Wang et al. 2008). Recent molecular evidence shows that there are at least two types of urea transporters in plants: the low-affinity channel-like major intrinsic proteins (MIPs) and the high-affinity active transporters (DUR3), the latter or which has a K_m for urea of 3 µM. Transport with DUR3 involves the co-transport of urea with protons driving secondary active transport even against a urea concentration gradient (Liu et al. 2003). Orthologs to the *Arabidopsis* gene for AtDUR3 are present in algae and most, if not all, higher plants. Urea transporters are particularly important in cotyledons of young seedlings where they function to transport urea from mitochondria, which is the location where arginine is converted into ornithine and urea by arginase (Wang et al. 2008). The high-affinity DUR urea transporters also appear to be important in vascular loading into phloem cells of mature leaves (Liu et al. 2015). Such urea transporters are of particular importance in agricultural systems where N fertilizer is applied in the form of urea in which case urea is taken up by roots or foliage. At least in the case of *Zea mays*, expression of the high-affinity urea transport gene (*Zmdur3*) is induced by low N conditions and repressed by restoration of N after N starvation (Liu et al. 2015). The protein is present in the plasma membrane of root epidermal cells, especially in N-starved cells. In addition, urea can activate the promoter for the transporter gene in the absence of other sources of N, an observation that is consistent with the role of DUR3 proteins as a urea transporter (Kojima et al. 2007; Merigout et al. 2008). The presence of ammonium nitrate reduces urea uptake. Transcriptomic analysis revealed large differences (960 genes in the roots and 474 in shoots) between urea-grown plants and those grown with ammonium nitrate, suggesting that urea is a potent regulator of many genes other than just urea transporters (Merigout et al. 2008).

Passive urea transport is conducted by MIPs that also function as aquaporins whose primary function is to facilitate the transport of water molecules across biological membranes. *Arabidopsis* contains 35 MIPs, some of which have been shown to function as urea transporters based on their ability to complement a yeast DUR3 mutant in a screen of an *Arabidopsis* seedling cDNA library (Liu et al. 2003; Witte 2011). The MIP passive urea transporters are of several classes including plasma membrane intrinsic proteins (PIPs), tonoplast intrinsic proteins (TIPs), and nodulin-26-like intrinsic proteins (NIPs). PIPs allow at best only modest levels of urea transport whereas probably all TIPs and NIPs are efficient in allowing urea transport (Witte 2011). It is clear that in addition to their role as water channels, MIPs also play a role in controlling urea concentrations between different cellular compartments. The ubiquitous nature of both urease and urea transport proteins provides further evidence that urea plays a vital role in the N metabolism of plants in general.

1.6 Ni Processing by Accessory Proteins

Several activation proteins are required to guide the insertion of Ni into the metallocenter of urease. This process occurs in bacterial, fungal, and plant ureases, though there are subtle differences between the plant and microbial systems. Most bacterial ureases are a "trimer of trimers"—$(\alpha\beta\gamma)_3$—and are encoded by the genes *ureC*, *ureB*, and *ureA*. Plant ureases are homo-oligomers of a single peptide (Eu1 for the embryo-specific urease and Eu4 for the ubiquitous urease) that is homologous to the three peptides of the bacterial ureases (Polacco et al. 2013). In bacterial ureases, three peptides are required for nickel insertion: UreD, UreF, and UreG. UreD acts as a chaperone that binds to the UreABC trimer (Park et al. 1994). This initial binding is followed by binding of the UreF protein to the UreABC–UreD complex, forming UreABC–UreDF. This is followed by the binding of UreG to form a UreABC–UreDFG complex (Polacco et al. 2013). UreG acts as a GTPase and nickel is initially bound to UreG in this final complex. This binding is facilitated by a Ni chaperone UreE, which is responsible for transferring Ni atoms to the UreABC–UreDFG complex in a process that Polacco et al. (2013) call a "bucket brigade." The sequence of Ni binding proceeds from UreE–UreG to UreF, then to UreD, and ultimately to a carbamate on a modified active-site lysine residue (Jabri et al. 1995). The formation of the carbamate involves carboxylation of the epsilon amino group of lysine. Each active site contains two Ni atoms.

The activation of plant ureases also involves the formation of the UreDFG complex with apo-urease, but there is no chaperone protein that acts similar to UreE. Polacco et al. (2013) speculate that UreG has an N-terminal domain that is His-rich and resembles the C-terminus of UreE, thus allowing plant UreG to provide the chaperone role played by UreE in bacteria. Numerous plant mutants in Ni-processing accessory proteins have been described. The simplest case is that of *Arabidopsis*, which has only a single copy of each component of the system such that mutants in genes encoding UreD, UreF, UreG, or urease itself have a urease-negative phenotype (Witte et al. 2005). Soybean has a paleopolyploid genome that contains two two paralogs for *UreD* and for *UreF*, a single copy of *UreG*, and three genes for urease (Polacco et al. 2011). UreD, UreF, and UreG are responsible for activating the two active forms of urease in soybean: the embryo-specific urease and the so-called ubiquitous urease of leaves. The third urease gene is apparently non-ureolytic (Polacco et al. 2013). Readers are referred to this reference for a detailed discussion of soybean mutants in genes for Ni accessory proteins.

1.7 Agricultural Issues Related to Urease

Plants encounter considerable urea in their environment because it is a major form of recyclable N excreted by animals. Thus urea is ubiquitous in natural environments. Urea also accounts for about 50% of the total nitrogen fertilizer used worldwide in agriculture (Wang et al. 2008). As discussed above, plants may take up urea directly from the soil using urea transporters that are present on the plasma membrane of root epidermal cells. Subsequent utilization of urea as a source of N for plant growth and metabolism requires the action of urease to convert the urea into NH_4^+ followed by incorporation of NH_4^+ into organic form by GS/GOGAT. There is no alternative means of assimilating urea in plants. Plants in which urease has been blocked fail to utilize the urea (Polacco et al. 2013). Foliar application of urea fertilizer can lead to foliar burn that can be alleviated by the application of Ni to maximize urease activity (Kutman et al. 2013). Alternatively, microbial urease, which is present in most soils, can convert the urea into NH_4^+, which can then be taken up by plants. This especially applies to agricultural settings where large amounts of urea are applied.

Nickel is the fifth most abundant element on earth and is present at moderate levels in most soils. For instance, a survey of soils from Oregon showed Ni levels between 22 and 96 parts per million (ppm) (Dalton et al. 1985). Worldwide, the range of Ni concentration in soils typically ranges from 0.2 to 450 ppm with an average value of approximately 20 ppm (Ahmad and Ashraf 2011). Even though the amount of Ni needed to prevent Ni deficiency in plants is small, there is still a potential for Ni-deficiency symptoms in the field, especially if urea fertilizer is used, although most fertilizers contain sufficient contaminating Ni to meet plant needs (Harasim and Filipek 2015).

Few studies have examined the potential for natural soils to provide insufficient Ni to meet the needs of plants largely because nearly all soils contain adequate amounts of Ni and the amount of Ni required by plants is very slight. An early exception to the lack of Ni deficiencies in the field is the work of Bertrand and de Wolff (1973), which showed increases in nodulation and yield in legume crops in response to application of Ni at a rate of up to 40 g of Ni per hectare. Oliveira et al. (2013) grew lettuce (*Lactuca sativa*) in soils amended with Ni and found no difference in yield in response to Ni, though leaf urease activity was increased by Ni application. Perhaps the most profound example of Ni deficiency in plants outside the lab or greenhouse is the case of pecan (*Carya illinoinensis*), which develops a wide range of Ni-related deficiency symptoms including early-season leaf chlorosis, dwarfing of foliage, blunting of leaf tips, necrosis of leaf tips, curled leaf margins, dwarfed internodes, distorted bud shape, brittle shoots, death of over-wintering shoots, diminished root system with dead fibrous roots, failure of foliar lamina to develop, loss of apical dominance, overall stunting, and even tree death (Wood et al. 2005). These symptoms are referred to as "mouse ear" due to the malformation of young leaves as shown in Figure 1.4 (Wood et al. 2004). These symptoms are also present in other woody perennials and are apparently caused by interference of Ni uptake caused by the accumulation of light metals (Zu, Cu, Mn, Fe, Ca, or Mg) that arises from long-term fertilizer usage. Foliar application of Ni salts completely alleviates these deficiency symptoms.

Several recent studies have addressed the potential benefits of Ni treatments under field conditions. Canola (*B. napus*) grown on two calcareous soils showed a positive response of growth and yield from Ni application (Moosavi et al. 2014). Data from long-term field trials indicated that 77% of samples of wheat, barley, and oats grown in Swedish soils appears to contain a level of Ni below the proposed critical concentration but no yield differences or symptoms were reported (Hamner et al. 2013). Ni amendments to soybeans grown on a low-Ni soil resulted in an increase in leaf urease activity but not in yield. No deficiency symptoms were evident (Dalton et al. 1985). Ni added directly to this soil also resulted in an increase in soil microbial urease activity,

FIGURE 1.4
Nickel deficiency symptom known as mouse ear (right) as compared to unaffected shoot (left) in pecan. (Courtesy of B. W. Wood, USDA–Agricultural Research Service.)

an observation that has implications to the use of urea fertilizers in agriculture. As with other aspects of Ni action, the amount of Ni required is low and levels too high are inhibitory. This is reflected in a dose–response curve showing that activity of microbial urease in a low-Ni soil could be increased threefold by the addition of 5 μmol Ni.g soil^{-1}, whereas 50 μmol Ni.g soil^{-1} was strongly inhibitory (Figure 1.5; Dalton et al. 1985).

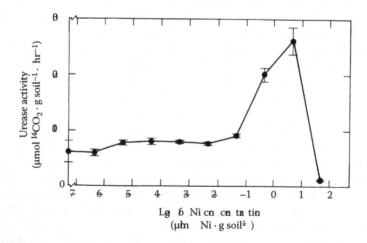

FIGURE 1.5
The effect of supplemental Ni on microbial urease activity in a low-Ni soil after 3 days of incubation with added glucose and urea. (From Dalton, D. A., H. J. Evans, and F. J. Hanus. 1985. Stimulation by nickel of soil microbial urease activity and urease and hydrogenase activities in soybean grown in a low-nickel soil. *Plant Soil* 88:245–258.)

1.8 Ni in Hydrogenase

Nitrogen-fixing plants such as soybeans and other legumes have a secondary requirement for Ni due to the importance of this element in activating hydrogenase that is present in the rhizobia of root nodules. Hydrogen gas is a by-product of the nitrogenase reaction: $N_2 + 8e^- + 8H+ + 16ATP \rightarrow 2NH_3 + H_2 + 16ADP + 16\ P_i$. This hydrogen product appears to be an unavoidable and undesirable reaction by nitrogenase as it represents the potential loss of energy. Hydrogenase acts to remove electrons from H_2 to drive ATP formation via the electron transport chain. The hydrogenase is strictly a product of the symbiotic bacteria and its action can save as much as 25% of the energy utilized by the nitrogenase reaction. Soybean plants with hydrogenase-containing rhizobia ("Hup$^+$") fix more nitrogen and have higher yields than those inoculated with strains lacking hydrogen-uptake capacity (Hup$^-$; Albrecht et al. 1979). This hydrogenase is also produced *ex planta* by rhizobia and allows the bacteria to be cultured chemolithotrophicly in a medium with H_2 as the sole energy source. This chemolithotrophic growth is dependent on the presence of Ni for hydrogenase to be active. Supplemental Ni significantly increased the activity of hydrogenase in symbiotic bacteroids isolated from root nodules of soybeans grown hydroponically in purified nutrient solution (Klucas et al. 1983) and from soybeans grown in soil with low Ni content (Dalton et al. 1985). In an early field experiment, Bertrand and DeWolff (1973) showed that the addition of up to 40 g of Ni per hectare increased nodulation and grain yield significantly.

Ni is also a component of several other bacterial enzymes: CO dehydrogenase, acireductone dioxygenase, Ni-superoxide dismutase, methyl coenzyme M reductase, and acetyl-CoA synthase, but none of these Ni-containing proteins are in higher plants and thus they are outside the scope of this review (Ragsdale 2009; Walsh and Orme-Johnson 1987).

1.9 Glyoxylase and Plant Stress

Recent studies have reported that Ni is required for the activation of glyoxalase I in plants (Mustafiz et al. 2014). Glyoxalase I is responsible for the detoxification of methylglyoxal (MG), a cytotoxic by-product of glycolysis and photosynthesis. MG is a mutagenic alpha-ketoaldehyde that may kill cells (Fabiano et al. 2015). MG formation begins with dihydroxyacetone phosphate, which reacts to form an enediol intermediate followed by an enol and then MG. The toxicity of MG is due to its tendency to produce reactive oxygen species (Kalapos 2008; Maeta et al. 2005), to generate glycation end products and tissue damage (Thornalley 2003), to inactivate antioxidant defenses

(Martins et al. 2001), and to disrupt electron flow through the mitochondrial respiratory chain (Ray et al. 1994). In the detoxification process, MG combines with reduced glutathione (GSH) and to form S-D-lactoylglutathione through catalysis by the Ni-containing glyoxalase I. In a subsequent reaction, the S-D-lactoylglutathione is converted in D-lactic acid by glyoxalase II and GSH is regenerated. The production of MG increases in plants under abiotic stress and expression of glyoxalase I is induced by this stress. Overexpression of glyoxalase I in transgenic tobacco plants provides enhanced tolerance of osmotic, oxidative, exogenous MG, and salinity stresses (Mustafiz et al. 2014). These observations indicate that Ni plays a role in plant stress responses. The involvement of GSH in the detoxification of MG also provides a link between Ni and antioxidant defenses since glyoxalase I plays a role in regulating the pool of GSH in cells (Fabiano et al. 2015).

1.10 Fungicidal Benefits of Ni

There is a long history of using moderate concentrations of Ni salts to act as a fungicide on crop plants (Demaree and Cole 1927; Polacco et al. 2013). For instance, Ni salts are an effective means of reducing damage from the disease pecan scab due to nickel's ability to inhibit the growth of the fungus *Fusicladium effusum* (Wood et al. 2012). Ni salts have been used to prevent other plant fungal diseases including cereal rusts, rice blast, rice spot disease, flax rust, and cotton wilt (Mishra and Kar 1974). The benefits are systemic and persistent but care must be taken to avoid excess levels of Ni as these can be toxic to the plant. There is a wide range of concentrations that have been reported to be effective fungicidal agents, but most studies report benefits at concentrations less than 100 ppm of Ni. Despite the proven efficacy of Ni as a fungicide, it is not used extensively for this purpose because of the risk of toxicity to plants. There is a large range of values reported to be toxic but some crop plants appear to be damaged by Ni levels as low as 1 ppm (Mishra and Kar 1974). So clearly, Ni levels that are reported to be beneficial against fungi can actually cause harm to the host plant due to direct toxicity.

Once the role of Ni in urease was established, it became evident that the fungicidal properties of Ni were related at least in part to its role in activation of urease. The interaction is complex and does not appear to be related to the normal catalytic activity of urease but rather to an enzymatic-independent mechanism involving fungal membrane permeabilization (Carlini and Ligabue-Braun 2016). The antifungal properties persist even if urease is treated with an irreversible inhibitor of ureolytic activity (Becker-Ritt et al. 2007). Soybean that was genetically engineered to be deficient in ubiquitous leaf urease had decreased fungal resistance (Wiebke-Ströhm et al. 2012).

Similarly, urease-negative soybean seedlings were more prone to damping off fungi than urease-positive seedlings (Polacco et al. 2013). Urease is also toxic to insects through a poorly understood mechanism. Since these toxic properties are not dependent on the ureolytic properties of urease, it appears that the fungicidal properties of Ni might be due in part to stabilization of urease and less turnover of the enzyme.

1.11 Conclusions

Nickel is an essential element for all plants due to its role in urease, the enzyme that catalyzes the conversion of urea into NH_4^+ and CO_2. Plants deficient in Ni are unable to metabolize urea, and this can lead to the toxic accumulation of urea in leaves. The amounts of Ni required are exceedingly slight such that deficiencies of Ni in the field are rare. Urease-negative mutants of plants—or Ni-deficient plants—generally grow normally except for the formation of necrotic lesions on leaves. The main role of urease is to process N stored in the form of arginase in seeds. Arginine is often the most abundant form of stored N in seeds. Arginine is degraded into urea by arginase, which is abundant in cotyledons of some plants, especially legumes. Subsequent incorporation of the urea into amino acids requires the action of urease. The transport of urea in plants involves abundant membrane-intrinsic aquaporins that co-transport urea along with water. Several accessory proteins are known to be essential for the insertion of Ni into the active site of urease. The ability to metabolize urea is especially critical in crops because N fertilizers very often consist of urea. Ni is also a component of hydrogenase, a beneficial enzyme present in rhizobia that are symbiotic with legumes. The plant enzyme glyoxylase, a component of plant response to stress, is also a Ni metalloenzyme. Ni may also be beneficial to plants because of its direct role as a fungicide, but this property is of limited practical importance because of the risks of Ni toxicity to plants at concentrations at or below those shown to have fungicidal properties.

References

Ahmad, M. S. A. and M. Ashraf. 2011. Essential roles and hazardous effects of nickel in plants. *Rev. Environ. Contam. Toxicol.* 214:125–167.

Albrecht, S. L., R. J. Maier, F. J. Hanus, S. A. Russell, D. W. Emerich, and H. J. Evans. 1979. Hydrogenase in *Rhizobium japonicum* increases nitrogen fixation by nodulated soybeans. *Science* 203:1255–1257.

Alibakhshi, M. and A. H. Khoshgoftarmanesh. 2016. The effect of nickel supply on bulb yield, urease and glutamine synthetase activity and concentrations of urea, amino acids and nitrogen of urea-fed plants. *Arch. Agronomy Soil Sci.* 62:37–51.

Arkoun, M., L. Jannin, P. Laine, P. Etienne, C. Masclaux-Daubresse, S. Citerne, M. Garnica, J. M. Garcia-Mina, J. C. Yvin, and A. Ourry. 2013. A physiological and molecular study of the effects of nickel deficiency and phenylphosphorodiamidate (PPD) application on urea metabolism in oilseed rape (*Brassica napus* L.). *Plant Soil* 362:79–92.

Arnon, D. I. and P. R. Stout. 1939. The essentiality of certain elements in minute quantity for plants with special reference to copper. *Plant Physiol.* 14:371–374.

Becker-Ritt, A. B., A. H. S. Martinelli, S. Mitidieri, V. Feder, G. E. Wassermann, L. Santi, M. H. Vainstein, J. T. A. Oliveira, L. M. Fiuza, G. Pasquali, and C. R. Carlini. 2007. Antifungal activity of plant and bacterial ureases. *Toxicon.* 50:971–983.

Bertrand, D. and A. De Wolff. 1973. Importance de nickel, comme oligoélément pour les *Rhizobium* des nodosites des legumimeuses. *C. R. Acad. Sci. Paris.* 276:1855–1858.

Brown, P. H., R. M. Welch, and E. E. Cary. 1987. Nickel: A micronutrient essential for higher plants. *Plant Physiol.* 85:801–803.

Broyer, T. C., A. B. Carlton, and C. M. Johnson. 1954. Chlorine—A micronutrient element for higher plants. *Plant Physiol.* 29:526–532.

Carlini, C. R. and R. Ligabue-Braun. 2016. Ureases as multifunctional toxic proteins: A review. *Toxicon.* 110:90–109.

Checkai, R. T., W. A. Norvell, and R. M. Welch. 1986. Investigation of nickel essentiality in higher plants using a recirculating resin-buffered hydroponic system. *Agron. Abstr.* 195.

Dalton, D. A., H. J. Evans, and F. J. Hanus. 1985. Stimulation by nickel of soil microbial urease activity and urease and hydrogenase activities in soybean grown in a low-nickel soil. *Plant Soil* 88:245–258.

Dalton, D. A., S. A. Russell, and H. J. Evans. 1988. Nickel as a micronutrient element for plants. *BioFactors* 1:11–16.

Demaree, J. B. and J. R. Cole. 1927. Dusting with monohydrated copper sulphate and lime for control of pecan scab. USDA Circular no. 412.

Dixon, N. E., C. Gazzola, R. L. Blakeley, and R. Zerner. 1975. Jack bean urease (E.C. 3.5.1.5). A metalloenzyme. A simple biological role for nickel. *J. Am. Chem. Soc.* 97:4131–4133.

Eskew, D. L., R. M. Welch, and E. E. Cary. 1983. Nickel: An essential micronutrient for legumes and possibly all higher plants. *Science* 222:621–623.

Eskew, D. L., R. M. Welch, and W. A. Norvell. 1984. Nickel in higher plants. Further evidence for an essential role. *Plant Physiol.* 76:691–693.

Fabiano, C. C., T. Tezotto, J. L. Favarin, J. C. Polacco, and P. Massafeera. 2015. Essentiality of nickel in plants: A role in plant stress. *Front. Plant Sci.* 6:1–4.

Gerendás, J. and B. Sattelmacher. 1997a. Significance of Ni supply for growth, urease activity and the contents of urea, amino acids and mineral nutrients of urea-grown plants. *Plant Soil* 190:153–162.

Gerendás, J. and B. Sattelmacher. 1997b. Significance of N source (urea vs. NH_4NO_3) and Ni supply for growth, urease activity and nitrogen metabolism of zucchini (*Curcubita pepo* convar. *giromontiina. Plant Soil* 196:217–222.

Gerendás, J., J. C. Polacco, S. K. Freymuth, and B. Sattelmacher. 1998a. Co does not replace Ni with respect to urease activity in zucchini (*Cucurbita pepo* convar. *giromontiina*) and soybean (*Glycine max*). *Plant Soil* 203:127–135.

Gerendás, J., Z. Zhu, and B. Sattelmacher. 1998b. Influence of N and Ni supply on nitrogen metabolism and urease activity in rice (*Oryza sativa* L.). *J. Exp. Botany.* 49:1545–1554.

Gerendás, J. and B. Sattelmacher. 1999. Influence of Ni supply on growth, urease activity and nitrogen metabolites of *Brassica napus* grown with NH_4NO_3 or urea as N source. *Ann. Bot.* 83:65–71.

Gerendás, J., J. C. Polacco, S. K. Freyermuth, and B. Sattelmacher. 1999. Significance of nickel for plant growth and metabolism. *J. Plant Nutr. Soil Sci.* 162:241–256.

Gheibi, M. N., M. J. Malakouti, B. Kholdebarin, F. Ghanati, S. Teimouri, and R. Sayadi. 2009. Significance of nickel supply for growth and chlorophyll content of wheat supplied with urea or ammonium nitrate. *J. Plant Nutr.* 32:1440–1450.

Gheibi, M. N., B. Kholdebarin, M. J. Malakouti, F. Ghanati, S. Teimouri, and R. Sayadi. 2011. Effect of various nickel levels on growth and chlorophyll content of corn plants supplied with urea and ammonium nitrate. *J. Food Agric. Environ.* 9:583–587.

Goldraij, A. and J. C. Polacco. 2000. Arginine degradation by arginase in mitochondria of soybean seedling cotyledons. *Planta* 210:652–658.

Gordon, W. R., S. S. Schwemmer, and W. S. Hillman. 1978. Nickel and the metabolism of urea by *Lemna paucicostata* Heggelm. 6746. *Planta* 140:265–268.

Hamner, K, J. Eriksson, and H. Kirchmann. 2013. Nickel in Swedish soils and cereal grain in relation to soil properties, fertilization, and seed quality. *Acta Agric. Scand. Sect. B—Soil Plant Sci.* 63:712–722.

Harasim, P. and T. Filipek. 2015. Nickel in the environment. *J. Elementol.* 20:525–534.

Hogan, M. E., I. E. Swift, and J. Done. 1983. Urease assay and ammonia release from leaf tissues. *Phytochem.* 22:663–667.

Holland, M. A. and J. C. Polacco. 1992. Urease-null and hydrogenase-null phenotypes of a phylloplane bacterium reveal altered nickel metabolism in two soybean mutants. *Plant Physiol.* 98:942–948.

Jabri, E., M. B. Carr, R. P. Hausinger, and P. A. Karplus. 1995. The crystal structure of urease from *Klebsiella aerogenes*. *Science* 268:998–1004.

Kalapos, M. P. 2008. The tandem of free radicals and methylglyoxal. *Chem. Biol. Interact.* 171:251–271.

Khosgoftarmanesh, A. H. and H. Bahmaniziari. 2012. Stimulating and toxicity effects of nickel on growth, yield, and fruit quality of cucumber supplied with different nitrogen sources. *J. Plant Nutr. Soil Sci.* 175:474–481.

Khosgoftarmanesh, A. H., F. Hosseini, and M. Afyuni. 2011. Nickel supplementation effect on growth, urease activity and urea and nitrate concentrations in lettuce supplied with different nitrogen sources. *Scientia Horticulturae* 130:381–385.

Klucas, R. V., F. J. Hanus, S. A. Russell, and H. J. Evans. 1983. Nickel: A micronutrient element for the hydrogen-dependent growth of *Rhizobium japonicum* and for expression of urease activity in soybean leaves. *Proc. Natl. Acad. Sci. U.S.A.* 80:2253–2257.

Kojima, S., A. Bohner, B. Gassert, L. X. Yuan, and N. von Wiren. 2007. AtDUR3 represents the major transporter for high-affinity urea transport across the plasma membrane of nitrogen-deficient *Arabidopsis* roots. *Plant J.* 52:30–40.

Krogmeier, M. J., G. W. McCarty, D. R. Shogren, and J. D. Bremmer. 1991. Effect of nickel deficiency in soybeans on the phytotoxicity of foliar-applied urea. *Plant Soil*. 135:283–286.

Kutman, B. Y., U. B. Kutman, and I. Cakmak. 2013. Nickel-enriched seed and externally supplied nickel improve growth and alleviate foliar urea damage in soybean. *Plant Soil* 363:61–75.

Liu, L. H., U. Ludewig, B. Gassert, W. B. Frommer, and N. von Wirén. 2003. Urea transport by nitrogen-regulated tonoplast intrinsic proteins in *Arabidopsis*. *Plant Physiol*. 133:1220–1228.

Liu, G. W., A. L. Sun, D. Q. Li, A. Athman, M. Gilliham, and L. H. Liu. 2015. Molecular identification and functional analysis of a maize (*Zea mays*) DUR3 homolog that transports urea with high affinity. *Planta* 241:861–874.

Maeta, K., S. Izawa, and Y. Inoue. 2005. Methylglyoxal, a metabolite derived from glycolysis, functions as a signal initiator of high osmolarity glycerol-mitogen-activated protein kinase cascade and calcineurin/Crz-mediated pathway in *Saccharomyces cerevisiae*. *J. Biol. Chem*. 280:253–260.

Martins, A. M. T. B. S., C. A. A. Cordeiro, and A. M. J. Ponces Freire. 2001. *In situ* analysis of methylglyoxal metabolism in *Saccharomyces cerevisiae*. *FEBS Lett*. 499:41–44.

Merigout, P., M. Lelandais, F. Bitton, J. P. Renou, X. Briand, C. Meyer, and F. Daniel-Vedele. 2008. Physiological and transcriptomic aspects of urea uptake and assimilation in *Arabidopsis* plants. *Plant Physiol*. 147:1225–1238.

Mishra, D. and M. Kar. 1974. Nickel in plant growth and metabolism. *Bot. Rev.* 40:395–452.

Moosavi, A. A., S. Mannsouri, M. Zahedifar, and M. R. Sadikhani. 2014. Effect of water stress and nickel application on yield components and agronomic characteristics of canola grown on two calcareous soils. *Arch. Agronomy Soil Sci*. 60:1747–1764.

Moraes, M. F., A. R. Reia, L. A. C. Moraes, J. Lavres, R. Vivian, C. P. Cabral, and E. Malavolta. 2009. Effects of molybdenum, nickel, and nitrogen sources on the mineral nutrition and growth of rice plants. *Comm. Soil Sci. Plant Analysis* 40:21–22.

Mustafiz, A., A. Ghosh, A. K. Tripathi, C. Kaur, A. K. Ganguly, N. S. Bhavesh, J. K. Tripathi, A. Pareek, S. K. Sopory, and S. L. Singla-Pareek. 2014. A unique Ni^{2+}-dependent and methylglyoxal-inducible rice glyoxalase I posses a single active site and functions in abiotic stress response. *Plant J*. 78:951–963.

Oliveira, T. C., R. L. F. Fontes, S. T. de Rezende, and V. H. Alvarez. 2013. Effects of nickel and nitrogen fertilization on lettuce growth and urease activity. *Revista Brasileira De Ciencia Do Solo*. 37:698–706.

Park, L. S., M. B. Carr, and R. P. Hausinger. 1994. In vitro activation of urease apoprotein and role of UreD as a chaperone required for nickel metallocenter assembly. *Proc. Natl. Acad. Sci. USA*. 91:3233–3237.

Polacco, J. C. 1977a. Nitrogen metabolism in soybean tissue culture. II. Urea utilization and urease synthesis require Ni^{2+}. *Plant Physiol*. 59:827–830.

Polacco, J. C. 1977b. Is nickel a universal component of plant ureases? *Plant Sci. Lett*. 10:249–255.

Polacco J. C. and E. A. Havir. 1979. Comparison of soybean urease isolated from seed and tissue culture. *J. Biol. Chem.* 254:1707–1715.

Polacco, J. C. and M. A. Holland. 1993. Roles of urease in plant cells. *Int. Rev. Cytol.* 145:65–103.

Polacco, J. C., D. L. Hyten, M. Medeiros-Silva, D. A. Sleper, and K. D. Bilyeu. 2011. Mutational analysis of the major soybean UreF paralogue involved in urease activation. *J. Exp. Bot.* 62:3599–3608.

Polacco, J. C., P. Mazzafera, and T. Tezotto. 2013. Opinion—Nickel and urease in plants; still many knowledge gaps. *Plant Sci.* 100–200:79–90.

Ragsdale, S. W. 2009. Nickel-based enzyme systems. *J. Biol. Chem.* 284:18571–18575.

Ray, S., S. Dutta, J. Halder, and M. Ray. 1994. Inhibition of electron flow through complex I of the mitochondrial respiratory chain of Ehrlich ascites carcinoma cells by methylglyoxal. *Biochem. J.* 303:69–72.

Shimada N. and T. Ando. 1980. Role of nickel in plant nutrition II. Effect of nickel on the assimilation of urea by plants. *Nippon Dojo Hiryogaku Zasshi* 51:493–496.

Shimada N., T. Ando, M. Tomiyama, and H. Kaku. 1980. Role of nickel in plant nutrition. I. Effects of nickel on growth of tomato and soybean. *Nippon Dojo Hiryogaku Zasshi* 51:487–492.

Shimada, N. and A. Matsuo. 1985. Role of nickel in plant nutrition (3). Effects of nickel on the growth of plants and the assimilation of urea by cucumber and barley. *Jpn. J. Soil Sci. Plant Nutr.* 51:257–263.

Stebbins, N., M. A. Holland, S. R. Cianzio, and J. C. Polacco. 1991. Genetic tests of the roles of embryonic ureases of soybean. *Plant Physiol.* 97:1004–1010.

Sumner, J. B. 1926. The isolation and crystallization of the enzyme urease. *J. Biol. Chem.* 69:435–441.

Tan, X. W., H. Ikeda, and M. Oda. 2000. Effects of nickel concentration in the nutrient solution on nitrogen assimilation and growth of tomato seedlings in hydroponic culture supplied with urea or nitrate as the sole nitrogen source. *Scientia Horticulturae* 84:265–273.

Thornalley, P. J. 2003. Glyoxalase I—Structure, function and a critical role in the enzymatic defence against glycation. *Biochem. Soc. Trans.* 31:1343–1348.

Vanetten, C. H., W. F. Kwolek, J. E. Peters, and A. S. Barclay. 1967. Plant seeds as protein sources for food or feed. Evaluation based on amino acid composition of 379 species. *J. Agric. Food Chem.* 15:1077–1089.

Walker, C. D., R. D. Graham, J. T. Madison, E. E. Cary, and R. M. Welch. 1985. Effects of Ni deficiency on some nitrogen metabolites in cowpeas (*Vigna unguiculata* L. Walp.). *Plant Physiol.* 79:474–479.

Walsh, C. T. and W. H. Orme-Johnson. 1987. Nickel enzymes. *Biochemistry* 26: 4901–4906.

Wang, W. H., B. Kohler, F. Q. Cao, and L. H. Liu. 2008. Molecular and physiological aspects of urea transport in higher plants. *Plant Sci.* 175:467–477.

Welch, R. M. 1981. The biological significance of nickel. *J. Plant Nutrition* 3:345–356.

Wiebke-Ströhm, B. and 16 others. 2012. Ubiquitous urease affects soybean susceptibility to fungi. *Plant Mol. Biol.* 79:75–87.

Witte, C.-P. 2011. Urea metabolism in plants. *Plant Sci.* 180:431–438.

Witte, C.-P., M. G. Rosso, and T. Romeis. 2005. Identification of three urease accessory proteins that are required for urease activation in *Arabidopsis*. *Plant Physiol.* 139:1155–1162.

Wood, B. W., C. C. Reilly, and A. P. Nyczepir. 2004. Mouse-ear of pecan: A nickel deficiency. *HortScience* 39:1238–2004.

Wood, B. W., C. C. Reilly, and A. P. Nyczepir. 2005. Field deficiency in nickel in trees: Symptoms and causes. Proceedings of the Vth International Symposium on Mineral Nutrition of Fruit Plants Book Series: *Acta Horticulturae* 721:83–97.

Wood, B. W., C. C. Reilly, C. H. Bock, and M. W. Hotchkiss. 2012. Suppression of pecan scab by nickel. *HortScience* 47:503–508.

2

Nickel Biogeochemistry at the Soil–Plant Interface

Walter W. Wenzel, Petra S. Kidd, Markus Puschenreiter, and Theresa Rosenkranz

CONTENTS

2.1 Introduction

Nickel, a metallic element with an atomic mass of 58.6934, is accommodated together with Fe and Co in the first d-block series of the transition metals.

The anthropogenic contribution to the global Ni cycle is estimated to exceed 50% (Sen and Peucker-Ehrenbrinck 2012). Total Ni concentrations in bedrocks range from <2 (sandstones) to >400 mg kg^{-1} (basaltic igneous rocks) and can reach 3600 mg kg^{-1} in ultramafic rocks such as serpentine (Sparks 2003). Total Ni concentrations in soils globally range between 2 and 750 mg kg^{-1} (Sparks 2002) but can exceed 2500 mg kg^{-1} in soils on ultramafic parent materials (Wenzel et al. 2003; Wenzel and Jockwer 1999). Background values of 1 M NH$_4$NO$_3$-extractable Ni increase from 250 µg kg^{-1} at pH 8 to >360 µg kg^{-1} below pH 5.5 (Lombi et al. 1998).

Nickel concentrations in plant tissues and common crops typically range between 0.01 and 5 mg kg^{-1} dry matter (Adriano 2001). However, Ni concentrations in shoots of hyperaccumulator plants may reach 12,500 mg kg^{-1} dry matter (Reeves and Baker 1984; Wenzel and Jockwer 1999). Nickel is classified as an essential plant micronutrient as, among other functions, it is a constituent of the enzymes urease, methyl coenzyme M reductase, hydrogenase, and carbon monoxide dehydrogenase (Adriano 2001). On the other hand, Ni toxicity has been observed in soils with Ni concentrations >50 mg kg^{-1} and, depending on the sensitivity of plants, is indicated at Ni concentrations in foliar tissues exceeding 20–150 mg kg^{-1} dry matter (Adriano 2001).

The plant rhizosphere is defined as the microenvironment influenced by root activities, characterized by physical, chemical, and biological changes compared to the bulk soil. Apart from removal of elements by root uptake, passive and active release of root products including, among others, organic compounds, protons, and CO$_2$ can modify the soil chemical milieu near roots substantially in terms of pH, redox potential, and presence of inorganic and organic ligands (e.g., Hinsinger and Courchesne 2008; Hinsinger et al. 2005). Concurrently, microbial communities and their activities in the rhizosphere can also strongly deviate from those in bulk soils (Buée et al. 2009). Consequently, the fate of metals such as Ni is expected to be affected by the activities of roots and their associated microbial communities.

Such rhizosphere effects are considered important for controlling metal release during weathering, metal solubility, mobility and availability, uptake in plants, leaching toward groundwater, and transfer of metals in the food chain. Information about the mechanism involved and the direction and extent of changes are crucial for improving the predictive power of mathematical plant–soil models, soil tests for assessing metal availability, deficiency and/or toxicity to plants, and leaching rates. Moreover, plant–microbe-based technologies for ecological engineering and environmental restoration/remediation are deemed to benefit from sound knowledge and informed use of rhizosphere processes controlling Ni biogeochemistry at the soil–root interface (Wenzel 2009).

While the fate of Ni in soils is fairly well understood, only a limited number of studies of Ni biogeochemistry in plant rhizospheres are available. Here, we review relevant literature to derive a comprehensive picture of the fate of Ni in rhizospheres of accumulator and non-accumulator plants and its interaction with microbial strains and communities, and explore possible applications of plant–microbe–soil interactions in biotechnology and plant-assisted soil remediation.

2.2 Nickel in Plants

Nickel is an essential element for plants. It is the co-factor of urease (Bai et al. 2006) and nickel-metallochaperones (Freyermuth et al. 2000). The typical concentration in plant tissue ranges from 0.05 to 5 mg kg^{-1} (Broadley et al. 2012). Kabata-Pendias (2011) has listed typical Ni concentrations in various food plants, varying from 0.06 to 2 mg kg^{-1}. Because of the very low demand, Ni deficiency is a very rare phenomenon. Nickel toxicity has been observed for Ni concentrations >10 mg kg^{-1} in sensitive plants and >50 mg kg^{-1} in moderately tolerant species (Gonnelli and Renella 2013). A detailed overview on the role of Ni as an essential micronutrient as well as on toxicity effects can be found in Yusuf et al. (2009).

Tolerance to high Ni concentrations is a common feature of plants growing on Ni-rich serpentine soils. As outlined by Brooks (1998), Ni is an important but not the only factor responsible for the unique flora on serpentine soils. Next to Ni, excessive amounts of magnesium, chromium, and cobalt as well as the low contents of calcium, molybdenum, nitrogen, phosphorus, and potassium result in highly specialized vegetation with a high degree of endemism. A further interesting phenomenon is the so-called serpentinomorphoses of plants grown on serpentine soils, which is characterized by dwarfism, glabrescence or pubescence, plagiotropism, and xeromorphic foliage (Gonnelli and Renella 2013). Since serpentine soils have typically a very low fertility, they are hardly used for agriculture. A significant transfer of Ni into the human food chain via harvested crop plants has not been reported yet. However, grazing animals may take up substantial amounts of nickel by inadvertently ingesting soil. As a result of that, Ni concentrations in the kidneys of 20% of cattle grazing on serpentine soils in the northwest of Spain reached toxic levels; however, nickel levels in muscles were still well below detectable levels (Miranda et al. 2005).

Most of the plants growing on serpentine soils are excluders, which are strongly limiting the root to shoot translocation of nickel (Gonnelli and Renella 2013). However, some specific plants, so-called hyperaccumulators, developed a strategy to accumulate Ni in their aboveground biomass, reaching Ni concentrations >1000 mg kg^{-1} and sometimes >10,000 mg kg^{-1}. In some extreme cases, concentrations of >20% nickel was found in the sap of *Sebertia acuminata* (Jaffré et al. 1976). So far, more than 400 Ni hyperaccumulating plant species have been identified (Krämer et al. 2010; van der Ent et al. 2012). Hyperaccumulators are characterized by (1) a shoot/root metal concentration ratio higher than 1, (2) a shoot/soil metal concentration ratio higher than 1, and (3) shoot concentrations exceeding a metal-specific concentration limit in its natural habitat (Baker and Whiting 2002). The current knowledge on Ni hyperaccumulation mechanisms has been summarized by Merlot et al. (2018).

2.3 Processes Controlling Nickel Biogeochemistry in the Rhizosphere

2.3.1 Nickel Uptake and Its Controls in the Rhizosphere

The large variation of Ni concentrations in plants and the corresponding rates of uptake in non-accumulator and accumulator species relate to differential controls of uptake and resupply from the soil solid phase.

In a rhizobox experiment with the Ni hyperaccumulator *Leptoplax emarginata*, Bartoli et al. (2012) determined Intact Plant Transpiration Stream Concentration Factors (TSCF = xylem/solution solute concentration ratio) of Ni. Based on TSCF $\gg$ 1 (5.2 ± 0.9) independent of the Ni availability in the growth medium (sand or topsoil), they concluded that Ni uptake by the hyperaccumulator was actively controlled (Bartoli et al. 2012).

Using laser ablation ICP-MS and a staining method, Moradi et al. (2010) imaged the distribution of total Ni in cross sections of *Berkheya coddii* roots grown on Ni-spiked soil and a low-Ni control soil. They found larger Ni concentrations in the cortex compared to the stele of roots from the Ni-spiked soil while the pattern was reversed on the control soil. From these results, they concluded that active uptake or ion selection mechanisms control Ni uptake in soils with low available Ni (Moradi et al. 2010).

In a study with Ni-spiked soil (10 to 100 mg kg^{-1}), using different measures of Ni availability (total Ni added, Ni in pore water, DGT-measured Ni concentration), Luo et al. (2010) showed that Ni uptake in hyperaccumulator *Noccaea caerulescens* is diffusion limited while this is not the case for the closely related non-accumulator species *Thlaspi arvense*.

2.3.2 Effects of Root-Induced Changes in pH and Redox

Based on the plant's nutrient uptake in the form of cations or anions and the release of protons or hydroxyl ions to maintain the cell internal charge balance, plants contribute to the acidification or alkalization of the rhizosphere (Hinsinger et al. 2003; Marschner 2012). Rhizosphere acidification can result in protonation of mineral surfaces and subsequent desorption of Ni attached to the soil solid phase, leading to increased Ni solubility with decreasing pH in soil solution (Kabata-Pendias 2011). Contrary to the general perception that rhizosphere acidification is actively induced by Ni-hyperaccumulating plants, less than half of the reviewed publications on pH changes in the rhizosphere of hyperaccumulator plants showed a decrease in pH of the soil solution in the rhizosphere (Wenzel et al. 2004). In this context, instead of acidification, a slight alkalization in the rhizosphere of *Alyssum murale* has been reported, whereas the acidification and reducing capacities of roots were greater in the non-accumulator *Raphanus sativus* compared to the Ni hyperaccumulator (Bernal et al. 1994). Moreover, Ni uptake in *A. murale* decreased with decreasing soil pH, and accumulation rates were higher at neutral to slightly alkaline pH (Kukier et al. 2004; Li et al. 2003b; Nkrumah et al. 2016). In a study comparing rhizosphere characteristics of *Noccaea goesingense* with that of the non-accumulator species *Silene vulgaris* and *Rumex acetosella*, an increase in rhizosphere pH compared to bulk soil was detected, probably related to the release of hydroxyl ions during mineral dissolution (Wenzel et al. 2003).

The dominant inorganic Ni species in soil solution is Ni^{2+}, which is stable over a range of pH and redox conditions (Adriano 2001). However, the reductive dissolution of Ni bearing Fe- and Mn-oxides can indirectly affect Ni speciation and solubility in soil. Differences in the amount of Ni found in the amorphous and crystalline Fe-oxide fractions in two hydromorphic soils were explained by the changes in redox conditions under the prevailing water regime, which influence the solubility of Fe-oxides and therefore the behavior of Ni bound to those oxides (Rinklebe and Shaheen 2014). A clear negative correlation between redox potential on Ni solubility was shown by Antić-Mladenović et al. (2011). Upon re-oxidation, precipitation with and sorption to newly formed Fe- and Mn-oxides could explain the decrease in Ni solubility in soil (Antić-Mladenović et al. 2011).

2.3.3 Complexation and Chelation of Ni in the Rhizosphere

About 40% of net carbon fixed by a plant is going belowground, with 19% being incorporated in the root biomass, 12% being released by root respiration, and about 5% being allocated to rhizodeposition (Jones et al. 2009). This rhizodeposition occurs in the form of secretion, plant mucilage, root lysates, and root exudates (Hirsch et al. 2013). Besides sugars and amino acids, root exudates are composed of low-molecular-weight organic acids (Jones 1998). In a plant cell, organic acids are fully dissociated at the neutral to slightly alkaline pH in the cytoplasm and are therefore excreted as organic anion (Jones 1998; Ryan et al. 2001). Therefore, organic acids do not actually contribute to the acidification of the rhizosphere. The solubility of Ni in the rhizosphere can be influenced by root exudation of such organic ligands, as Ni solubility can increase depending on the quality and quantity of root exudates and the Ni–organic complex that is formed (Clemens et al. 2002; Molas 2002). Some complexes can have a higher affinity to the soil solid phase, while others remain in solution. As free Ni^{2+} is removed from the soil solution by a complexing agent, the decrease in free metal activity could lead to ligand-induced dissolution of Ni that was previously unavailable until equilibrium solubility is reached (Agnello et al. 2014). Comparisons of the rhizosphere biogeochemistry of different plants showed that Ni solubility was increased in the rhizosphere of hyperaccumulators. The labile pool of Ni was depleted by plant uptake, indicating that Ni was mobilized from less soluble phases, possibly by ligand-induced co-dissolution of Ni-bearing minerals by root exudates (Puschenreiter et al. 2005; Wenzel et al. 2003). On the other hand, a decrease in free Ni²⁺ in soil solution can reduce Ni toxicity. Exudation of low-molecular-weight organic acids is known as a detoxifying mechanism in response to heavy metal stress with the effect of trace element immobilization and reduced toxicity to the plants (Ma et al. 2001; Montiel-Rozas et al. 2016). Early publications have suggested that citrate (Lee et al. 1978) and histidine (Krämer et al. 1996) are among the substances involved in Ni uptake by hyperaccumulator species, possibly acting as Ni chelators in the rhizosphere. However, upon increasing Ni exposure, root exudates of the hyperaccumulator *N. goesingense* and the non-accumulating plant *T. arvense* showed that exudation rates of citrate and histidine remained unchanged in the hyperaccumulator, while histidine and citrate accumulated in the exudates of *T. arvense* (Salt et al. 2000). In another study characterizing the composition of the dissolved organic carbon (DOC) from soil-grown *N. goesingense*, no high-affinity Ni chelators were found in the root exudates (Puschenreiter et al. 2003). Besides the effect of root exudates on the solubility of Ni in the rhizosphere, they also act as a valuable carbon source for rhizosphere-inhabiting bacteria (Doornbos et al. 2012). Certain bacterial strains can influence the Ni biogeochemistry in the rhizosphere that has been reported for phytoremediation-related experiments (Álvarez-López et al. 2016a; Becerra-Castro et al. 2009, 2011; Cabello-Conejo et al. 2014).

2.3.4 Enhanced Weathering of Ni-Rich Minerals in the Rhizosphere

Ni concentrations in soil greatly depend on the composition of the parent material (Kabata-Pendias 2011). Weathering of Ni-containing bedrock largely depends on moisture and temperature regimes of the respective region, on the composition of the bedrock as well as on biological activity acting on the mineral phase (Baumeister et al. 2015; Massoura et al. 2006). In upper soil horizons, weathering of Ni-rich minerals is influenced by biotic processes mediated by root activity. Such has been investigated in batch experiments using the Ni hyperaccumulator *L. emarginata* (Chardot-Jacques et al. 2013). Root activity of *L. emarginata* resulted in a more than twofold higher dissolution of the Ni-bearing mineral chrysotile that was supplied to the plant as the sole Ni source and 88% of the solubilized Ni was taken up by the plant compared to an unplanted treatment. It is suggested that the strong depletion of Ni by the plant and the resulting gradient in Ni concentration induced chrysotile dissolution. Interestingly, the pH of the leachates was higher than the initial pH of the nutrient solution that was supplied to the plant, indicating that OH⁻ ions released with the chrysotile dissolution buffered the solution. Root-induced acidification targeting Ni mobilization could have been obscured by this process. Similarly, it is suggested that the release of OH⁻ ions along with mineral weathering controls the pH in the rhizosphere of *T. goesingense* (Wenzel et al. 2003). Chardot-Jacques et al. (2013) also tested the influence of two different bacterial inoculants on the weathering of chrysotile, but did not find a significant contribution to the weathering induced by the plant roots. In contrast, the weathering of ultramafic rock was increased by the activity of a rhizobacterial strain, solubilizing Ni associated with Mn-oxides through oxalate exudation (Becerra-Castro et al. 2013).

2.4 Nickel Fractionation and Speciation in the Soil Solid Phase

The characterization of Ni in the solid phase, as well as Ni speciation in soil solution, is essential for the understanding of Ni behavior in soils. Fractionation and speciation of Ni determine Ni bioavailability in soil and therefore play a major role in the development of management strategies for Ni-rich soils, especially in regard to Ni immobilization, or phytoextraction and phytomining applications. The biogeochemical processes discussed in Section 2.3 greatly influence the speciation of Ni in the rhizosphere and the overall fractionation of Ni in soil.

Ni in the soil solid phase can be bound to exchange sites, specifically adsorbed, adsorbed or occluded into sesquioxides, bound to the clay lattice, or bound to organic matter (Adriano 2001). Ni bound to the surface of the soil solid phase in inner-sphere complexes is characterized by low availability, whereas Ni bound to outer-sphere complexes is easily soluble, existing as hydrated Ni species bound to the surface of those minerals (Ma and Hooda 2010; Uren 1984). Sorption to the soil mineral phase can be influenced by competing metal species, organic compounds, or complex formation. Depending on the surface charge of an organometallic complex, either attraction or repulsion to a charged soil mineral surface can occur (Neubauer et al. 2002). Nickel sorption to surfaces of goethite and montmorillonite in the presence of citrate in a pH range of 4–7.5 has been investigated (Marcussen et al. 2009). It is suggested that Ni is nonspecifically adsorbed to the negatively charged surface of montmorillonite and, in the presence of citrate anionic Ni–citrate complexes, reduced the sorption to negatively charged surfaces of montmorillonite. In a sorption experiment with sandy loam topsoil at pH 6, citrate prevented Ni from sorption as negatively charged Ni–citrate complexes were formed. Another complexing agent, arginine, did not influence Ni sorption, as cationic $NiH_3arg_2^{3+}$ complexes bound to cation exchange sites (Poulsen and Hansen 2000). In this context, the effect of root exudates on Ni behavior in soil greatly depends on the quality of the root exudates.

Surface precipitation of Ni has been demonstrated to control the partitioning of Ni between soil solution and soil solid phase in Ni-contaminated soils, clay mineral, and oxide surfaces at neutral to basic pH (Scheckel and Sparks 2001; Scheidegger et al. 1996; Shi et al. 2012). McNear et al. (2007) determined the speciation of Ni in Ni-contaminated mineral and muck soils collected in the vicinity of a historic Ni refinery using synchrotron-based micro-XAFS and micro-X-ray fluorescence spectroscopy. The prevalent species in the mineral soil was Ni incorporated into mixed hydroxide surface precipitates of layered double hydroxides (LDHs). Following a pH increase from 7.0 to 7.5 and a higher Si solubility after liming the soil, those precipitates can transform to more stable Ni phyllosilicates. In the organic soil, Ni speciation was dominated by complexes of Ni with fulvic acid. Liming leads to more stable Ni–organic acid complexes, also shown in a decreased solubility of Ni in the stirred-flow experiment and a reduction of Ni uptake in plants (Everhart et al. 2006; Kukier and Chaney 2001). Ni–organic complexes have been shown to be the major fraction in the short term, while with time and aging of a soil, the formation of more stable Ni-LDH precipitates becomes more important (Shi et al. 2012).

Sequential extractions are commonly used to characterize metal fractionation in soil (Krishnamurti and Naidu 2007). The fractions of a sequential extraction are operationally defined and sample alteration and redistribution of Ni from one fraction to the other cannot be excluded during extraction steps. This should be kept in mind when interpreting such results. Fractions targeted by different extraction solutions, from the most soluble fraction to more recalcitrant fractions, include soluble and exchangeable Ni, carbonate-associated, associated with Mn-oxides, bound to soil organic matter, bound to amorphous Fe-oxides, and occluded into crystalline Fe-oxides and the residual fraction associated with silicates (Tessier et al. 1979). In general, Ni was found to be present mainly in the residual fraction and to a lesser extent associated with crystalline and amorphous Fe-oxides, which has been shown in hydromorphic soils in Germany (Rinklebe and Shaheen 2014), a serpentine soil in Serbia (Antić-Mladenović et al. 2011), and serpentine soils of Taiwan and Austria (Hseu et al. 2017).

Factors controlling Ni availability and partitioning in Ni-rich soils were extensively studied using XRD analyses. The authors showed that Ni was present in primary minerals (such as serpentines, chlorites, and talc), secondary clay minerals, and Fe-Mn oxyhydroxides (Echevarria et al. 2006; Massoura et al. 2006). The mineral fraction contributing to the availability of Ni in ultramafic soil using the isotopic exchange method showed that the exchangeable fraction of Ni in ultramafic soils was correlated with concentrations of amorphous Fe-oxyhydroxides and amorphous Fe-oxides (Massoura et al. 2006). Using synchrotron radiation methods to determine Ni fractionation in ultramafic soils, it was shown that primary serpentine minerals were the most important Ni-bearing mineral fraction. Moreover, Ni was also found to be associated with inosilicate minerals and iron oxides such as goethite, hematite, and ferrihydrite (Siebecker et al. 2017). The characterization of Ni isotope fractionation can give information on processes involved in weathering of Ni-rich parent rock and partitioning of Ni within a soil. Estrade et al. (2015) investigated isotopic fractionation in soils developed on serpentinite bedrock and the degree and control of Ni fractionation during weathering. Weathering of the ultramafic rock resulted in loss of heavy isotopes by leaching processes and enrichment in light Ni isotopes in the soil compared to the bedrock. Moreover, it was shown that the isotopic composition of the labile Ni fraction (DTPA-extractable) was heavier than the Ni in the soil solid phase (Estrade et al. 2015; Ratié et al. 2015). Moreover, the isotopic composition of the plant litter was heavier than in the respective rhizosphere soil, which explains the enrichment of heavier Ni isotopes in the topsoil of an investigated Cambisol compared to the underlying horizons (Estrade et al. 2015).

2.5 Nickel Solubility and Resupply

2.5.1 Nickel Concentrations and Speciation in Soil Solution

Nickel concentrations in field-collected soil solutions from mineral and organic horizons of an Austrian forest region were found to range between <10 and >1000 μg L^{-1} (Wenzel et al. 1997), with the higher concentrations being influenced by Ni dissolution from serpentine. The main inorganic species of Ni in soil solutions include Ni^{2+}, $NiSO_4^0$, and $NiHCO_3^+$ in acidic soils, and $NiCO_3^0$, $NiHCO_3^+$, and Ni^{2+} in alkaline soils (Kirk 2004). Organic Ni complexes can contribute to speciation substantially especially in non-alkaline soils (Kirk 2004). Ni speciation largely determines Ni bioavailability and therefore possibly toxicity. Ni in soil solution is relatively stable in the form of free Ni^{2+}, but can be also associated with organic and inorganic ligands (Uren 1984). In a study of 13 long-term contaminated soils, Nolan et al. (2009) determined Ni^{2+} in pore waters using a Donnan dialysis membrane technique. As a percentage of total Ni in pore water, Ni^{2+} ranged between 21% and 80% (average, 58%). The results indicated that a relevant proportion of total soluble Ni was present in complexed forms. The contribution of Ni bound to the dissolved organic matter (DOM) was estimated to account for ≤17% using the Windermere Humic Aqueous Model VI (WHAM VI). Inorganic complexes of Ni were also calculated using WHAM VI; their contribution to total dissolved Ni ranged between 9.96% and 70.5% for SO_4^{2-}, between 0.01% and 17.7% for Cl$^-$, up to 37.4% for CO_3^{2-}, and up to 13.0% for HCO_3^-. Using isotopic exchange, they determined the contribution of non-labile complexes to the total dissolved Ni to range between 0% and 9%. The labile pool of Ni in the soil solid phase was determined as the resin E value ranged between 0.9% and 32.4%, indicating a relatively small proportion relative to the total Ni concentration in soil (Nolan et al. 2009).

Even though the predicted fraction of Ni bound to DOM in soil may be generally small, Ni–organic complexes can contribute substantially to Ni speciation in rhizosphere soil solutions as plant roots typically increase the quantity and change the quality of DOM in soil solution. When the DOM in the rhizosphere increases, more Ni becomes complexed and this shift from free Ni^{2+} to DOM–Ni complexes can induce solubilization of Ni from the soil solid phase (Wenzel et al. 2003).

2.5.2 Controls of Ni Retention and Resupply from the Soil Solid Phase

Ni sorption in soils is most effective at pH >7.5 and decreases to virtually zero below pH 4.5 (Adriano 2001). In highly weathered tropical soils, Ni sorption envelopes reach the plateau even between pH 5 and 6 while some sorption occurs even below pH 4, and in some soils at pH 2 (Moreira et al. 2008). In aerobic soil environments, Ni, together with Cd, Co, and Zn, belongs to the most soluble, labile metals with average linear ion distribution coefficients ranging from 10^5 (pH 10) to <1 L kg^{-1} (pH 2) (Sauvé et al. 2000). Compared to Pb and Cu, Ni solubility clearly increases more pronouncedly in response to lowering soil pH (Sauvé et al. 2000).

Pérez and Anderson (2009) employed the diffusive gradients in thin films (DGT) technique to assess labile Ni concentrations (c_{DGT}) considered to reflect Ni bioavailability in fertilized field soils. They also used the ratio between a quasi-total Ni fraction (acid digestible Ni) and c_{DGT} as a distribution coefficient to monitor changes in Ni bioavailability in response to fertilization. Distribution coefficients (K_d) of Ni generally increase with pH (Degryse et al. 2009; Staunton 2004), but in some soils, they decrease again in the alkaline range of pH (Staunton 2004). Decreased K_d is indicative of a decrease of the buffer power and resupply from the soil solid phase.

2.5.3 Effect of Root Activities on Ni Solubility, Solution Speciation, and Resupply

Total water-soluble Ni concentrations (solution/soil ratio, 10 mL g^{-1}, <0.5 mm sieve fraction of soil) in the rhizosphere of eight different plant species cultivated for 3 months in pots varied between 10 and 30 µg kg^{-1} (Nguyen et al. 2017). The experimental plants included graminaceous (*Festuca arundinacea*, *Panicum virgatum*, and *Phalaris arundinacea*) as well as herbaceous (*Trifolium repens*) and woody (*Alnus crispa*, *Populus* sp., *Cornus stolonifera*, and *Salix nigra*) dicotyledonous species. Total Ni concentrations in solution linearly increased in response to the increase of DOC from about 4.7 in the unplanted control to 6.7 mg L^{-1} in the rhizosphere of *T. repens*. This clearly indicates an effect of root activities on Ni solubility in the rhizosphere of non-accumulator plants.

In their rhizobox experiment with hyperacccumulator *L. emarginata*, Bartoli et al. (2012) measured steep concentration gradients of Ni in soil solution declining from the bulk soil (100%) toward the root compartment (<25%). In another rhizobox experiment (system of Wenzel et al. 2001) with *B. coddii*, Moradi et al. (2010) imaged (magnetic resonance imaging) Ni concentrations in soil solution perpendicular to the root plane after supplying a Ni^{2+} solution of 10 mg L^{-1}. They found increasing Ni concentrations toward the root plane, indicating Ni accumulation in the rhizosphere. Numerical simulations suggested that the Ni distribution was a result of transpiration-driven advective water flow toward the root plane and Ni diffusion toward the bulk soil. Numerical simulations also showed that at initial Ni concentrations between 1.0 and 0.1 mg L^{-1}, the distribution pattern of Ni in the rhizosphere changed from accumulation to depletion. Further simulations demonstrated that the Ni concentration pattern also switched from accumulation to depletion if the uptake rate coefficient or transpiration rate were decreased (Moradi et al. 2010).

Chemical imaging of Ni fluxes around individual root axes of the Cd/Zn accumulating willow *Salix smithiana* in a rhizotron experiment using DGT and subsequent laser ablation ICP-MS revealed a bimodal distribution of Ni fluxes (= labile Ni), with a narrow zone of Ni depletion close to the root and accumulation in the adjacent zone (Hoefer et al. 2017; Figure 2.1). The authors related this finding to the concurrent decrease in soil pH (imaged using planar optodes) and possibly co-dissolution of Ni associated with Mn-oxides (Hoefer et al. 2017).

Using the same approach, Williams et al. (2014) imaged the Ni distribution around rice roots grown in submerged conditions. For Ni, they show varying patterns, including Ni depletion next to the root and accumulation in the adjacent zone, or a similar pattern but with an additional tiny zone of Ni flux increase at the soil–root interface. Generally, the distribution of Ni and other metals was found to be controlled by the complex interaction of oxygen release to rice root aerenchyma and subsequent interacting changes in redox, pH, and Fe speciation (Williams et al. 2014).

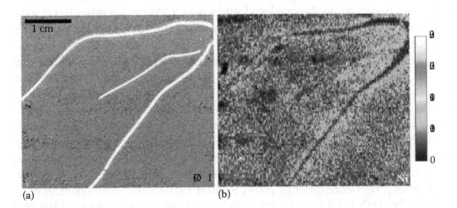

(a)   (b)

FIGURE 2.1
(See color insert.) Photograph of a willow (*S. smithiana*) root (a) and related nickel depletion zone chemically imaged by combining diffusive gradients in thin films (DGT) deployment with subsequent laser ablation (LA)–ICP-MS. The scale is showing nickel fluxes in pg m^{-2} s^{-2} (b). (Modified with permission from Hoefer C., J. Santner, S. M. Borisov et al. 2017. Integrating chemical imaging of cationic trace metal solutes and pH into a single hydrogel layer. *Analytica Chimica Acta* 950:88–97.)

FIGURE 2.2
Nickel hyperaccumulator *N. goesingensis* (formerly *T. goesingense* Halaczy) growing on a serpentine site near Redlschlag, Eastern Austria. (Photos by Walter W. Wenzel.)

In a field study on the rhizosphere characteristics of Ni hyperaccumulator *Noccaea goesingensis* (syn. *Thlaspi goesingense*; Figure 2.2) and the non-accumulator species *S. vulgaris* and *R. acetosella* natively growing on serpentine soil, Wenzel et al. (2003) found increased total Ni concentrations in water extracts of the root zones compared with those of bulk soils while the labile (NH_4NO_3-exchangeable) Ni fraction decreased. The increase in soluble Ni was most pronounced in the root zone of the hyperaccumulator. The observed trend of soluble Ni (hyperaccumulator > non-accumulators > bulk soil) was associated with a similar pattern of DOC, with a factor of approximately 3 between the hyperaccumulator root zone and the bulk soil for both. Biogeochemical modeling suggested that the observed increase of soluble Ni in rhizosphere soils was closely linked with the formation of dissolved organic Ni complexes (Figure 2.3). Further, close correlations between Ni on the one hand and Ca and Mg on the other hand indicated co-dissolution of Ni from Ni-bearing minerals such as forsterite, likely enhanced by the presence of organic ligands in the rhizosphere.

The results of the field study were confirmed in a rhizobox experiment with *N. goesingensis* grown on serpentine soil from the same site except that the correlation between soluble Ni and DOC was rather weak in the latter (Puschenreiter et al. 2005; Figure 2.4). While ion competition between Ni and other cations (Ca, Mg) could be ruled out as explanation for the increased soluble Ni in the rhizosphere, the authors observed decreased Vanselow coefficients of Ni toward the root plane, indicating decreased selectivity for Ni in the rhizosphere, possibly related to sorption of oxalate released by *N. goesingensis* roots (Puschenreiter et al. 2005). Similar to the field study (Wenzel et al. 2003), the labile (NH_4NO_3-exchangeable) Ni fraction decreased toward the root plane, indicating depletion due to excessive uptake by the hyperaccumulator (Puschenreiter et al. 2005). The opposing trends of the labile and soluble fractions indicate decreased distribution coefficients near the root compared to the bulk soil. This is consistent with the commonly observed decrease of distribution coefficients in 13 soils with varied properties but Ni concentrations in the normal range upon the addition of organic ligands (Staunton 2004).

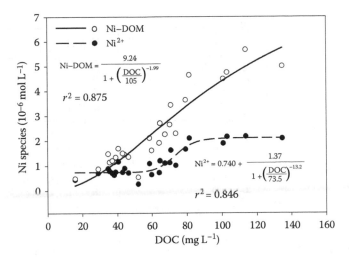

FIGURE 2.3
Nickel speciation modeled using MINTEQ in water extracts from rhizosphere soils of nickel hyperaccumulator *N. goesingensis* formerly *T. goesingense*) and two non-accumulator species (*S. vulgaris* and *R. acetosella*). The soils were collected from field-grown specimens. The graph shows increased total nickel concentrations in solution in response to increased DOC due to the formation of organic nickel complexes. The largest DOC and soluble nickel concentrations were observed in rhizospheres of the hyperaccumulator, the lowest in bulk soils, with intermediate values for the non-accumulator species. (With permission from Wenzel, W. W., M. Bunkowski, M. Puschenreiter et al. 2003. Rhizosphere characteristics of indigenously growing nickel hyperaccumulator and excluder plants on serpentine soil. *Environ. Pollut.* 123:131–138.)

The bulk of information presented in this section suggests that diffusion and resupply from soil are generally not limiting Ni uptake in non-accumulator plants species. However, chemical imaging recently showed that narrow depletion zones of DGT-labile Ni can be observed along the root axes of rice and a willow, both not known to accumulate Ni in plant tissues above normal concentrations. It appears that the formation of Ni concentration gradients in the rhizosphere can be more complex (bimodal) than initially expected. In the rhizosphere of Ni hyperaccumulator plants, labile Ni fractions are typically decreased while it has been repeatedly observed that soluble Ni is concomitantly increased. There is evidence for root-induced modification of Ni speciation in soil solutions (increased complexation by DOM), changes of the exchange complex (occupation of sorption sites by DOM and corresponding loss of selectivity for Ni sorption), and enhanced resupply of Ni from mineral phases, probably triggered by organic compounds released by plant roots and associated microorganisms. The observed decrease in labile Ni is consistent with other evidence suggesting active and diffusion-limited uptake by Ni hyperaccumulators. Modifications of Ni pore water concentrations and speciation in the rhizosphere of such plants can be considered as response to the limited resupply.

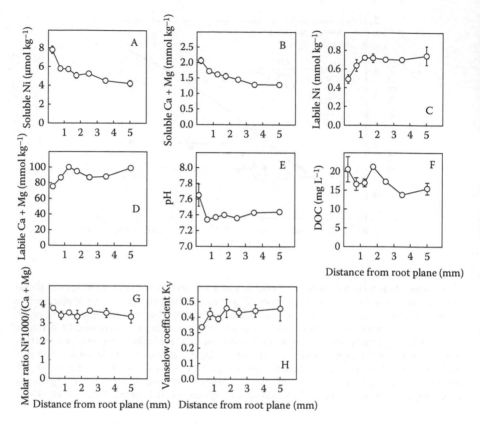

FIGURE 2.4
Gradients of soluble and labile nickel and other properties across the rhizosphere of *N. goesingensis* (formerly *T. goesingense*) grown in a rhizobox experiment. The plots show depletion of labile (exchangeable) nickel in the rhizosphere whereas soluble nickel increases toward the root plane. This apparent contradiction is explained by the decreased selectivity of the exchange complex for nickel as indicated by the decreased Vanselow coefficients calculated from calcium, magnesium, and nickel concentrations in solution and in the labile (exchangeable) fraction. (Modified with permission from Puschenreiter, M., A. Schnepf, I. M. Millán et al. 2005. Changes of Ni biogeochemistry in the rhizosphere of the hyperaccumulator *Thlaspi goesingense*. *Plant Soil* 271:205–218.)

2.6 Microbial Interactions

Ultramafic (serpentine) soils represent model environments for studying interactions between plant species and their associated microorganisms and Ni biogeochemistry. The extreme environments of serpentine soils (high concentrations of Co, Cr, and particularly Ni, nutrient deficiency and adverse Ca:Mg quotient, lack of soil structure and water retention capacity, etc.) have often been attributed the term "serpentine syndrome" and associated with the development of the particular plant communities (metallophytes) inhabiting these environments. However, like plants, these soils also host unique soil microbial communities. Early microbiological studies, based on traditional culture-dependent methods, led to the common belief that the physicochemical conditions of serpentine soils, and particularly the toxic metal concentrations, significantly reduced soil microbial density, biomass, and activity (Acea and Carballas 1986; DeGrood et al. 2005; Lipman 1926; Pal et al. 2005). Indeed, long-term exposure to elevated concentrations of trace metals has been shown to adversely affect soil biodiversity and functionality in soils contaminated through anthropogenic activities (Lorenz et al. 2006; Renella et al. 2006). However, more recent molecular-based studies have demonstrated that naturally metal-enriched soils, such as Ni-rich serpentine soils, host a diverse microbial community (Bordez et al. 2016; Gourmelon et al. 2016; Lopez et al. 2017; Mengoni et al. 2004; Oline 2006). Moreover, the plant metallophytes growing in serpentine areas strongly influence the soil microorganisms in their surroundings and have been shown to host plant species-specific microbial communities (as a result of differences in the composition of root exudates, secretions, and lysates or in litter quality). Excess concentrations of trace metals typically induce a shift in species composition and lead to the selection of metal-tolerant bacteria (Becerra-Castro et al. 2012; Giller et al. 1998; Mertens et al. 2006; Pereira and Castro 2014). Bacterial strains isolated from serpentine soils have been found to tolerate Ni concentrations in their culture medium up to one order of magnitude higher than strains from other soil types (Schlegel et al. 1991; Turgay et al. 2012). Furthermore, nickel hyperaccumulating plant species growing in these soils have been shown to select for Ni-tolerant bacterial strains in their rhizosphere. The critical role played by soil microbes, including bacteria, archaea, and fungi, in soil nutrient cycling is well recognized (Philippot et al. 2013), but more recent studies indicate that the microbial communities associated with hyperaccumulators may also play a role in plant metal resistance and the hyperaccumulation process itself (Glick 2010; Lebeau et al. 2008; Sessitsch et al. 2013). Plant-induced selective enrichment of Ni-tolerant bacteria in the rhizosphere has been documented in the Ni-hyperaccumulating tree *Sebertia* (= *Pycnandra*) *acuminata* or members of the Mediterranean hyperaccumulators of the genus *Alyssum* (= *Odontarrhena*), such as *A. bertolonii* and *A. serpyllifolium* (Álvarez-López et al. 2016a; Becerra-Castro et al. 2009; Mengoni et al. 2001). Becerra-Castro et al. (2009) found that the densities of Ni-tolerant bacteria in the rhizosphere were positively correlated with soil Ni availability. These Ni-tolerant bacterial isolates often present plant growth-promoting traits and are considered as promising biotechnological tools to enhance growth of plants in both naturally metal-rich or metal-contaminated soils (see Section 2.7).

Rhizosphere bacterial communities associated with these plants differ significantly from those of non-accumulating plant species growing in the same soil or from nonvegetated (bulk) soil. Moreover, differences observed in the composition of the rhizosphere bacterial community have been shown to be population-specific in the case of some hyperaccumulators, such as that observed among different populations of *A. serpyllifolium* across the Iberian Peninsula (Álvarez-López et al. 2016a). Culturable rhizobacterial communities associated with temperate Ni hyperaccumulators (including *A. murale*, *A. bertolonii*, and *A. serpyllifolium*) have commonly shown a dominance of bacteria belonging to the phyla *Proteobacteria* and, in particular, *Actinobacteria* (Abou-Shanab et al. 2010; Álvarez-López et al. 2016a; Idris et al. 2004; Mengoni et al. 2001). The predominance of actinobacterial strains has been related to the high adaptability of such Gram-positive bacteria to toxic concentrations of trace metals (DeGrood et al. 2005). Such a plant-driven microorganism selection has also been observed in the phenotype of the rhizobacterial strains isolated from either hyperaccumulators or non-accumulating plants. For example, Álvarez-López et al. (2016a) observed a greater number of P-solubilizing and indoleacetic acid (IAA)-producing bacterial strains associated with the hyperaccumulator *A. serpyllifolium*, while siderophore producers were more common to the Ni excluder *Dactylis glomerata*.

The growing interest in the hyperaccumulator plant microbiome has led to a move from studies based on culture-dependent techniques toward more novel metagenomic approaches (Benizri and Kidd 2018; Thijs et al. 2017). These techniques provide a more in-depth characterization of the phylogenetic structure of soil microbial communities (considering that culture-dependent techniques reflect <1% of living microorganisms present) and can enhance our understanding of the diversity and ecology of microorganisms and their interactions with hyperaccumulating plants and edaphic conditions. A few pioneer studies can be found in which metagenomic techniques have been used to assess soil microbial diversity under different plant covers in ultramafic soils of both temperate and tropical or subtropical regions. Lopez et al. (2017) used a high-throughput sequencing technique (454-pyrosequencing) to relate changes in the rhizosphere bacterial community of *A. murale* to edaphic factors (altitude and Ni bioavailability). Although *Chloroflexi* was the major phylum present (with >50% relative abundance), its abundance decreased with increasing soil Ni availability, while the relative abundance of *Proteobacteria* (particularly *Alphaproteobacteria*) and *Actinobacteria* were found to increase. Multivariate statistical analyses associated the phyla *Actinobacteria, Gemmatimonadetes, Acidobacteria, Bacteroidetes,* and *Proteobacteria* with the highest concentrations of available (DTPA-extractable) and exchangeable Ni fractions. Touceda-González (2017), using the same approach, characterized the rhizobacterial communities associated with two populations of the Ni hyperaccumulator *A. sepryllifolium* endemic to the Iberian Peninsula and compared these with nonvegetated soil at the same sites. All the soils were dominated by *Actinobacteria, Proteobacteria, Acidobacteria, Bacteroidetes, Gemmatimonadetes,* and *Firmicutes*. The relative abundance of *Actinobacteria* and *Gemmatimonadetes* phyla in the ultramafic soils was higher than that usually found in other soils of similar characteristics but not enriched in trace metals, thus corroborating the dominance of

Actinobacteria previously found in ultramafic soils using culture-dependent techniques. Significant differences were found in the abundance of bacterial taxa (at all taxonomic levels) between the hyperaccumulators, and among hyperaccumulators and nonvegetated soil. At the family and genus levels, the rhizosphere of both populations of *Alyssum* was enriched in members of *Blastococcus* (Geodermatophilaceae), *Pseudonocardia* (*Actinobacteria*), and *Methylobacterium*, and of the Rhodobacteraceae family (*Alphaproteobacteria*). *Blastococcus* strains are among the most common bacteria associated with stone biodeterioration (Urzi et al. 2001) and have been described as metal(loid)-resistant (Gtari et al. 2012). Touceda-González (2017) suggested that the relative enrichment of *Blastococcus* strains in the rhizosphere of *Alyssum* could reflect their role in mineral alteration and nutrient and Ni mobilization at the root–soil interface of this hyperaccumulator.

Gourmelon et al. (2016) assessed both soil bacterial and fungal diversity associated with four different plant formations (shrubland to rainforest) in two ultramafic complexes (Kopéto Massif and Massif du Grand Sud) of New Caledonia using high-throughput sequencing (Illumina MiSeq). Bacterial and fungal communities were both structured by the type of plant formation and were significantly related to the dominant plant species (e.g., the forest pioneer tree *Nothofagus aequilateralis*). The bacterial communities were dominated by *Proteobacteria, Planctomycetes, Acidobacteria,* and *Actinobacteria*, which (with the exception of *Planctomycetes*) coincides with other studies carried out in New Caledonian ultramafic soils (Bordez et al. 2016). On the other hand, *Basidiomycota* were the most dominant phyla in fungal communities associated with rainforests. However, in these studies, the authors did not find soil Ni availability to be a significant soil property influencing the microbial community composition.

2.7 Applications and Management of Nickel Biogeochemistry in the Rhizosphere in Phytoremediation

The typical background concentration of nickel in soils is 13–37 mg kg^{-1} (Kabata-Pendias 2011). For arable soils, a maximum permissible limit of 50 mg kg^{-1} was suggested (Herselman et al. 2005). In naturally Ni-enriched soils but also in anthropogenically contaminated soils, Ni concentrations may exceed 1000 mg kg^{-1} (Kabata-Pendias 2011). Natural Ni enrichment in soils is typically found in serpentine soils that developed on ultramafic bedrocks. On these soils, which typically contain 1400–2000 mg Ni kg^{-1}, adapted vegetation and microflora have been evolved; thus, no remediation attempts are required for those naturally Ni-enriched soils. In contrast, anthropogenic Ni contamination in soils may occur near metal smelters or close to the galvanic/plating industry. On these soils, high Ni transfer into plants may cause plant toxicity and contamination of the human food chain and thus may require efforts for reducing Ni mobility or reducing Ni content in soil.

Phytoremediation comprises a set of soil remediation technologies, which are based on the use of plants and associated microbes and, in some cases, the application of soil conditioners as well. Phytoremediation technologies have been developed for the treatment of soils polluted by organic contaminants (e.g., petroleum hydrocarbons, PAHs, etc.) or by metals and/or metalloids. For the latter, phytoextraction (removal of contaminants), phytostabilization (immobilization of contaminants), or phytovolatilization (transformation and release into the atmosphere) may be applied (Chaney et al. 2010).

Ni hyperaccumulating plants may be used for the phytoextraction of Ni on polluted soils or for phytomining of Ni on serpentine soils. Technologically, phytomining and phytoextraction are the same, but they differ in their target: phytoextraction is typically referred to when targeting the removal of contaminants from the growth substrate (i.e., soil, water), whereas phytomining refers to the recovery of the accumulated metal. Over the last decade, the concept of microbial-assisted phytoremediation (MAP) has arisen, in which microbes associated with metal-tolerant and (hyper)accumulating plants can be used to enhance plant growth, health, and stress resistance (Thijs et al. 2016). Plant-associated microorganisms that have received much attention for these applications are the plant growth-promoting bacteria (PGPB) and mycorrhizal and endophytic fungi (Coninx et al. 2017; Mendes et al. 2013).

Li et al. (2003b) tested the Ni phytoextraction efficiency of *Alyssum murale* and *A. corsicum* and found similar phytoextraction efficiency. If perennial plants are used, the question arises which harvesting time is optimal for achieving the largest phytoextraction efficiency. Adamidis et al. (2017) reported that for *Alyssum lesbiacum*, harvesting in the third or fourth year would reveal the highest Ni removal rates. However, for the most successful Ni phytoextraction crop, that is, *A. murale*, mostly an annual harvest has been implemented (Bani et al. 2015).

As for conventional agriculture, agronomic measures may significantly influence the growth and metal accumulation rates and thus the phytoextraction efficiency (Kidd et al. 2015). Bani et al. (2015) showed that fertilization and herbicide application can dramatically increase the biomass and the quantity of phytoextracted nickel. The same effect was observed upon compost application (Álvarez-López et al. 2016b). Also, inoculation with plant-associated microbes has been shown to increase growth and shoot Ni concentration and, consequently, the total amount of nickel in the harvested biomass (described further below). Furthermore, the selection of high-performance populations and cultivars may provide more efficient plants (Li et al. 2003a). Another agronomic measure is the use of compounds that increase the metal bioavailability, that is, enhanced phytoextraction. In this approach, plants with lower Ni accumulation efficiency, but higher biomass production, are typically used. Metal-mobilizing compounds are applied to the metal-rich soil shortly before the phytoextraction crops are harvested for increasing the metal bioavailability in soil and thus the accumulation in shoots. Hsiao et al. (2007) tested the efficiency of DTPA, EDTA, and low-molecular-weight organic acids on Ni phytoextraction in *Brassica juncea*. Although an increase of Ni concentration in shoot biomass was found upon EDTA and DTPA application, the biomass and thus the amount of phytoextracted Ni were decreased. Despite the potential increase of the phytoextraction efficiency, the application of metal-mobilizing compounds may cause secondary problems such as groundwater pollution (Nowack et al. 2006).

Phytomining of Ni has gained increasing interest in the last 20 years. Pioneer field work on Ni phytomining has been carried out 1995 in Oregon, USA (Li et al. 2003a), and later in Albania (Bani et al. 2015). Nickel phytomining is based on the use of Ni hyperaccumulating plants for recovering Ni that accumulated in the shoot biomass. Up to now, mostly *A. murale* has been used; further tested species were *A. corsicum*, *Leoptoplax emarginata*, and *Bornmuellera tymphaea*. In field experiments in Albania, 0.3–9 t/ha of *A. murale* biomass was harvested, resulting in a maximum Ni yield of 105 kg Ni ha^{-1} (Bani et al. 2015). The harvested shoot biomass could be used for recovering Ni in its metallic form (Barbaroux et al. 2011) or as a Ni ammonium sulfate salt (Barbaroux et al. 2012). To date, phytomining has mostly been tested and developed on serpentine soils. Metal phytomining has also been tested on waste incineration ashes, but its efficiency was strongly limited due to the strongly limited growth of hyperaccumulator plants on these substrates and due to limited metal availability (Rosenkranz et al. 2017). A detailed overview on different aspects of phytomining can be found in a recently published book (van der Ent et al. 2018).

Since the early 2000s, numerous authors have focused on the identification of useful bioinoculants for application in phytoremediation of metal-contaminated soils (Benizri and Kidd 2018; Kidd et al. 2017; Sessitsch et al. 2013; Thijs et al. 2016). In phytoextraction or phytomining scenarios, MAP aims to enhance metal removal by inoculating metal (hyper)accumulating plants with plant-associated microbes, which are able to increase the metal accumulation capacity of the phytoextracting plants, by (i) improving plant biomass and thus metal yield and/or (ii) increasing metal bioavailability in the soil and thus plant uptake and bioaccumulation. Microorganisms can increase the availability of essential plant nutrients, such as nitrogen (N_2-fixing organisms), phosphorus (by solubilization or mineralization through the production of organic acids and/or phosphatases), or iron [by releasing Fe(III)-specific chelating agents or siderophores] (Glick 2010). PGPB can also directly influence plant growth and physiology through the production of phytohormones (e.g., IAA) or by reducing stress ethylene levels in plants through the production of the enzyme 1-aminocyclopropane-1-carboxylate deaminase. Some bacteria act as biocontrol agents, inhibiting or reducing plant diseases indirectly by competing for nutrients and space (niche exclusion), producing antimicrobial compounds or through the induction of plant defense mechanisms (Compant et al. 2005). Rhizobacterial strains with the capacity to mobilize metals, such as Ni, from less labile soil fractions are of great interest since they contribute toward efficient replenishment of soil labile metal pools, leading to enhanced plant metal uptake and an overall improvement in the efficiency of the phytoextraction process. Several rhizobacterial strains associated with the hyperaccumulators *A. murale* and *A. sepryllifolium* have been described in the literature as efficient Ni mobilizers (Abou-Shanab et al. 2006; Becerra-Castro et al. 2013). These were successfully applied as bacterial inoculants to (hyper)accumulating plant species at a bench scale, improving both the plant growth and Ni uptake and accumulation. Stimulation in plant biomass is the most common outcome of the inoculation process, while enhanced Ni uptake is less documented (Sessitsch et al. 2013). Much research has been dedicated to evaluating the plant species-specific nature of these plant–microbial associations (Durand et al. 2016; Ghasemi et al. 2018). The expected promotion in plant growth or Ni accumulation is not always observed, and plant response has been found to vary according to the soil type and properties, and abiotic factors (Benizri and Kidd 2018; Cabello-Conejo et al. 2014; Ghasemi et al. 2018). Inoculation methods differ greatly among studies, and few studies have been able to successfully monitor the proliferation and survival of the inoculant used. The survival of introduced inoculants can be challenged by the strong competition encountered from the soil indigenous microbes (Thijs et al. 2016). Although promising results can be found in the literature, further studies are required to fully elucidate the complexity of these plant–microbial associations and molecular techniques will be required to enable efficient tracking of the introduced inoculant(s) in terms of their abundance and viability. Microorganisms possess various mechanisms of metal resistance, such as exclusion by intra- and extracellular sequestration by chelating compounds, sorption by structural components of the cell envelope; active efflux transport systems that excrete toxic or overconcentrated metals from the cell, or enzymatic transformation and detoxification (Ndeddy Aka and Babalola 2017). These mechanisms appear to be plasmid mediated and are highly specific to a particular cation or anion. The introduction of PGP microbes with appropriate metal-tolerant and detoxification systems on plasmids has been proposed as a potential strategy to adapt the local microbial community for metal tolerance, which in turn may result in increased metal phytoextraction (Thijs et al. 2017). When discussing microbial-assisted remediation of soils contaminated with organic pollutants, Garbisu et al. (2017) differentiated cell bioaugmentation, which relies on the survival and growth of the inoculated strains, from genetic bioaugmentation, which is based on the spread of catabolic genes, located in mobile genetic elements, into native microbial populations. Thus, in the case of metal-tolerant PGP microbial inoculants, the genes encoding metal resistance mechanisms that are found on plasmids could be spread by horizontal gene transfer into the established, indigenous soil bacterial community.

Plant-associated microbes clearly play an important role in soil metal bio-geochemistry and in plant metal accumulation. However, before their use as bioinoculants can be fully exploited, it will be necessary to also assess their potential benefits on a field scale.

2.8 Conclusions

Nickel biogeochemistry and microbial communities and their activities vary largely among the rhizospheres of non-accumulator and (hyper-) accumulator plant species. This offers opportunities to employ and manage rhizosphere features in biotechnology and the management and remediation of polluted soils by selecting appropriate plant–microbial associations. The differential uptake strategies are generally associated with opposing pattern of root activities and their effect on Ni bioavailability. High uptake rates commonly result in depletion of labile Ni while Ni solubility is often maintained by accelerated root activities, in particular release of organic ligands and subsequent complexation and solubilization of Ni. Such plants and their rhizospheres can be explored for phytoextraction of Ni-polluted soils and phytomining technologies, even though the latter is currently limited by poor plan growth on waste materials with adverse growth conditions. Selection of Ni excluder crops and cultivars on the other hand can serve as a means to minimize pollutant transfer into the food chain in agriculture and forestry. Highly tolerant excluder species can be used in phytostabilization of Ni-polluted areas. Specific microbial strains and communities are candidates for either enhancing Ni solubility and uptake or immobilization of Ni in the abovementioned technological applications.

References

Abou-Shanab, R., J. Angle, and R. Chaney. 2006. Bacterial inoculants affecting nickel uptake by *Alyssum murale* from low, moderate and high Ni soils. *Soil Biol. Biochem.* 38:2882–2889.

Abou-Shanab, R., P. van Berkum, J. S. Angle et al. 2010. Characterization of Ni-resistant bacteria in the rhizosphere of the hyperaccumulator *Alyssum murale* by 16S rRNA gene sequence analysis. *World J. Microbiol. Biotechnol.* 26:101–108.

Acea, M. and T. Carballas. 1986. Estudio de la población microbiana de diversos tipos de suelos de zona húmeda (N. O. de España). *An. Edafol. Agrobiol.* 45:381–398.

Adamidis, G. C., M. Aloupi, P. Mastoras et al. 2017. Is annual or perennial harvesting more efficient in Ni phytoextraction? *Plant Soil* 418:205–218.

Adriano, D. C. 2001. *Trace Elements in Terrestrial Environments, Biogeochemistry, Bioavailiability, and Risk of Metals.* 2nd ed. New York: Springer-Verlag.

Agnello, A. C., D. Huguenot, E. D. Van Hullebusch et al. 2014. Enhanced phytoremediation: A review of low molecular weight organic acids and surfactants used as amendments. *Crit. Rev. Environ. Sci. Technol.* 44:2531–2576.

Álvarez-López, V., A. Prieto-Fernández, C. Becerra-Castro et al. 2016a. Rhizobacterial communities associated with the flora of three serpentine outcrops of the Iberian Peninsula. *Plant Soil* 403:233–252.

Álvarez-López, V., A. Prieto-Fernández, and X. Cabello-Conejo. 2016b. Organic amendments for improving biomass production and metal yield of Ni hyperaccumulating plants. *Sci. Total Environ.* 548–549:370–379.

Antić-Mladenović, S., Rinklebe, J., Frohne, T. et al. 2011. Impact of controlled redox conditions on nickel in a serpentine soil. *J. Soils Sediments* 11:406–415.

Bai, C., C. C. Reilly, and B. W. Wood. 2006. Nickel deficiency disrupts metabolism of ureides, amino acids, and organic acids of young pecan foliage. *Plant Physiol.* 140:433–443.

Baker, A.J.M., S.N. Whiting. 2002. In search of the holy grail—A further step in understanding metal hyperaccumulation. *New Phytologist* 155:1– 4.

Bani, A., G. Echevarria, S. Sulçe et al. 2015. Improving the agronomy of *Alyssum murale* for extensive phytomining: A five-year field study. *Int. J. Phytoremediation* 17:117–127.

Barbaroux, R., G. Mercier, J. F. Blais et al. 2011. A new method for obtaining nickel metal from the hyperaccumulator plant *Alyssum murale*. *Sep. Purif. Technol.* 83:57–65.

Barbaroux, R., E. Plasari, G. Mercier et al. 2012. A new process for nickel ammonium disulfate production from ash of the hyperaccumulating plant *Alyssum murale*. *Sci. Total Environ.* 423:111–119.

Bartoli, F., D. Coinchelin, C. Robin et al. 2012. Impact of active transport and transpiration on nickel and cadmium accumulation in the leaves of the Ni-hyperaccumulator *Leptoplax emarginata*: A biophysical approach. *Plant Soil* 350:99–115.

Baumeister, J. L., E. M. Hausrath, A. A. Olsen et al. 2015. Biogeochemical weathering of serpentinites: An examination of incipient dissolution affecting serpentine soil formation. *Appl. Geochem.* 54:74–84.

Becerra-Castro, C., P. Kidd, M. Kuffner et al. 2013. Bacterially induced weathering of ultramafic rock and its implications for phytoextraction. *Appl. Environ. Microbiol.* 79:5094–5103.

Becerra-Castro, C., C. Monterroso, M. García-Lestón et al. 2009. Rhizosphere microbial densities and trace metal tolerance of the nickel hyperaccumulator *Alyssum serpyllifolium* subsp. *lusitanicum*. *Int. J. Phytoremediation* 11:525–541.

Becerra-Castro, C., C. Monterroso, A. Prieto-Fernández et al. 2012. Pseudometallophytes colonizing Pb/Zn mine tailings: A description of the plant–microorganism–rhizosphere soil system and isolation of metal-tolerant bacteria *J. Hazard Mater.* 217–218:350–359.

Becerra-Castro, C., A. Prieto-Fernández, V. Alvarez-Lopez et al. 2011. Nickel solubilizing capacity and characterization of rhizobacteria isolated from hyperaccumulating and non-hyperaccumulating subspecies of *Alyssum serpyllifolium*. *Int. J. Phytoremediation* 13:229–244.

Benizri, E. and P. S. Kidd. 2018. The role of the rhizosphere and microbes associated with hyperaccumulator plants in metal accumulation. In: *Agromining: Farming for Metals. Extracting Unconventional Resources Using Plant*, eds. Van der Ent, A., G. Echevarria, A. J. M. Baker et al., Cham, Switzerland: Springer International Publishing.

Bernal, M. P., S. P. McGrath, A. J. Miller et al. 1994. Comparison of the chemical changes in the rhizosphere of the nickel hyperaccumulator *Alyssum murale* with the non-accumulator *Raphanus sativus*. *Plant Soil* 164:251–259.

Bordez, L., P. Jourand, M. Ducousso et al. 2016. Distribution patterns of microbial communities in ultramafic landscape: A metagenetic approach highlights the strong relationships between diversity and environmental traits. *Mol. Ecol.* 25:2258–2272.

Broadley, M., P. H. Brown, I. Cakmak et al. 2012. Function of nutrients: Micronutrients. In: *Marschner's Mineral Nutrition of Higher Plants*, 3rd ed., ed. Marschner, P., pp. 191–248, Amsterdam, NL: Elsevier.

Brooks, R. R. 1998. Geobotany and hyperaccumulators. In: *Plants that Hyperaccumulate Heavy Metals*, ed. R. R. Brooks, pp. 55–94, Oxon: CAB International.

Buée, M., W. De Boer, F. Martin et al. 2009. The rhizosphere zoo: An overview of plant-associated communities of microorganisms, including phages, bacteria, archaea, and fungi, and of some of their structuring factors. *Plant Soil* 321:189–212.

Cabello-Conejo, M. I., C. Becerra-Castro, A. Prieto-Fernández et al. 2014. Rhizobacterial inoculants can improve nickel phytoextraction by the hyperaccumulator *Alyssum pintodasilvae*. *Plant Soil* 379:35–50.

Chaney, R., C. L. Broadhurst, and T. Centofanti. 2010. Phytoremediation of soil trace elements. In: *Trace Elements in Soil*, ed. Hooda, P. S., Chichester, UK: Wiley.

Chardot-Jacques, V., C. Calvaruso, B. Simon et al. 2013. Chrysotile dissolution in the rhizosphere of the nickel hyperaccumulator *Leptoplax emarginata*. *Environ. Sci. Technol.* 47:2612–2620.

Clemens, S., M. G. Palmgren, and U. Krämer. 2002. A long way ahead: Understanding and engineering plant metal accumulation. *Trends Plant Sci.* 7:309–315.

Compant, S., B. Duffy, J. Nowak et al. 2005. Use of plant growth-promoting bacteria for biocontrol of plant diseases: Principles, mechanisms of action, and future prospects. *Appl. Environ. Microbiol.* 71:4951–4959.

Coninx, L., V. Martinova, and F. Rineau. 2017. Mycorrhiza-assisted phytoremediation. In: *Phytoremediation*, ed. Cuypers, A. and J. Vangronsveld, London: Academic Press.

DeGrood, S. H., V. P. Claassen, and K. M. Scow. 2005. Microbial community composition on native and drastically disturbed serpentine soils. *Soil Biol. Biochem.* 37:1427–1435.

Degryse, F., E. Smolders, and D. R. Parker. 2009. Partitioning of metals (Cd, Co, Cu, Ni, Pb, Zn) in soils: Concepts, methodologies, prediction and applications—A review. *Eur. J. Soil Sci.* 60:590–612.

Doornbos, R. F., L. C. Van Loon, and P. A. H. M. Bakker. 2012. Impact of root exudates and plant defense signaling on bacterial communities in the rhizosphere. A review. *Agron. Sustain. Dev.* 32:227–243.

Durand, A., S. Piutti, M. Rue et al. 2016. Improving nickel phytoextraction by co-cropping hyperaccumulator plants inoculated by plant growth promoting rhizobacteria. *Plant Soil* 399:179–192.

Echevarria, G., S. T. Massoura, T. Sterckeman et al. 2006. Assessment and control of the bioavailability of nickel in soils. *Environ. Toxicol. Chem.* 25:643.

Estrade, N., C. Cloquet, G. Echevarria et al. 2015. Weathering and vegetation controls on nickel isotope fractionation in surface ultramafic environments (Albania). *Earth Planet. Sci. Lett.* 423:24–35.

Everhart, J. L., D. McNear, E. Peltier et al. 2006. Assessing nickel bioavailability in smelter-contaminated soils. *Sci. Total Environ.* 367:732–744.

Freyermuth, S. K., M. Bacanamwo, J. C. Polacco. 2000. The soybean *Eu3* gene encodes an Ni-binding protein necessary for urease activity. *Plant J.* 21:53–60.

Garbisu, C., O. Garaiyurrebaso, L. Epelde et al. 2017. Plasmid-mediated bioaugmentation for the bioremediation of contaminated soils. *Front. Microbiol.* 8:1966.

Ghasemi, Z., S. M. Ghaderian, A. Prieto-Fernández et al. 2018. Plant species-specificity and effects of bioinoculants and fertilization on plant performance for nickel phytomining. *Plant Soil* (in press).

Giller, K. E., E. Witter, and S. P. McGrath. 1998. Toxicity of heavy metals to microorganisms and microbial processes in agricultural soils: A review. *Soil Biol. Biochem.* 30:1389–1414.

Glick, B. R. 2010. Using soil bacteria to facilitate phytoremediation. *Biotechnol. Adv.* 28:367–374.

Gonnelli, C. and G. Renella. 2013. Chromium and nickel. In: *Heavy Metals in Soils*, 3rd ed., ed. Alloway, B. J., Dordrecht, Heidelberg, New York, London: Springer.

Gourmelon, V., L. Maggia, J. R. Powell et al. 2016. Environmental and geographical factors of structure and soil microbial diversity in New Caledonian ultramafic substrates: A metagenomic approach. *PLOS ONE* 11.

Gtari, M., I. Essoussi, R. Maaoui et al. 2012. Contrasted resistance of stone-dwelling *Geodermatophilaceae* species to stresses known to give rise to reactive oxygen species. *FEMS Microbiol. Ecol.* 80:566–577.

Herselman, J. E., C. E. Steyn, and M. V. Fey. 2005. Baseline concentration of Cd, Co, Cr, Cu, Pb, Ni and Zn in surface soils of South Africa. *South African J. Sci.* 101:509–512.

Hinsinger, P. and F. Courchesne. 2008. Biogeochemistry of metals and metalloids at the soil–root interface. In: *Biophysico-Chemical Processes of Heavy Metals and Metalloids in Soil Environments,* eds. Violante, A., P. M. Huang, and G. M. Gadd, pp. 267–310, New York: John Wiley & Sons.

Hinsinger, P., G. R. Gobran, P. J. Gregory et al. 2005. Rhizosphere geometry and heterogeneity arising from root-mediated physical and chemical processes. *New Phytol.* 168:293–303.

Hinsinger, P., C. Plassard, C. Tang et al. 2003. Origins of root-mediated pH changes in the rhizosphere and their responses to environmental constraints: A review. *Plant Soil* 248:43–59.

Hirsch, P. R., A. J. Miller, and P. G. Dennis. 2013. Do root exudates exert more influence on rhizosphere bacterial community structure than other rhizodeposits? *Mol. Microb. Ecol. Rhizosph.* 1:229–242.

Hoefer, C., J. Santner, S. M. Borisov et al. 2017. Integrating chemical imaging of cationic trace metal solutes and pH into a single hydrogel layer. *Anal. Chim. Acta* 950:88–97.

Hseu, Z.-Y., Y. C. Su, F. Zehetner et al. 2017. Leaching potential of geogenic nickel in serpentine soils from Taiwan and Austria. *J. Environ. Manage.* 186:151–157.

Hsiao, K.-H., P.-H. Kao, and Z.-Y. Hseu. 2007. Effects of chelators on chromium and nickel uptake by *Brassica juncea* on serpentine-mine tailings for phytoextraction. *J. Hazard. Mater.* 148:366–376.

Idris, R., R. Trifonova, M. Puschenreiter et al. 2004. Bacterial communities associated with flowering plants of the Ni hyperaccumulator *Thlaspi goesingense*. *Appl. Environ. Microbiol.* 70:2667–2677.

Jaffré, T., R. R. Brooks, J. Lee et al. 1976. *Sebertia acuminata*: A hyperaccumulator of nickel from New Caledonia. *Science* 193:579–580.

Jones, D. 1998. Organic acids in the rhizosphere—A critical review. *Plant Soil* 205:25–44.

Jones, D. L., C. Nguyen, and R. D. Finlay. 2009. Carbon flow in the rhizosphere: Carbon trading at the soil-root interface. *Plant Soil* 321:5–33. doi:10.1007/s11104 -009-9925-0

Kabata-Pendias, A. 2011. *Trace Elements in Soils and Plants*, 4th ed. Boca Raton: CRC Press.

Kidd, P., M. Mench, V. Álvarez-López et al. 2015. Agronomic practices for improving gentle remediation of trace element-contaminated soils. *Int. J. Phytoremediation* 17:1005–1037.

Kidd, P., V. Álvarez-López, C. Becerra-Castro et al. 2017. Potential role of plant-associated bacteria in plant metal uptake and implications in phytotechnologies. In: *Phytoremediation*, ed. Cuypers, A. and J. Vangronsveld, London: Academic Press.

Kirk, G. 2004. *The Biogeochemistry of Submerged Soils*. Chichester, England: John Wiley & Sons.

Krämer, U., J. D. Cotter-Howells, J. M. Charnock et al. 1996. Free histidine as a metal chelator in plants that accumulate nickel. *Nature* 379:635–638.

Krämer, U. 2010. Metal hyperaccumulation in plants. *Annu. Rev. Plant Biol.* 61:517–534.

Krishnamurti, G. S. R. and R. Naidu. 2007. Chemical speciation and bioavailability of trace metals. In: *Biophysico-Chemical Processes of Heavy Metals and Metalloids in Soil Environments*, pp. 419–466.

Kukier, U. and R. L. Chaney. 2001. Amelioration of nickel phytotoxicity in muck and mineral soils. *J. Environ. Qual.* 30:1949–1960.

Kukier, U., C. A. Peters, R. L. Chaney et al. 2004. The effect of pH on metal accumulation in two species. *J. Environ. Qual.* 33:2090.

Lebeau, T., A. Braud, and K. Jezequel. 2008. Performance of bioaugmentation-assisted phytoextraction applied to metal contaminated soils: A review. *Environ. Pollut.* 153:497–522.

Lee, J., R. D. Reeves, R. R. Brooks et al. 1978. The relation between nickel and citric acid in some nickel-accumulating plants. *Phytochemistry* 17:1033–1035.

Li, Y.-M., R. L. Chaney, E. Brewer et al. 2003a. Development of a technology for commercial phytoextraction of nickel: Economic and technical considerations. *Plant Soil* 249:107–115.

Li, Y.-M., R. L. Chaney, E. P. Brewer et al. 2003b. Phytoextraction of nickel and cobalt by hyperaccumulator *Alyssum* species grown on nickel-contaminated soils. *Environ. Sci. Technol.* 37:1463–1468.

Lipman, C. B. 1926. The bacterial flora of serpentine soils. *J. Bacteriol.* 12:315–318.

Lombi, E., W. W. Wenzel, and D. C. Adriano. 1998. Soil contamination, risk reduction and remediation. *Land Contam. Reclam.* 6:183–197.

Lopez, S., S. Piutti, J. Vallance et al. 2017. Nickel drives bacterial community diversity in the rhizosphere of the hyperaccumulator *Alyssum murale*. *Soil Biol. Biochem.* 114:121–130.

Lorenz, N., T. Hintemann, T. Kramarewa et al. 2006. Response of microbial activity and microbial community composition in soils to long-term arsenic and cadmium exposure. *Soil Biol. Biochem.* 38:1430–1437.

Luo, U., H. Zhang, F.-J. Zhao et al. 2010. Distinguishing diffusional and plant control of Cd and Ni uptake by hyperaccumulator and nonhyperaccumulator plants. *Environ. Sci. Technol.* 44: 6636–6641.

Ma, J. F., P. R. Ryan, and E. Delhaize. 2001. Aluminium tolerance in plants and the complexing role of organic acids. *Trends Plant Sci.* 6:273–278.

Ma, Y. and P. S. Hooda. 2010. Chromium, nickel and cobalt. In: *Trace Elements in Soils*, pp. 111–133, ed. Hooda, P. S., Wiley-Blackwell, Hoboken, NJ.

Marcussen, H., P. E. Holm, B. W. Strobel et al. 2009. Nickel sorption to goethite and montmorillonite in presence of citrate. *Environ. Sci. Technol.* 43:1122–1127.

Marschner, P. 2012. *Marschner's Mineral Nutrition of Higher Plants*, 3rd ed., Amsterdam Elsevier.

Massoura, S. T., G. Echevarria, T. Becquer et al. 2006. Control of nickel availability by nickel bearing minerals in natural and anthropogenic soils. *Geoderma* 136:28–37.

McNear, D. H., R. L. Chaney, and D. L. Sparks. 2007. The effects of soil type and chemical treatment on nickel speciation in refinery enriched soils: A multi-technique investigation. *Geochim. Cosmochim. Acta* 71:2190–2208.

Mendes, R., P. Garbeva, and J. M. Raaijmakers. 2013. The rhizosphere microbiome: Significance of plant beneficial, plant pathogenic, and human pathogenic microorganisms. *FEMS Microbiol. Rev.* 37:634–663.

Mengoni, A., R. Barzanti, C. Gonnelli et al. 2001. Characterization of nickel-resistant bacteria isolated from serpentine soil. *Environ. Microbiol.* 3:691–698.

Mengoni, A., E. Grassi, R. Barzanti et al. 2004. Genetic diversity of bacterial communities of serpentine soil and of rhizosphere of the nickel-hyperaccumulator plant *Alyssum bertolonii*. *Microb. Ecol.* 48:209–217.

Merlot, S., V. Sanchez Garcia de la Torre, and M. Hanikenne. 2018. Physiology and molecular biology of trace element hyperaccumulation. In: *Agromining: Farming for Metals*, eds. Van der Ent, A., G. Echevarria, A. J. M. Baker et al., Cham, Switzerland: Springer.

Miranda, M., M. López Alonso, C. Castillo et al. 2005. Effects of moderate pollution on toxic and trace metal levels in calves from a polluted area of northern Spain. *Environ. Int.* 31:543–548.

Mertens, J., P. Vervaeke, E. Meers et al. 2006. Seasonal changes of metals in willow (*Salix* sp.) stands for phytoremediation on dredged sediment. *Environ. Sci. Technol.* 40:1962–1968.

Molas, J. 2002. Changes of chloroplast ultrastructure and total chlorophyll concentration in cabbage leaves caused by excess of organic Ni(II) complexes. *Environ. Exp. Bot.* 47:115–126.

Montiel-Rozas, M. M., E. Madejón, and P. Madejón. 2016. Effect of heavy metals and organic matter on root exudates (low molecular weight organic acids) of herbaceous species: An assessment in sand and soil conditions under different levels of contamination. *Environ. Pollut.* 216:273–281.

Moradi, A. B., S. Swoboda, B. Robinson et al. 2010. Mapping of nickel in root cross-sections of the hyperaccumulator plant *Berkheya coddii* using laser ablation ICP-MS. *Environ. Exp. Botany* 69:24–31.

Moreira, C. S., J. C. Casagrande, L. R. Ferracciú Alleoni et al. 2008. Nickel adsorption in two Oxisols and an Alfisol as affected by pH, nature of the electrolyte, and ionic strength of soil solution. *J. Soils Sediments* 8:442–451.

Ndeddy Aka, R. J. and O. O. Babalola. 2017. Identification and characterization of Cr-, Cd-, and Ni-tolerant bacteria isolated from mine tailings. *Bioremediation J.* 21:1–19.

Neubauer, U., G. Furrer, and R. Schulin. 2002. Heavy metal sorption on soil minerals affected by the siderophore desferrioxamine B: The role of Fe(III) (hydr)oxides and dissolved Fe(III). *Eur. J. Soil Sci.* 53:45–55.

Nguyen, T. X. T., M. Amyot, and M. Labrecque. 2017. Differential effects of plant root systems on nickel, copper and silver bioavailability in contaminated soil. *Chemosphere* 168:131–138.

Nkrumah, P. N., A. J. M. Baker, R. L. Chaney et al. 2016. Current status and challenges in developing nickel phytomining: An agronomic perspective. *Plant Soil* 406:55–69.

Nolan, A. L., Y. Ma, E. Lombi et al. 2009. Speciation and isotopic exchangeability of nickel in soil solution. *J. Environ. Qual.* 38:485–492.

Nowack, B., R. Schulin, and B. Robinson. 2006. Critical assessment of chelant-enhanced metal phytoextraction. *Environ. Sci. Technol.* 40:5225–5232.

Oline, D. K. 2006. Phylogenetic comparisons of bacterial communities from serpentine and nonserpentine soils. *Appl. Environ. Microbiol.* 72:6965–6971.

Pal, A., S. Dutta, P. K. Mukherjee et al. 2005. Occurrence of heavy metal-resistance in microflora from serpentine soil of Andaman. *J. Basic Microbiol.* 45:207–218.

Pereira, S. I. A. and P. M. L. Castro. 2014. Diversity and characterization of culturable bacterial endophytes from *Zea mays* and their potential as plant growth-promoting agents in metal-degraded soils. *Environ. Sci. Pollut. Res.* 21:14110–14123.

Pérez, A. L. and K. A. Anderson. 2009. Soil-diffusive gradient in thin films partition coefficients estimate metal bioavailability to crops at fertilized field sites. *Environ. Toxicol. Chem.* 28:230–237.

Philippot, L., J. M. Raaijmakers, P. Lemanceau et al. 2013. Going back to the roots: The microbial ecology of the rhizosphere. *Nat. Rev. Microbiol.* 11:789–799.

Poulsen, I. F. and H. C. B. Hansen. 2000. Soil sorption of nickel in presence of citrate or arginine. *Water Air Soil Pollut.* 120:249–259.

Puschenreiter, M., A. Schnepf, I. M. Millán et al. 2005. Changes of Ni biogeochemistry in the rhizosphere of the hyperaccumulator *Thlaspi goesingense*. *Plant Soil* 271:205–218.

Puschenreiter, M., S. Wieczorek, O. Horak et al. 2003. Chemical changes in the rhizosphere of metal hyperaccumulator and excluder *Thlaspi* species. *J. Plant Nutr. Soil Sci.* 166:579–584.

Ratié, G., D. Jouvin, J. Garnier et al. 2015. Nickel isotope fractionation during tropical weathering of ultramafic rocks. *Chem. Geol.* 402:68–76.

Reeves, R. D. and A. J. M. Baker. 1984. Studies on metal uptake by plants from serpentine and non-serpentine populations of *Thlaspi goesingense* Hálácsy (*Cruciferae*). *New Phytologist* 98:191–204.

Renella, G., D. Egamberdiyeva, L. Landi et al. 2006. Microbial activity and hydrolase activities during decomposition of root exudates released by an artificial root surface in Cd-contaminated soils. *Soil Biol. Biochem.* 38:702–708.

Rinklebe, J. and S. M. Shaheen. 2014. Assessing the mobilization of cadmium, lead, and nickel using a seven-step sequential extraction technique in contaminated floodplain soil profiles along the Central Elbe River, Germany. *Water Air Soil Pollut.* 225.

Rosenkranz, T., J. Kisser, W. W. Wenzel et al. 2017. Waste or substrate for metal hyperaccumulating plants—The potential of phytomining on waste incineration bottom ash. *Sci. Total Environ.* 575:910–918.

Ryan, P., E. Delhaize, and D. Jones. 2001. Function and mechanism of organic anion exudation from plant roots. *Annu. Physiol. Plan Mol. Biol.* 52:527–560.

Salt, D. E., N. Kato, U. Krämer et al. 2000. *The Role of Root Exudates in Nickel Hyperaccumulation and Tolerance in Accumulator and Nonaccumulator Species of Thlaspi.* Boca Raton, FL: Lewis Publishers.

Sauvé, S., W. Hendershot, and H. E. Allen. 2000. Solid-solution partitioning of metals in contaminated soils: Dependence on pH, total metal burden, and organic matter. *Environ. Sci. Technol.* 34:1125–1131.

Scheckel, K. G. and D. L. Sparks. 2001. Dissolution kinetics of nickel surface precipitates on clay mineral and oxide surfaces. *Soil Sci. Soc. Am. J.* 65:685.

Scheidegger, A. M., G. M. Lamble, and D. L. Sparks. 1996. Investigation of Ni sorption on pyrophyllite: An XAFS study. *Environ. Sci. Technol.* 30:548–554.

Schlegel, H. G., J. P. Cosson, and A. J. M. Baker. 1991. Nickel-hyperaccumulating plants provide a niche for nickel-resistant bacteria. *Botanica Acta* 104:18–25.

Sen, I. S. and B. Peucker-Ehrenbrink. 2012. Anthropogenic disturbance of element cycles at the earth's surface. *Environ. Sci. Technol.* 46:8601–8609.

Sessitsch, A., M. Kuffner, P. Kidd et al. 2013. The role of plant-associated bacteria in the mobilization and phytoextraction of trace elements in contaminated soils. *Soil Biol. Biochem.* 60:182–194.

Shi, Z., E. Peltier, and D. L. Sparks. 2012. Kinetics of Ni sorption in soils: Roles of soil organic matter and Ni precipitation. *Environ. Sci. Technol.* 46:2212–2219.

Siebecker, M. G., R. L. Chaney, and D. L. Sparks. 2017. Nickel speciation in several serpentine (ultramafic) topsoils via bulk synchrotron-based techniques. *Geoderma* 298:35–45.

Sparks, D. L. 2003. *Environmental Soil Chemistry.* 2nd ed. Amsterdam: Academic Press.

Staunton, S. 2004. Sensitivity analysis of the distribution coefficient, K_d, of nickel with changing soil chemical properties. *Geoderma* 122:281–290.

Tessier, A., P. G. C. Campbell, and M. Bisson. 1979. Sequential extraction procedure for the speciation of particulate trace metals. *Anal. Chem.* 51:844–851.

Thijs, S., T. Langill, and J. Vangronsveld. 2017. The bacterial and fungal microbiota of hyperaccumulator plants: Small organisms, large influence. In: *Phytoremediation*, pp. 43–86, ed. A. Cuypers and J. Vangronsveld, London: Academic Press.

Thijs, S., W. Sillen, F. Rineau et al. 2016. Towards an enhanced understanding of plant–microbiome interactions to improve phytoremediation: Engineering the metaorganism. *Front. Microbiol.* 7:341.

Touceda-González, M. 2017. Molecular tools to characterize plant-associated bacterial communities in soils naturally enriched or contaminated with trace metals. PhD thesis, University of Santiago de Compostela, Spain.

Turgay, O. C., A. Gormez, and S. Bilen. 2012. Isolation and characterization of metal resistant-tolerant rhizosphere bacteria from the serpentine soils in Turkey. *Environ. Monit. Assess.* 184:515–526.

Uren, N. C. 1984. Forms, reactions and availability of iron in soils. *J. Plant Nutr.* 7:165–176.

Urzi, C., L. Brusetti, P. Salamone et al. 2001. Biodiversity of *Geodermatophilaceae* isolated from altered stones and monuments in the Mediterranean basin. *Environ. Microbiol.* 3:471–479.

Van der Ent, A., A. J. M. Baker, R. D. Reeves et al. 2012. Hyperaccumulators of metal and metalloid trace elements: Facts and fiction. *Plant Soil* 362:319–334.

Van der Ent, A., A. J. M. Baker, R. D. Reeves et al. 2015. "Agromining": Farming for metals in the future? *Environ. Sci. Technol.* 49:4773–4780.

Van der Ent, A., G. Echevarria, A. J. M. Baker et al. 2018. *Agromining: Farming for Metals Extracting Unconventional Resources Using Plants.* Cham, Switzerland: Springer International Publishing.

Wenzel, W. W. 2009. Rhizosphere processes and management in plant-assisted bioremediation (phytoremediation) of soils. *Plant Soil* 321:385–408.

Wenzel, W. W., M. Bunkowski, M. Puschenreiter et al. 2003. Rhizosphere characteristics of indigenously growing nickel hyperaccumulator and excluder plants on serpentine soil. *Environ. Pollut.* 123:131–138.

Wenzel, W. W. and F. Jockwer. 1999. Accumulation of heavy metals in plants grown on metalliferous soils of Austria. *Environ. Pollut.* 104:145–155.

Wenzel, W. W., R. S. Sletten, G. Wieshammer et al. 1997. Adsorption of trace metals by tension lysimeters: Nylon membrane vs. porous ceramic cups. *J. Environ. Qual.* 26:1430–1434.

Wenzel, W. W., G. Wieshammer, W. J. Fitz et al. 2001. Novel rhizobox design to assess rhizosphere characteristics at high spatial resolution. *Plant Soil* 237:37–45.

Wenzel, W. W., E. Lombi, and D. C. Adriano. 2004. Root and rhizosphere processes in metal hyperaccumulation and phytoremediation technology. In: *Heavy Metal Stress in Plants.* Heidelberg: Springer Verlag.

Williams, P. N., J. Santner, M. Larsen et al. 2014. Localized flux maxima of arsenic, lead, and iron around root apices in flooded lowland rice. *Environ. Sci. Technol.* 48:8498–8506.

Yusuf, M., Q. Fariduddin, S. Hayat et al. 2009. Nickel: An overview of uptake, essentiality and toxicity in plants. *Bull. Environ. Contam. Toxicol.* 86:1–17.

3

Biogeochemistry of Nickel in Soils, Plants, and the Rhizosphere

James W. Morris, Kirk G. Scheckel, and David H. McNear

CONTENTS

3.1 Introduction

3.1.1 Chemical Properties of Ni

Nickel (Ni) was first recognized as a unique element in 1751 upon its isolation by the Swedish mineralogist Axel Fredrik Cronstedt. It is a d-block transition metal with an electron structure of [Ar] $3d^8 4s^2$. The grand majority of available Ni occurs as one of five stable isotopes: ^{58}Ni, ^{60}Ni, ^{61}Ni, ^{62}Ni, and ^{64}Ni, with the majority occurring as ^{58}Ni (68%) and ^{60}Ni (26%) (Japan Atomic Energy Agency 2016). However, there are several natural and synthetic isotopes of Ni that are unstable, with many having short half-lives (Japan Atomic Energy Agency 2016). Ni is recognized as one of the primary constituents of Earth's core, making it the fifth most abundant element on Earth. Despite its overall abundance, however, Ni is a relatively minor constituent of the Earth's crust, having an average concentration of less than 0.01% by weight and ranking 24th in terms of abundance. Ni is very heterogeneously distributed among crustal rocks ranging from less than 0.0001% in sandstone and granite to 4% in coveted ore deposits (Duke 1980). Ni can be found in igneous, sedimentary, and metamorphic rocks as well as Ni ores. While there are many Ni-bearing minerals, most are considered rare. The most common Ni-bearing minerals occur in tandem with economically important ore deposits where Ni concentrations are relatively high (Table 3.1), with such deposits being globally distributed (Figure 3.1). In soils, Ni ranges from 5 to 500 mg kg^{-1} (Lindsay 1979). Serpentine clay-rich soils are noted for being geologically enriched in Ni and have been the focus for use of hyperaccumulating plants to phytomine Ni (Chaney et al. 1995).

3.1.2 Introduction of Ni into Soils and the Aquatic Environment

Industrial and agricultural activity, the burning of fossil fuels, and the weathering of geologic sources of Ni have led to the introduction of Ni into soil and aquatic environments (Duke 1980; Richter and Theis 1980). World production of Ni was 2.25 million metric tons in 2016, with the Philippines being the world's top producer of Ni ore followed by Russia and Canada (United States Geological Survey [USGS] 2017). World Ni reserves are estimated to be approximately 78 million metric tons, with Australia, Brazil, and Russia holding approximately 47% of the total (USGS 2017). It is further estimated that there may be as many as 130 million metric tons of available Ni worldwide, with 40% residing in sulfide deposits and the remaining 60% in laterite deposits (USGS 2017). Globally, the production of Ni-bearing stainless steel consumes 60% of the world's Ni, followed by nonferrous alloys at 12%, surface finishing at 11%, and alloy steels at 10%. The remaining 7% is consumed by a variety of chemical applications, such as the production of batteries, catalysts, and reagents (USGS 2016). According to the USGS and United States Environmental Protection Agency (USEPA), 210,000 metric tons of Ni were

TABLE 3.1

Ni-Bearing Minerals That Are Most Commonly Found in Economically Important Ni Deposits

Name	Group	Formula	Most Common Mode of Occurrence	Example Deposit
Pentlandite	Sulfide	$(Fe, Ni)_9S_8$	In mafic intrusions or remobilized phase after metamorphism	Noril'sk, Russia Bushveld, South Africa Voisey's Bay, Canada Kambalda, Western Australia
Ni replacement in pyrrhotite	Sulfide	$Fe_{1-x}S_x$	In mafic intrusions	Noril'sk, Russia Bushveld, South Africa Voisey's Bay, Canada Kambalda, Western Australia
Garnierite	Hydrous nickel silicate (serpentine)	$(Ni, Mg)_3Si_2O_5(OH)_4$	In laterites related to ultramafic rocks	New Caledonia Sulawesi, Indonesia Cerro Matoso, Columbia
Nickeliferous limonite	Hydroxide	$(Fe, Ni)O(OH)$	In laterites related to ultramafic rocks	New Caledonia Sulawesi, Indonesia Euboea, Greece
Millerite	Sulfide	NiS	Mafic intrusions where metasomatism has remobilized Ni and S from pentlandite; also from metamorphism of olivine	Silver Swan, Western Australia Sudbury, Canada
Niccolite	Nickel arsenide	$NiAs$	Hydrothermal replacement of pentlandite in mafic intrusions or metasomatisms of low-sulfur mafic rocks. Seafloor manganese nodules	Cobalt, Ontario Widgiemooltha Dome and Kambalda, Western Australia
Nickeliferous goethite	Hydrated oxide	$(Fe, Ni)O(OH)$	In laterites related to ultramafic rocks	Koniambo Massif, New Caledonia
Siegenite	Sulfide	$(Ni, Co)_3S_4$	Hydrothermal veins	Siegen, Germany Jachymov, Czech Republic

Source: Richter, R. O. and T. L. Theis. 1980. Nickel speciation in a soil/water system. In: *Nickel in the Environment*, ed. J. O. Nriagu, pp. 189–202, New York: John Wiley & Sons.

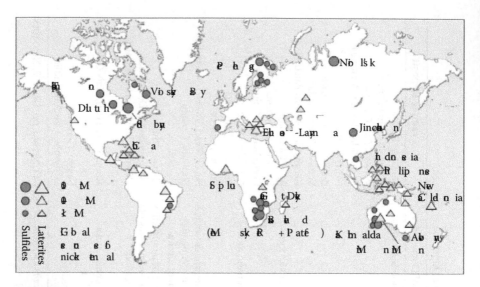

FIGURE 3.1
Global distribution of economically important sources of Ni ore as laterite and sulfide deposits. Note the high degree of complementarity between the above map and that displaying
the global distribution of ophiolites, which approximates the distribution of ultramafic soils.
(From Richter, R. O. and T. L. Theis. 1980. Nickel speciation in a soil/water system. In: *Nickel in
the Environment*, ed. J. O. Nriagu, pp. 189–202, New York: John Wiley & Sons.)

consumed and 18,282 metric tons of Ni and Ni compounds were released to
the environment in the United States in 2016 (USEPA 2017; USGS 2017). Over
the past 20 years, it is estimated that approximately 401,665 metric tons of
Ni and Ni compounds have been released to the environment in the United
States alone, with about 39% of those releases occurring between 1998 and
2002, while the amount of Ni and Ni compounds released from 2003 to 2016
has been relatively stable at 16,358 ∓ 1890 metric tons per year (Figure 3.2)
(USEPA 2017; USGS 1997–2017). The majority of Ni released into the environment in the United States is deposited onto land or into the atmosphere,
with surface waters and injection wells being the other important sinks for
anthropogenic Ni (USEPA 2017).

3.1.3 Biological Ni Requirements and Toxicity Thresholds

Ni is a micronutrient that is essential to the health of higher plants, microbes,
and some animals, possibly including humans, and so it maintains ecological
and agricultural, as well as industrial, significance (Brown et al. 1987; USEPA
2000). In plants, Ni is required in extremely small quantities (0.01–5 mg Ni/kg
dry plant weight) and so Ni deficiency is seldom an issue; however, Ni-enriched
soils such as ultramafic and Ni-polluted soils may impose strong selection
pressures on plant and microbial communities, where selection by Ni pollution may have uncertain ecological consequences. With respect to humans,

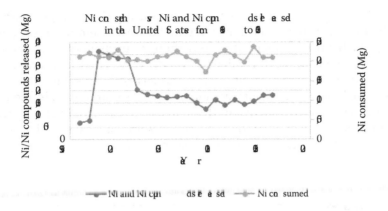

FIGURE 3.2
Ni consumed (imported and mined) and Ni released to the environment as Ni metal or Ni compounds in the United States from 1996–2016 (USEPA 2017; USGS 1997–2017). Data for both series are given in megagrams (Mg), which are equivalent to metric tons. In order to mitigate the loss of resolution due to scaling, the data are plotted on primary and secondary axes.

the USEPA has not established a reference concentration for Ni, and only has medium confidence in its reference dose of 0.02 mg per kg body weight per day for soluble Ni salts (USEPA 2000). Other thresholds for Ni exposure set by the USEPA include no more than 0.1 mg/L Ni in drinking water and exposure to no more than 1 mg of Ni per cubic meter air during a standard work week (8 hours per day, 40 hours per week) (Registry 2005). Humans are most frequently exposed via ingestion of food, where it is estimated that adults ingest an average of 100–300 µg Ni per day while the estimated daily requirement for a 70 kg adult is roughly 3.5 mg Ni per day, or 50 µg Ni per kg body mass (USEPA 2000). Other routes of exposure may include inhalation of Ni released into the atmosphere by fossil fuel combustion, biosolid incineration, manufacturing, occupational exposure where Ni is processed or used, the smoking of tobacco, and contact with household items such as cooking/eating utensils and jewelry, drinking of water containing Ni, and exposure to soil-containing Ni (Registry 2005). Such exposures may result in a wide variety of acute and chronic health effects that can range from mild dermatitis in the case of dermal exposure to cancer in the case of chronic inhalation of Ni refinery dust (USEPA 2000). Thus, the widespread contact, redistribution, and release of Ni by human activities is of increasing global concern, and along with it, the need to better understand Ni's place within terrestrial ecosystems.

The following chapter will provide a brief discussion of Ni as it relates to soils, soil microbiology, and plants with some perspective scattered throughout. While the information contained herein is exhaustive, it should be noted that there is very little research aimed at understanding the complex interactions that occur at the Ni–soil–plant–microbe interface. As such, the reader is cautioned to accept the following body of information as a snapshot of the current state of the science on Ni–soil–plant–microbe interactions—a science that is still in its infancy and must overcome the highly contingent nature of soils, biological systems, and the interplay between them in order to grow. The hope is that this chapter will inspire thoughtful creativity and embolden researchers to engage such challenges in new ways so as to strengthen the claim on knowledge regarding Ni and the terrestrial systems that control its fate, which may create consequential feedbacks that in turn influence the ecological fate of such systems. In that context, it seems only appropriate to begin the discussion of Ni in soils and plants with the most notorious of geo-genically Ni-enriched ecosystems—the serpentine soils.

3.2 Serpentine Soils

3.2.1 Nickel-Bearing Parent Materials

Serpentine soils are part of a larger, distinct group of soils known as ultramafic soils. Ultramafic soils form from the weathering of igneous or metamorphic rocks composed of 70% or greater ferromagnesian (mafic) minerals, thus the designation of "ultramafic" (Brady et al. 2005). Serpentinization occurs when peridotite, a rock composed mostly of the mineral olivine, undergoes anaerobic oxidation followed by hydrolysis within oceanic crust giving rise to serpentine minerals, a subgroup of the kaolinite–serpentine group of minerals. The serpentine subgroup minerals are 1:1 phyllosilicates, making them less susceptible to high rates of weathering compared to olivine. Thus, Ni is thought to be more readily available in soils formed from partially serpentinized peridotite as opposed to soils formed from more highly serpentinized parent materials (Kierczak et al. 2016; Pędziwiatr et al. 2017). There are several known serpentine subgroup minerals represented by the generic chemical formula $X_{2-3}[Si_2O_5]$ $(OH)_4$, where X may be represented by Mg^{2+}, Fe^{2+}, Fe^{3+}, Ni^{2+}, Mn^{2+}, Al, or Zn^{2+} in octahedral coordination. Chrysotile, lizardite, and antigorite are the most abundant of the serpentine subgroup minerals, with the grand majority of the rest being considered rare in soils. Serpentinite is a catch-all term referring to rocks composed of one or more serpentine minerals. Serpentinite is found within geological formations known as ophiolites (Tashakor et al. 2013). Ophiolites form at convergent tectonic plate boundaries where oceanic crust has been obducted, forcing the ultramafic rocks within it, such as serpentinite, to the continental surface where chemical and physical weathering shape them into some of the most interesting soils studied to date. Given the unique circumstances of their formation, the distribution of serpentine soils is global, but patchy, with potential ages ranging from 1 billion (start of Neoproterozoic) to about 34 million years (end of Eocene) old (Tashakor et al. 2013; Vaughan and Scarrow 2003). A map showing the global distribution of ophiolites and their predicted ages can be seen in Figure 3.3 (Vaughan and Scarrow 2003).

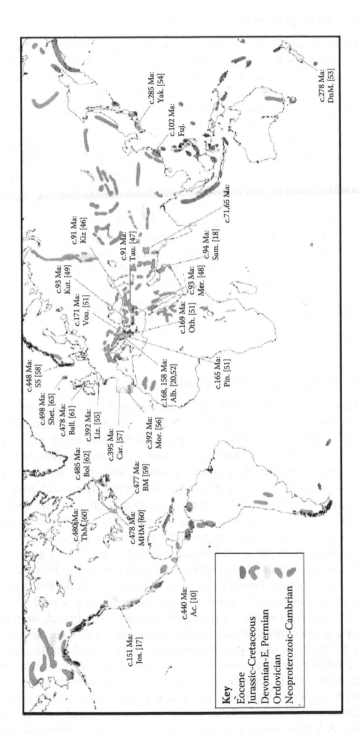

FIGURE 3.3

(See color insert.) The global distribution of ophilites with approximate ages provides a proxy for the patchy, global distribution of ultramafic soils. (From Vaughan, A. P. M. and J. H. Scarrow. 2003. Ophiolite obduction pulses as a proxy indicator of superplume events? *Earth Planet. Sci. Lett.* 213 (3–4):407–416. doi: http://dx.doi.org/10.1016/S0012-821X(03)00330-3.)

3.2.2 Properties of Serpentine-Derived Soils

The weathering of serpentinite results in a plethora of unique edaphic factors that impose strong selection pressures on plant life (Brady et al. 2005). This suite of properties is characteristic of serpentine soils and may include low Ca-to-Mg ratio; minimal clay and silt content resulting in low cation exchange capacity (CEC), reduced water retention, ready leaching of nutrient cations; low levels of fundamental macronutrients such as N, P, and K; increased temperature relative to surrounding soils; and high levels of geologically derived metals such as cobalt (Co), chromium (Cr), and Ni (Ni), with levels of Ni commonly being in excess of 1000 parts per million (ppm) (Brady et al. 2005). However, not all serpentine soils are created equal, and any given serpentine taxon may exhibit various combinations of the above characteristics with low Ca-to-Mg ratio and enrichment of trace metals being considered the common denominator among them. So effective are serpentine soils at withering the Darwinian fitness of plant life that Hans Jenny coined the term "serpentine syndrome" to describe the harsh chemical and physical impositions suffered by plant life attempting to stake a claim on serpentine territory. Appropriately, plant life on serpentine soils is often sparse and endemic. Thus, serpentine ecosystems are typically open and sharply delineated in contrast to the surrounding landscape.

3.2.3 Selective Pressure of Serpentine Soils

Serpentine-adapted plants are generally characterized by a few key morphological traits that separate them from closely related, non–serpentine-adapted relatives. Typically, serpentine-adapted species will exhibit reduced stature, a more developed root system, and xeromorphic foliage (Brady et al. 2005). Serpentine-associated molecular adaptations, however, are less well characterized. As serpentine syndrome is multifaceted, the exact properties of a serpentine system that select for, or against, a given plant species is likely to be specific to the species and site in question. Not all serpentine sites may exhibit each constituent of serpentine syndrome, nor are all plants sensitive to each constituent alone with equal intensity. Selection is generally thought to occur as a result of Mg toxicity, Ca or other nutrient deficiencies, drought sensitivity, trace metal toxicity, or a combination of these (Brady et al. 2005). Elevated quantities of exchangeable trace metals pose one of the most unique selection pressures within serpentine systems. The broad, nonspecific mechanisms of trace metal toxicity affect the most minute single-celled soil organisms to the most intricate evolutions of botany, and can therefore be significant modifiers of plant and microbial community structure and ecosystem dynamics. While the effects of elevated levels of trace metals on biological entities are well characterized in a general sense, less is known about the precise effects of specific trace metals such as Ni on plants and soil microorganisms. A logical starting point in the attempt to understand what such effects might be is with an examination of the chemical behavior of Ni in bulk and rhizosphere soils.

3.2.4 Nickel Chemistry in Soil

3.2.4.1 Ni Chemistry in Soils, Aquifers, and Sediments

Nickel may be immobilized through formation of pure Ni precipitates such as hydroxides, silicates, or sulfides (Mattigod et al. 1997; Merlen et al. 1995; Peltier et al. 2006; Scheidegger et al. 1998; Scheinost and Sparks 2000; Thoenen 1999) or through coprecipitation with other soil-forming minerals such as silicates, iron oxides/sulfides, or carbonates (Ford et al. 1999a; Hoffmann and Stipp 2001; Huerta-Diaz and Morse 1992; Manceau 1986; Manceau et al. 1985). Predicted Ni concentrations in the absence of sulfide for several potential pure Ni precipitates suggest that phyllosilicate and layered double hydroxide (LDH) precipitates (incorporating aluminum) may result in dissolved Ni concentrations below most relevant regulatory criteria over a pH range typical for groundwater. These data also point to the limited capability of pure Ni carbonates and hydroxides in controlling dissolved Ni concentrations to sufficiently low values except under very alkaline conditions. The Eh–pH conditions under which these solubility-limiting phases may form is shown in Figure 3.4. According to these data, Ni-bearing phyllosilicate and/or LDH precipitates possess large stability fields indicating their relative importance

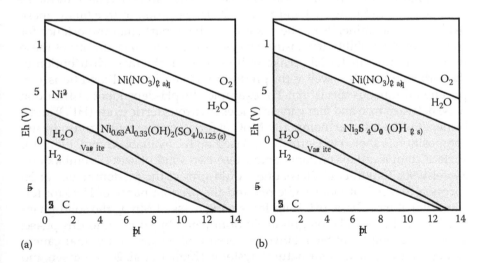

(a)   (b)

FIGURE 3.4
Eh–pH diagrams for Ni at 25°C. (a) System Ni-H_2O-Ca-Al-NO_3-HCO_3-SO_4 (2 mg Ni/L; 40 mg Ca/L; 3 mg Al/L; 6 mg NO_3/L; 60 mg HCO_3/L; 100 mg SO_4/L). Stability fields for solids are shaded green (Vaesite = NiS2). (b) Same system plus 3 mg Si/L. Thermodynamic data for $Ni_3Si_4O_{10}(OH)_2$ and $Ni_{0.63}Al_{0.33}(OH)_2(SO4)_{0.125}$ are from Peltier et al. (2006). [Note that the solubility of the Ni-Al-SO4 LDH was adjusted to correct for charge imbalance for the chemical structure published in Peltier et al. (2006).]

to controlling Ni solubility under a range of conditions. These calculations point to the importance of dissolved aluminum and silicon concentrations in soil water relative to the potential sequestration of Ni via precipitation (Ford et al. 1999b; Scheinost et al. 1999).

Retention of Ni may also occur via coprecipitation during the formation of (hydr)oxides or sulfides of iron. These minerals have been observed to form at the boundaries between oxidizing and reducing zones within ground-water plumes. There are numerous laboratory and field observations that demonstrate the capacity of these precipitates for Ni uptake (Coughlin and Stone 1995; Ford et al. 1997, 1999a; Huerta-Diaz and Morse 1992; Schultz et al. 1987). Under these circumstances, the solubility of Ni will depend on the stability of the host precipitate phase. For example, iron oxide precipitates may alternatively transform to more stable forms (Ford et al. 1997), stabiliz-ing coprecipitated Ni over the long term, or these precipitates may dissolve concurrent with changes in groundwater redox chemistry (Zachara et al. 2001).

Adsorption of Ni in soil environments is dependent on pH, temperature, and type of sorbent (minerals or organic matter), as well as the concentration of aqueous complexing agents, competition from other adsorbing cations, and the ionic strength in groundwater. Ni has been shown to adsorb onto many solid components encountered in aquifer sediments, components that may be analogous to those found in surface soils, including iron/manganese oxides, clay minerals (Bradbury and Baeyens 2005; Dähn et al. 2003), and solid organic matter (Nachtegaal and Sparks 2003). Sorption to iron/manganese oxides and clay minerals has been shown to be of particular importance for controlling Ni mobility in subsurface systems. The relative affinity of these individual minerals for Ni uptake will depend on the mass distribution of the sorbent minerals as well as the predominant geochemical conditions (e.g., pH and Ni aqueous speciation). For example, the pH-dependent distribution of Ni between iron and manganese oxides [hydrous ferric oxide (HFO) and a birnessite-like mineral (nominally MnO_2)] for a representative groundwater composition is shown in Figure 3.5a. Based on the available compilations for surface complexation constants onto these two solid phases (Dzombak and Morel 1990; Tonkin et al. 2004), one would project the predominance of Ni sorption to MnO_2 at more acidic pH and the predominance of HFO (or fer-rihydrite) at more basic pH. With increasing mass of MnO_2, the solid-phase speciation of Ni will be progressively dominated by sorption to this phase. There are examples of the relative preference of Ni sorption to manganese oxides over iron oxides for natural systems (Kjøller et al. 2004; Larsen and Postma 1997; Manceau et al. 2002, 2007). As shown in Figure 3.5b, Ni adsorp-tion may be inhibited (or Ni desorption enhanced) through the formation of solution complexes with organic ligands such as EDTA or natural organic matter (Bryce and Clark 1996; Nowack et al. 1997).

As previously noted, adsorption of Ni onto mineral surfaces may serve as a precursory step to the formation of trace precipitates that reduce the

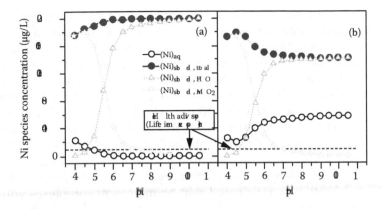

FIGURE 3.5
(a) Ni sorption as a function of pH in the presence of a hypothetical soil system with iron and manganese oxides reflective of the crustal abundance of these elements (Schulze 2002; assumed 30% porosity with 185.0 g HFO/L and 1.66 g MnO2/L). (b) Same conditions as in (a), but with 10 μM EDTA added. Nominal groundwater composition: 0.005 mol/L NaCl, 0.001 mol/L K2SO4, 0.001 mol/L MgNO3, 0.001 mol/L CaCO3, and 34 μ mol/L Ni (2 mg Ni/L). Model predictions using Visual MINTEQ Version 2.50 (based on MINTEQA2 described in Allison et al. 1991) with available surface complexation parameters derived from Dzombak and Morel (1990) and Tonkin et al. (2004); kaolinite set as an "infinite" solid for pH titration.

potential for desorption with changes in groundwater chemistry. This may be realized through the nucleation and growth of surface precipitates on clay mineral surfaces due to continued uptake of Ni (Dähn et al. 2002; Scheckel et al. 2000; Scheckel and Sparks 2000, 2001). This type of process may compete with other adsorption processes, such as ion exchange, depending on the prevailing groundwater chemistry and characteristics of the clay mineral (Elzinga and Sparks 2001).

Nickel is one of the most mobile of the heavy metals in the aquatic environment. The mobility of Ni in the aquatic environment is controlled largely by competition between various sorbents to scavenge it from solution and ligands to form non-sorptive complexes. Although data are limited, it appears that in pristine environments, hydrous oxides and phyllosilicates control Ni mobility via coprecipitation and sorption. In polluted environments, the more prevalent organic compounds will keep Ni soluble by ligand complexation. In reducing environments, insoluble Ni sulfide may form. Ni chloride is water-soluble and would be expected to release divalent Ni into the water. The atmosphere is a major conduit for Ni as particulate matter. Contributions to atmospheric loading come from both natural sources and anthropogenic activity, with input from both stationary and mobile sources. Various dry and wet precipitation processes remove particulate matter as washout or fallout from the atmosphere with transfer to soils and waters. Soil-borne Ni may enter waters by surface runoff or by percolation into groundwater. Once Ni is in surface and groundwater systems, physical and chemical interactions (complexation, precipitation/dissolution, adsorption/desorption, and oxidation/reduction) occur that will determine its fate and that of its constituents. In ambient aqueous systems, Ni exists in the divalent oxidation state and is not subject to oxidation-state transformations under typical conditions. Ni predominantly exists as a cationic species (Ni^{2+}) or various hydrolysis species (e.g., $NiOH^+$) at near-neutral pH (Baes and Mesmer 1986). However, Ni may also form dissolved complexes in the presence of high concentrations of inorganic ions such as carbonate/bicarbonate and sulfate (Chen et al. 2005; Hummel and Curti 2003) or organic ligands such as natural/synthetic carboxylic acids and dissolved humic compounds (Baeyens et al. 2003; Bryce and Clark 1996; Strathmann and Myneni 2004). It is anticipated that Ni may form complexes with dissolved sulfide under sulfate-reducing conditions, although the current state of knowledge is insufficient to ascertain the relative importance of these species in aqueous systems (Thoenen 1999). The formation of solution complexes, especially with organic ligands, may limit sorption of Ni to mineral surfaces in aquifer sediments.

3.2.4.2 Ni Chemistry in the Rhizosphere

While Ni chemistry in the rhizosphere of plants is poorly understood, the above discussion provides some meaningful insight. As in bulk surface soils, aquifers, and sediments, Ni mobility in the rhizosphere may be modulated by well-established edaphic factors such as clay mineralogy, CEC, soil texture, moisture content, redox potential, pH, soil organic matter (SOM) content and composition, and the species and concentrations of competing ions. However, the relative mobility of Ni in rhizosphere soil compared to other trace metals seems to show no distinct pattern when examined in the context of electronegativity, stability of hydrolysis products, soil type, and various sequential extractions (Antoniadis et al. 2017). Aside from edaphic factors, nickel mobility could be altered by rhizosphere microorganisms and root exudates; however, such effects are difficult to differentiate from each other and from abiotic factors, as the activity of roots and associated microorganisms may alter the physical, chemical, and biological state of the soil environment. Root exudates can be defined by three general categories: ions, such as CO_3^{2-}, H^+, NH_4^+, PO_4^{3-}, and SO_4^{2-}; low-molecular-weight organics such as amino acids, fatty acids, flavonoids, organic acid phenolics, and sugars; and high-molecular-weight organics such as enzymes, mucilage, and polysaccharides (Antoniadis et al. 2017). In addition to root exudates, the rhizosphere is likely to contain a multitude of microbial exudates as well. Such compounds may foster Ni–plant–microbe interactions by altering the pH, redox state, and ion exchange dynamics in the rhizosphere and via the formation of Ni–organic complexes (Antoniadis et al. 2017). In many cases, the mechanisms by which exudates such as low-molecular-weight organic acids play a role in Ni chemistry are difficult to elucidate, as such molecules are capable of serving a dual role in Ni–plant interactions: they may serve as a source of acidity resulting in the solubilization of Ni or as a chelator, or potentially both. As mentioned in Section 3.2.4.1, a prevalence of organic ligands may keep Ni in solution via complexation, which might explain why increases in bioavailable Ni have been noted in the rhizosphere despite variability in pH and presumably reducing conditions (due to plant and microbial respiration) that may otherwise favor the formation of insoluble species. Much more research is needed in this area, and powerful methodologies for investigating in situ Ni speciation, distribution, and mobility along with controlling factors like pH and redox conditions in the rhizosphere are available. For example, by combining planar optode sensors that measure pH and O_2 with density gradient thin films infused with a cation binding agent, the distribution of phytoavailable Ni and variable that may influence it can be measured using rhizotrons (Hoefer et al. 2017; Santner et al. 2015). Likewise, highly tunable X-rays produced by synchrotrons provide an excellent means to produce two-dimensional maps of Ni distribution while allowing for the determination of Ni speciation across the bulk-rhizosphere continuum and within plant tissues (McNear et al. 2005a,b).

3.2.5 Ni Uptake

The uptake of Ni by plant roots is modulated at two course levels—the rhizosphere level and the plant soma level. Thus, uptake may be controlled, in part, by edaphic factors in the rhizosphere such as clay mineralogy, CEC, soil texture, moisture content, redox potential, pH, SOM content and composition, and the species and concentrations of competing ions; the latter five properties can be directly influenced by plant biology and that of the rhizosphere microbiome. Likewise, it is possible that root exudates effect microbial ecology so as to alter the mobility and subsequent uptake of Ni. At the plant soma level, uptake is presumed to occur via facilitated diffusion or active transport, where the mechanism responsible depends largely on individual genetics, plant species, the concentration of phytoavailable Ni, and the species of phytoavailable Ni present in the rhizosphere (Chen et al. 2009; Seregin and Kozhevnikova 2006). Antoniadis et al. (2017) provide an interesting list of soil-to-plant transfer coefficients for Ni across a variety of plant species that exemplifies this point. Currently, it is generally thought that the free Ni^{2+} ion is the most phytoavailable form in soils (Deng et al. 2017). When free, ionic Ni^{2+} is available in high quantities, in the context of the biology of a given plant species, passive diffusion is likely to be the mechanism of entry into the root body. Uptake of Ni via passive diffusion may occur through nonselective, low-affinity cation channels through which plants absorb other elements such as Ca^{2+}, Co^{2+}, Cu^{2+}, Fe^{2+}, Mn^{2+}, and Zn^{2+}. Thus, it is intuitive that the uptake of Ni may be competitively inhibited by the presence of other divalent cations and vice versa. However, the modulation of Ni uptake via competition between Ni and other divalent cations has been observed to be largely nonspecific in nature and may be highly contingent on the plant species in question. Generally speaking, Ca^{2+}, Cu^{2+}, Mg^{2+}, and Zn^{2+} have been shown to inhibit Ni^{2+}. Additionally, studies in *Arabidopsis thaliana* have demonstrated that Ni^{2+} may be transported into the root through IRT1, a known Fe^{3+} transporter, and Ni^{2+} uptake increases under times of Fe^{3+} deficiency (Nishida et al. 2011, 2012). It has also been suggested that Ni^{2+} may be absorbed by plant roots by way of Mg^{2+} transport systems due to chemical similarities between the Ni^{2+} and Mg^{2+} ions. However, experimental verification of Ni absorption into roots via Mg^{2+}-specific transporters is lacking, and much of what is known about Ni uptake in plants is inferred from the results of experiments focusing on competitive uptake, where reductions in the uptake of other cations is correlated with increased uptake of Ni and the mechanism presumed to be as described above.

When phytoavailable Ni exists in the rhizosphere primarily as Ni-organic complexes, especially when the ligand is of lower molecular weight, it is possible that Ni may enter the root body via active transport (Chen et al. 2009). As Ni is a required micronutrient that typically exists in low concentrations in the soil environment, active transport may play a more important role in Ni acquisition than is currently recognized; however, experimental evidence for the active transport of Ni into root cells is extremely scarce. Like other nutrients, when Ni is scarce, plants may rely on phytosiderophores or low-molecular-weight organic acids for the release and/or uptake of Ni^{2+}. Likewise, other chelators such as nicotianomine, histidine, cysteine, and metallothionein may play an important role in Ni acquisition by plants. As an example, the uptake of Ni^{2+} along with other metals in conjunction with nicotianamine and phytosiderophores was demonstrated to occur in yeast via ZmYS1, a transmembrane transport protein in maize (Schaaf et al. 2004). However, it is suspected that in plants chelators may play a more direct role in vascular transport of Ni once it is loaded into the xylem.

3.2.6 Physiological Role of Ni in Plants

In plants, the only widely accepted physiological role of Ni is as an essential component of the active sight in urease—a protein that catalyzes the hydrolysis of urea, resulting in the release of ammonia and hydroxamic acid, which then spontaneously hydrolyzes to release another ammonia along with bicarbonate (Carlini and Ligabue-Braun 2016; Fabiano et al. 2015; Polacco et al. 2013). There are two major forms of plant urease currently recognized by the plant science community—one is found in the seeds of some plants and the other is a ubiquitous form thought to be present in the vegetative tissues of all plants (Fabiano et al. 2015; Seregin and Kozhevnikova 2006). Two Ni atoms are necessary at the active sight for proper catalytic activity of the ubiquitous form of urease (Seregin and Kozhevnikova 2006). Urease may be active in intracellular and extracellular environments. While Ni is typically present in plant vegetative tissues in low concentrations, and the ubiquitous form of urease is thought to have low activity in such tissues, the N released from urea hydrolysis may serve as an important source of N for the synthesis of N containing biomolecules under times of N stress. Active, extracellular ureases are relatively stable and may affect rhizosphere conditions by increasing soil pH and facilitating the precipitation of carbonates, which in turn has the potential to alter nutrient dynamics (Carlini and Ligabue-Braun 2016). Further, extracellular ureases may decrease the efficiency of urea fertilization in agroecosystems resulting in release of ammonia to the atmosphere (Carlini and Ligabue-Braun 2016).

A wide variety of additional roles for urease in plants and plant–microbe–soil interactions are present in the literature. However, such phenomenon do not constitute additional, unique roles for Ni in plant biology, as the relationship between Ni and the urease enzyme does not change from one scenario to the next. Nevertheless, such scenarios may provide further insight into the necessity of Ni in plant biology, as Ni may be a necessary structural component of urease with respect to both urealytic and non-urealytic functions. Thus, some of these additional urease functions are noteworthy and are briefly mentioned here. Urease and urease-derived peptides have been observed to exhibit antifungal activity against filamentous fungi and yeasts, and have also been observed to have insecticidal properties in natural environments, suggesting potential roles for urease and urease products in defense (Carlini and Ligabue-Braun 2016). A variant of jack bean urease, known as canatoxin, utilizes one Ni atom and one Zn in its active site, and is proposed to play a primarily non-enzymatic, defensive role against phytopathogenic fungi and herbivorous insects (Carlini and Ligabue-Braun 2016). Additionally, ureases may be involved in microbe–microbe signaling as well as plant–microbe signaling in important processes such as nodulation (Polacco et al. 2013).

It has also been reported that Ni may be involved in the activation of malate dehydrogenase in the cytoplasm of plant cells (Polacco et al. 2013). Malate dehydrogenase has been implicated in the synthesis of NADH that is subsequently used by nitrate reductase, potentially providing another link between Ni and N metabolism in plants (Polacco et al. 2013). Ni has also been reported to possibly play a role in glyoxalase I activation in rice (Fabiano et al. 2015). While a first for plants, Ni and Zn requirements for glyoxalase activation is well known in microorganisms. In both plants and microorganisms, glyoxalase enzymes are responsible for degrading the toxic compound methylglyoxal, which may increase oxidative stress and interfere with cellular division when present at elevated levels (Fabiano et al. 2015). Furthermore, the process by which methylglyoxal is degraded requires the consumption of reduced glutathione (glyoxalase I) and its subsequent regeneration (glyoxalase II) (Fabiano et al. 2015). Other reports suggest that glyoxalase I enzymes may be induced in response to a variety of biotic and abiotic stressors that are often associated with reactive oxygen species (ROS) generation and oxidative stress (Fabiano et al. 2015). These relationships—between glyoxalase I,

glutathione, and oxidative stress response—suggest that Ni may also function in glutathione homeostasis and oxidative stress tolerance in some plants
(Fabiano et al. 2015).

3.2.7 Ni Toxicity and Tolerance in Plants

Symptoms of Ni toxicity in plants may be elicited by concentrations as low
as 10–15 µg of Ni per gram of plant tissue (Rascio and Navari-Izzo 2011).
Physiologically excess Ni may have a wide variety of toxic effects on plants
depending on the intensity of the exposure, plant species exposed, stage of
plant development at the time of exposure, the overall condition of the plant
at the time of exposure, and growth conditions. In general, plants challenged
with Ni stress experience nutrient imbalances; perturbations in photosynthesis; wilting, chlorosis, and/or necrosis of leaf tissue; and the inhibition
of normal growth and developmental patterns (Chen et al. 2009). Likewise,
plants may exhibit a diverse array of responses to such challenges that are
difficult to generalize. Thus far, information regarding plant responses
to Ni stress, and the investigation thereof, is largely centralized around a
theory of detoxification by chelation. Such studies report elevated levels of
metallothioneins, phytochelatin precursors such as glutathione and thiol
glutathione, O-acetyl-ʟ-serine, nicotianamine, nicotianamine synthase, free
histidine, and Ni^{2+}–NA and Ni^{2+}–organic acid complexes in response to Ni^{2+}
exposure—all responses commonly associated with detoxification of heavy
metals in a general sense (Chen et al. 2009). However, it is unclear at this
time what, if any, mechanisms plants may have for coping with Ni^{2+} stress
specifically.

3.2.8 Ni Hyperaccumulation in Plants

3.2.8.1 Ni Hyperaccumulation Defined

One of the most interesting outcomes of ultramafic selection pressure on
plant life has been the evolution of species that hyperaccumulate Ni within
their leaves to normally toxic concentrations. These hyperaccumulator plants
are generally defined as wild varieties that accumulate Ni and other trace
metals within their leaf tissues to concentrations at least two to three orders
of magnitude more than that of other plants growing on non-ultramafic soils,
and at least one order of magnitude more than plants growing on ultramafic
soils (Pollard et al. 2014; van der Ent et al. 2013). Generally accepted threshold
values for Ni hyperaccumulation (µg Ni/g dry plant tissue) are >1000 ppm
Ni; however, concentrations >10,000 ppm have been observed, with the term
"hypernickelophore" having been suggested as a special designation for such
plants (van der Ent et al. 2013). However, Ni hyperaccumulation is simply

considered to occur on a spectrum across which no special categorization scheme has been applied at this time. Although there is some debate as to whether there are "obligate" and "facultative" hyperaccumulators, where facultative hyperaccumulators may occur on metal-enriched and non-enriched soils but only hyperaccumulate when growing on metal enriched soils (Pollard et al. 2014; van der Ent et al. 2013). Plants categorized as Ni hyperaccumulators should exhibit hyperaccumulation within wild, self-sustaining populations, not excluding those discovered on sites anthropogenically enriched with Ni, in order to avoid assigning hyperaccumulation status to species in which it occurs only as an artifact of experimentation (van der Ent et al. 2013). Ni hyperaccumulation was first discovered in the serpentine endemic species *Alyssum bertolonii* in 1948 (Kramer 2010). Now, there are roughly 500 species of hyperaccumulator plants described to date, with approximately 400 of those described as Ni hyperaccumulators, occurring across 42 different plant families. The Brassicaceae and Euphorbiaceae families together comprise the majority of known Ni hyperaccumulators, with each containing 25% of known species (Kramer 2010). Hyperaccumulation has evolved in several distantly related plant families. Regardless of the evidence suggesting that hyperaccumulation has evolved multiple times independently, the study of hyperaccumulation has been largely relegated to two species within the Brassicaceae family, *Arabidopsis halleri* and *Noccaea caerulescens*, which are now widely regarded as model species for the study of hyperaccumulation, with *N. caerulescens* serving as a model Ni hyperaccumulator (Rascio and Navari-Izzo 2011). Thus, evidence supporting the accepted mechanisms of hyperaccumulation should be interpreted and extrapolated to other species with caution.

Most plant species that are capable of tolerating elevated levels of trace metals to some extent do so by complexing them with cell wall components within the apoplastic spaces of the root. Plants with this phenotype are often referred to as "excluders" of Ni and other trace metals, as the complexation to cell wall components within the epidermis and cortex of the root excludes metals from entering the vascular system. However, hyperaccumulators differ from most other plant species in four distinguishing features: enhanced absorption of heavy metals from the soil into the root system; efficient root-to-shoot translocation; enhanced tolerance and sequestration of heavy metals within the shoot tissue, usually in the cell walls or vacuoles of leaves; and efficient systems for cell-to-cell distribution of hyperaccumulated metals in shoots (Kramer 2010; Rascio and Navari-Izzo 2011). The current consensus is that these three processes are not mediated by novel genes, but rather the overexpression of commonly occurring genes involved in metal homeostasis as a result of mutated promoters and duplicate copies (Kramer 2010; Rascio and Navari-Izzo 2011). Most of these genes are involved in transport of cations across the plasma membrane or tonoplast.

3.2.8.2 Physiology of Ni Hyperaccumulation

Following is a brief summary of the widely accepted information regarding the physiology of Ni hyperaccumulation in plants. It is hypothesized that trace metals such as Ni may enter the plant through transporters for physiologically important metal cations, such as Mg^{2+} and Zn^{2+}. Constitutive overexpression of ZIP (zinc-regulated transporter iron-regulated transporter) genes may serve an important role in trace metal uptake into the roots of hyperaccumulators, especially as this overexpression indicates the loss of metal regulated, or induced, metal transport into the root system (Rascio and Navari-Izzo 2011). Thus, it is suggested that the feedback mechanisms for regulating the entry of trace metal ions into the root system has been compromised in hyperaccumulator plants. Once in the roots, Ni must be efficiently translocated across the casparian strip and into the xylem. The overexpression of HMAs (heavy metal transporting ATPases) and MATE (multidrug and toxin efflux) proteins such as FDR3 (overexpressed in the pericycle of *N. caerulescens* and *A. halleri*) has been implicated as important to efficient xylem loading of some trace metals (Rascio and Navari-Izzo 2011). In the case of Ni, some evidence suggests that along with the overexpression of transporters, complexation to histidine or nicotinamine may be important for keeping Ni in the cytosol, thus allowing for efficient passage from root cells into the xylem. In contrast to HMAs and MATE proteins, the overexpression of YSL (Yellow Strip1-Like) family of proteins in the roots and shoots of hyperaccumulators may be involved in the both the loading and unloading of the xylem nicotinamine–metal (Ni-nicotinamine complexes in *N. caerulescens*) (Kramer 2010; Rascio and Navari-Izzo 2011). Additionally, there is some evidence that suggests that like uptake, Zn^{2+} and Fe^{3+} efflux proteins may be important in the xylem loading of Ni (Deng et al. 2017). It is not yet clear whether Ni undergoes translocation primarily as hydrated metal ions or complexed with organic ligands, and the evidence suggests that it may be plant species specific rather than a generalizable phenomenon. Regardless, Ni has been observed to be translocated in complex with an organic ligand, such as citrate, histidine, nicotinamine, or phytate, as hydrous Ni^{2+}, and as a mixture of free and complexed forms (Deng et al. 2017; McNear et al. 2005, 2010). While the specific ligands used to transport Ni within the vascular system of plants, in a general sense it can be expected that Ni will be bound to those with reactive groups containing nitrogen and oxygen (McNear et al. 2010).

Once in the leaf tissues, hyperaccumulator strategies consist of complexing Ni with ligands or storing them in compartments with low physiological activity, such as the cell wall and vacuole, in order to protect sensitive biochemical processes (Kramer 2010; Rascio and Navari-Izzo 2011). In the case of Ni, hyperaccumulators primarily sequester Ni in the vacuoles of epidermal cells as Ni–organic or Ni–carboxylic acid complexes; however, the cuticle and trichomes could possibly serve as Ni sinks, especially when exceptionally large quantities of Ni are present in the leaf (Deng et al. 2017; Rascio and Navari-Izzo 2011). CDF (cation diffusion facilitator) proteins, also named MTPs (metal transport proteins) may mediate the transport of divalent cations from the cytosol into the vacuole (Rascio and Navari-Izzo 2011). As an example, MTP1 (localized at tonoplast) has been observed to be overexpressed in leaves of some Zn/Ni hyperaccumulators (Deng et al. 2017; Rascio and Navari-Izzo 2011). Additionally, V-ATPase may facilitate transport of Ni^{2+} from the cytoplasm into the vacuole via N^{2+}/H^+ antiporter activity (Deng et al. 2017; Rascio and Navari-Izzo 2011). Ni has also been observed to be translocated by the phloem, where the speciation in the phloem may differ from that of the xylem. Nickel has been observed to be translocated from old to young leaves, from shoots to roots, and to reproductive organs and seeds; however, very little is known about these processes (Deng et al. 2017). Figure 3.6 provides a snapshot of how Ni, along with Fe and Zn,

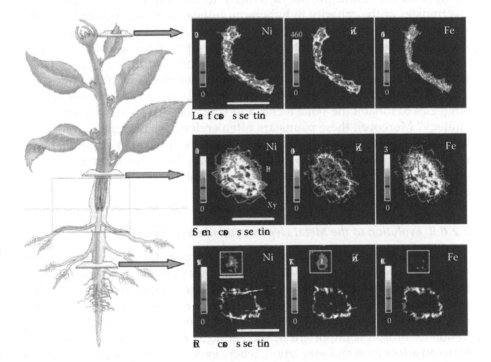

FIGURE 3.6
(See color insert.) Ni, Fe, and Zn fluorescence computed microtomography images of a leaf, stem, coarse and fine root cross sections from *A. murale* "Kotodesh." Inset in root tomogram is of a finer root. The colorimetric scale maps region-specific relative metal concentrations ($\mu g\ g^{-1}$) for each element, with brighter colors indicating areas of higher enrichment. The yellow scale bar represents ~500 µm; the white scale bar (root inset) represents ~100 µm. [Plant figure adapted from *Plant Physiology*, 3rd ed., Tiaz and Zeiger (eds.) with permission from Sinauer Associates, Inc., Publishers.] (From McNear, D. H., E. Peltier, J. Everhart, R. L. Chaney, S. Sutton, M. Newville, M. Rivers, and D. L. Sparks. 2005a. Application of quantitative fluorescence and absorption-edge computed microtomography to image metal compartmentalization in *Alyssum murale*. *Environ. Sci. Technol.* 39(7):2210–2218. doi: 10.1021/es0492034.)

is distributed within the Ni hyperaccumulator *Alyssum murale*, where Ni can easily be seen to be concentrated within the vascular system and the leaves (McNear et al. 2005a).

Ni hyperaccumulating plants have garnered interest over the years for their potential use as a means to remediate Ni-contaminated soils. Theoretically, such plants could be used to "mine" Ni from Ni-contaminated soils. The Ni-rich aerial tissues could be harvested and the Ni could be retrieved via incineration or other means, and recycled. However, given the slow rate of growth, small stature, ultramafic endemism, and/or climate tolerances of many known hyperaccumulators, the economic feasibility of such methods of remediating soil has not yet panned out. As we learn more about the molecular mechanisms behind Ni hyperaccumulation, biotechnological applications could provide a remedy to this problem by allowing for the transfer of the capacity to hyperaccumulate Ni to faster-growing, more robust, and easier-to-harvest plants. However, caution is needed, as fostering Ni-rich crops could result in the transfer of Ni from soils into the food chain. Interestingly, many Ni hyperaccumulators and their closely related, non-hyperaccumulator counterparts often inhabit the same ultramafic soils. Thus, it is difficult not to ask, why do closely related species utilize opposite strategies to inhabit the same niche, and what has driven this divergence in strategy? Moreover, these plants are often endemic to ultramafic soils, or at least have a high affinity for them. As ultramafic soils often impose a range of abiotic challenges to plants growing on them, Ni hyperaccumulators and other ultramafic specialists could be good plant systems for studying plant adaptation to abiotic stress.

3.2.8.3 Evolution of the Metal Hyperaccumulating Phenotype

There are several hypotheses as to how heavy metal hyperaccumulation increases plant fitness. However, as Boyd aptly points out, when assigning function to a trait in the context of evolution, one must be keen to remember that the currently observed function of a trait may not be the function for which the trait was originally selected for millions of years ago (Boyd 2004). In other words, it is important in such endeavors to be able to differentiate between adaptation and exaptation, where exaptation may occur as a means of acclimating to new challenges without evolving new traits (Boyd 2004). Additionally, it would be wise to keep in mind that evolution is not goal oriented—in order for genes associated with survival on ultramafic soils to increase in frequency within the population of a given species, they had to be present in the population of their ancestor in the first place. In other words, plants did not evolve to inhabit ultramafic soils, but rather the ultramafic environment imposed selection pressures that shaped the genetic makeup of the plant populations that inhabit them. Taking the prior two points together, the differentiation between what adaptations were originally selected for when land plants first encountered Ni-enriched, ultramafic soils and how such traits have since been enhanced, diminished, or put to novel uses is a daunting task. However, as previously mentioned, it has been suggested that Ni hyperaccumulation may be the result of constitutive overexpression of common genes, rather than the accumulation of novel ones, suggesting that the adaptations required for Ni hyperaccumulation may have been exacted on the regulation of the metal homeostasis system, rather than the alteration of the system itself, which could simplify the problem considerably and help explain the independent evolution of hyperaccumulation across distantly related plant families.

The tolerance hypothesis suggests that heavy metal hyperaccumulation evolved simply as a means of dealing with high concentrations of metallic elements in the rhizosphere by translocating them to aerial tissues and sequestering them away from critical physiological processes, such as photosynthetic centers (Boyd 2004; Boyd and Martens 1998). This hypothesis is certainly plausible, as an abundance of heavy metal cations may outcompete comparatively scarce nutrient cations for binding and uptake sites along roots, exacerbating problems of nutrient deficiency in soils already low in plant nutrients. Additionally, this strategy may mitigate any negative effects that free metal ions impose directly on the roots or on plant growth-promoting rhizobacteria communities. The drought resistance hypothesis suggests that the accumulation of heavy metals within leaf tissues may reduce transpiration across the cuticle, resulting in greater water retention within the plant body in landscapes that are often drought prone and subject to reduced water holding capacity (Boyd and Martens 1998). The hypothesis of elementary allelopathy puts forth the idea that metal-rich leaf tissue further enriches the topsoil in metals after senescence and decay, reducing competition within the hyperaccumulator's zone of influence (Boyd and Martens 1998).

The defense hypothesis suggests that by sequestering toxic levels of metallic trace elements within leaf tissues, hyperaccumulators are able to discourage herbivorous insects from feeding on leaves, which are particularly important to a plant of small stature (Boyd 2012). This is the most widely researched hypothesis and consequently the most widely accepted, but it requires further investigation. Additionally, it is also posited that use of metals as a defense mechanism is beneficial to the plant, as it reduces the need to synthesize defense compounds. Metals such as Ni may also provide constant protection, whereas defense compounds are often synthesized in response to an attack. However, it is unclear as to whether or not the cost of synthesizing defense compounds is greater than the cost of constitutively overexpressing the genes needed in order hyperaccumulate and sequester toxic levels of metals, especially if a given plant is never preyed upon to any significant measure. This requires special attention, especially in the case where overexpressed transporters are ATPases or symporters, as overexpression of transporters that require ATP for function is metabolically costly. The overexpression of symporters is also costly, as symport usually involves the disruption of electrical potential or the H^+ gradient across the membrane, which requires energy to correct. Furthermore, there is no evidence to suggest that herbivorous insects posed such a great selection pressure on serpentine plant communities, as excluders of heavy metals are more prevalent on the landscape than hyperaccumulators. Last, the joint effects hypothesis claims that elemental defense and organic defense mechanisms act in tandem, reducing the metabolic costs of defense to the plant overall; however, this argument requires close consideration of the previously mentioned points regarding the defense hypothesis (Boyd 2012).

One difficulty shared across all of the above hypotheses is that hyperaccumulated Ni may exhibit variations in temporal and spatial distribution through plant tissues over the duration of the plant's life. The benefits described by any given hypothesis may only increase fitness if the event they are proposed to defend against coincided regularly with certain patterns of Ni hyperaccumulation (Boyd et al. 2008). While these hypotheses seem to address the aforementioned question of why closely related species may have adapted different strategies for inhabiting similar niches, a key element—the rhizosphere—seems to absent from all of them. The physicochemical properties of the rhizosphere, along with the genetic capacity and biochemical activity of the soil microbiota, can differ greatly from that of bulk soil. How might the relationship between rhizosphere microorganisms and soil Ni have influenced plant biology with respect to Ni? As microorganisms often serve as the gatekeepers between geochemistry and biochemistry, the consortium of microorganisms that are active in the rhizosphere of plants growing in Ni-enriched soil may hold the key to Ni cycling and bioavailability—the front lines of Ni hyperaccumulation. Last, Boyd points out that an understanding of hyperaccumulator ecology is only in its infancy. While it is important to understand how hyperaccumulation takes place at the plant molecular level, and how it is effected by rhizosphere processes, that understanding must be placed in the context of the ecology of Ni hyperaccumulating species. Only then will we be able to speculate with success as to why plants may have developed this interesting trait.

3.3 Ni–Microbe Relationships

Like plants, certain metals are required by prokaryotic microorganism in order for them to carry out critical physiological processes. As microorganisms have presumably always lived in environments containing metal ions and have always needed them, it is suggested that genetic mechanisms for regulating metal ions evolved at the same time as those involved in carbon metabolism. Over their 3.5-billion-year history, bacteria in particular have honed their ability to protect themselves against environmental toxins. Thus, regulation and tolerance of both essential and nonessential metals have an extensive evolutionary history, resulting in multiple mechanisms of regulation and tolerance along with the partitioning of these mechanisms among genetic compartments. Essential metals are largely regulated by chromosomal genes, while nonessential metals are often regulated by genes contained within plasmids and other mobile genetic elements, allowing for such traits to be passed among bacteria via transduction, conjugation, or transformation.

Essential metals serve critical functions in osmoregulation, cell wall and protein stability, the catalytic activity of metalloenzymes, and biologically mediated redox chemistry (Macomber and Hausinger 2011). Ni in particular serves as a cofactor in nine enzymes, eight of which are exclusive to microorganisms, with the exception being urease, which is also found in plants and some invertebrates (Boer et al. 2014). Within these enzymes, Ni is often coordinated with histidine or cysteine, and sometimes with aspartate, glutamate, and lysine carbamate (urease) (Boer et al. 2014; Nies 1999). Like all elements, physiologically excessive levels of Ni can be toxic to microorganisms. Evidence regarding the mechanisms of Ni toxicity in microorganisms is sparse; however, some evidence exists for the competitive (Zn and Fe metalloenzymes) and noncompetitive inhibition of enzymes, oxidative stress, and the disruption of electron transport by the replacement of redox-reactive elements such as Fe and Cu in key enzymes with Ni^{2+}, which is relatively redox stable (Macomber and Hausinger 2011). As Ni is an essential micronutrient for many microorganisms, they have developed tools for sensing Ni and regulating its intake and efflux, some of which depend on Ni-regulated gene expression.

3.3.1 Ni Influx

Ni influx may occur across transmembrane proteins that are either specific or nonspecific to Ni. When Ni concentrations are relatively high, Ni may enter microbial cells nonspecifically via the passive Mg^{2+} transport protein, CorA (Macomber and Hausinger 2011; Nies 1999). As exceedingly high concentrations of Mg and Ni may occur together in serpentine systems, it is possible that regulation of the CorA transport system in order to mitigate the influx of Mg may serve a dual role by also limiting Ni uptake. There are three known systems for the specific uptake of Ni in microorganisms. The most well known and understood is the *nik* operon, which encodes a Ni-specific ATP-binding cassette transport system and a protein, NikR, which represses the expression of the operon when Ni^{2+} concentrations are adequate (Macomber and Hausinger 2011). Interestingly, the NikR protein contains two Ni binding sites, a high-affinity site in the C-terminal domain and a low-affinity site in the N-terminal domain (Macomber and Hausinger 2011). Satisfying the high-affinity sites is enough to facilitate repression; however, the affinity between NikR and its DNA target is increased by as much as a 1000-fold when both the high- and low-affinity sites are occupied, with the low-affinity N-terminal being the DNA binding domain (Macomber and Hausinger 2011). NikR homologs are widely distributed in archaea and bacteria and the exact mechanisms of action vary; notably, NikR has been shown to regulate urease expression in some microorganisms (Macomber and Hausinger 2011).

Uptake also occurs via the Ni-cobalt transporter family proteins (NiCoT), namely, HoxN, a Ni permease with high affinity for Ni, and NhlF, a permease that transports Ni and Co (Macomber and Hausinger 2011; Nies 1999). Additionally, transport of Ni across the outer membrane of Gram-negative organisms may be modulated by two proteins—FecA3 and FrpB4 of the TonB-dependent transport family of proteins (Macomber and Hausinger 2011). Aside from Ni and Ni complexes, these proteins are known to be involved in the transport of siderophores, carbohydrates, and vitamin B(12) (Macomber and Hausinger 2011).

3.3.2 Ni Efflux

As Ni is needed in minute quantities and can be toxic when these quantities are surpassed, it is no surprise that Ni efflux systems are more numerous than influx systems. There are six to seven known Ni efflux systems, three of which are known to be plasmid encoded (Macomber and Hausinger 2011). Among the plasmid-encoded Ni efflux systems are *cnrCBA* (cobalt-Ni resistance), *nccCBA* (Ni-cadmium-cobalt resistance), and *nreB*. Both *cnrCBA* and *nccCBA* are members of the resistance–nodulation–division family (RDN) and are common among bacteria resistant to Ni. Another member of the RDN family, *cznCBA* (cadmium-zinc-Ni resistance cluster), occurs on the bacterial chromosome and is also commonly associated with metal tolerance (Macomber and Hausinger 2011; Nies 1999). *nreB* and related proteins are part of the major facilitator superfamily of efflux proteins and have been noted to pump Ni out of the cytoplasm and into the periplasm of many bacteria using the proton motive force (Macomber and Hausinger 2011).

One of the more recently discovered systems for Ni efflux is that facilitated by the protein RcnA. Discovered in *Escherichia coli*, RcnA is an efflux pump that moves Ni and cobalt out of the cytoplasm (Macomber and Hausinger 2011). The expression of RcnA is governed by a transcriptional repressor known as RcnR. In this case, transcription is de-repressed in the presence of Ni, meaning that RcnR is released from its DNA target once it is bound with Ni, allowing transcription and translation of the efflux pump (Macomber and Hausinger 2011). Potential homologs of RcnA have been noted in a wide variety of prokaryotes, including archaea; alpha, beta, gamma proteobacteria; and cyanobacteria (Macomber and Hausinger 2011). Additionally, high levels of expression via plasmids containing many copies of RcnA have been demonstrated to dramatically increase bacterial resilience when challenged with Ni, while deletion of RcnA has the reverse effect—a two- to threefold increase in Ni sensitization (Macomber and Hausinger 2011).

Other chromosomal Ni transporters can be found in the cation diffusion facilitator (Cdf) family and include DmeF (Divalent metal efflux), CzcD, and NepA. DmeF was discovered within the chromosome of *Cupriavidus metallidurans* and is capable of removing iron, zinc, cobalt, cadmium, and Ni from the cytoplasm and appears to be constitutively expressed in *C. metallidurans*. CzcD and NepA were discovered in *Bacillus subtilis* and *Rhizobium etli*, respectively. Members of the Cdf family have been noted in archaea, bacteria, and eukaryotes. Last, two transcriptional repressors, NmtR and KmtR, were observed to sense Ni and cobalt in *Mycobacterium tuberculosis* and may regulate the transcription of *nmtA*, a gene encoding a Ni efflux pump (Macomber and Hausinger 2011). Like RcnR, these appear to be repressors in their native form, releasing from their DNA targets and allowing transcription once Ni concentrations within the cell are great enough.

3.3.3 Ni Tolerance

There are six generally accepted mechanisms by which microorganisms tolerate excessive quantities of metals: (1) avoidance, (2) exclusion by a physical barrier with reduced permeability to metals, (3) active transport (efflux) of the metal out of the cell, (4) sequestration (intra- or extracellular), (5) enzymatic detoxification, and (6) desensitization of cellular targets of metallic elements (Bruins et al. 2000). A given microorganism may utilize one or more of these strategies depending on its particular suite of genes.

Active transport, or efflux, of metal ions out of the cell is the most commonly observed and well-understood strategy, especially in reference to Ni homeostasis and tolerance. The primary genes involved in this strategy are described above. However, there is some evidence that prokaryotic organisms may employ sequestration and enzymatic detoxification strategies. One method of detoxification, reduction of ionic Ni to metallic Ni via H^+ metabolism, has been observed in *Thiocapsa roseopersicina* (Bruins et al. 2000). Thus, it is reasonable to assume that other anaerobic microorganisms (obligate or facultative) may be able to mitigate Ni toxicity via reduction. However, the redox potential of Ni^{2+} (−678 mV) is beyond the range of most organisms growing in aerobic conditions (between the proton–hydrogen redox couple [−421 mV] and the oxygen–hydrogen redox couple [+808 mV]), and so is not likely to be a widely utilized mechanism (Nies 1999). Some sulfur-reducing bacteria have been observed to form extracellular Ni precipitates by forming Ni–protein complexes and Ni–sulfides (Bruins et al. 2000). Interestingly, *Pseudomonas aeruginosa* is able to sequester Ni in the periplasm by transforming it into crystalline forms of Ni carbide and Ni phosphide (Bruins et al. 2000). Several other instances of Ni-inducible sequestration via protein complexes have been observed and may occur in the cytoplasm and the periplasm, including the possible formation of Ni-containing inclusion bodies in *Streptomyces coelicolor* (Bruins et al. 2000). On this note, many have speculated that Ni chaperones that are involved in the transport of Ni and its placement within Ni-metalloenzymes may aid in Ni tolerance by transiently sequestering Ni away from sensitive components; however, there is no strong evidence to support such claims. Last, motile bacteria may avoid Ni exposure via a negative chemotactic response they sense Ni in their environment, as has been observed in *E. coli* and *Helicobacter pylori* (Macomber and Hausinger 2011).

Interestingly, Ni toxicity may be intertwined with Fe metabolism and toxicity. Some prokaryotes exhibit a Ni-dependent increase in Fe uptake when challenged with excessive quantities of Ni (Macomber and Hausinger 2011). It is proposed that Ni may replace Fe in important metalloenzymes, thus the greater need for Fe under Ni stress (Macomber and Hausinger 2011). This is an interesting prospect, as serpentine soils are formed from Fe- and Mg-rich minerals. An abundance of Fe may aid in mitigating Ni toxicity in serpentine systems; however, Fe is also more redox reactive. Thus, a Ni-dependent increase in Fe uptake may also result in significant ROS production. As mentioned before, the exceedingly high levels of Mg in serpentine may also result in the downregulation of CorA, and so mitigate nonspecific influx of Ni. Together, these two edaphic factors (excess Mg leading to less nonspecific uptake and excess Fe needed to mitigate Ni toxicity) may play an important role in Ni homeostasis and tolerance for prokaryotes living in serpentine systems.

3.4 The Serpentine Microbiome and Hyperaccumulator Rhizobiome

While our understanding of the genetic and molecular mechanisms of Ni hyperaccumulation remains minimal, our understanding of how the rhizobiome contributes to the hyperaccumulator phenotype is even less so. Furthermore, the effects of serpentine syndrome on the serpentine microbiome has been largely ignored, and the rhizobiome of hyperaccumulator plants have only been tentatively investigated. The rhizobiome of only about 10% of hyperaccumulator plants have been investigated, with majority of these being Cd and Zn hyperaccumulators despite the fact that Ni hyperaccumulators are the most abundant representatives of the hyperaccumulator community (Visioli et al. 2014). However, there is consensus among the hyperaccumulator and serpentine communities about the increase in available metal ions and the selection of metal-tolerant microorganisms within the rhizosphere of hyperaccumulator plants, including Ni. Additionally, it is commonly accepted that metal-tolerant rhizosphere microorganisms may facilitate plant growth in metal-enriched soils by secretion of phytohormones and beneficial enzymes, increasing the solubility of metals, and increasing root surface area by inducing hairy root development (Visioli et al. 2014).

Non–metal-specific genera that seem to be common occupants of the hyperaccumulator rhizobiome include members of *Arthrobacter, Bacillus, Curtobacterium, Microbacterium, Pseudomonas, Sphingomonas,* and *Variovorax,* with *Arthrobacter, Bacillus,* and *Microbacterium* appearing to be common members of the Ni hyperaccumulator rhizobiome (Visioli et al. 2014), although the supporting evidence is mainly derived from culture-dependent studies carried out in metal-contaminated soils, which may be biased, especially considering the stark differences between serpentine conditions and compromised soil conditions, as well as disparities between soil conditions and culture conditions. Also, it is exceptionally important to note that simply because a microorganism is culturable from the rhizosphere does not mean that it plays an important role in the rhizosphere. In the same vein, culture-dependent studies tend to focus only on fast-growing heterotrophic organisms, which represent only a portion of the rhizosphere community and may have less influence in serpentine systems where carbon and nutrients are often limiting.

Ultramafic soil microbial communities are even less well characterized. However, there is some evidence to suggest some interesting relationships between ultramafic soils, the microbial communities living within them, and ultramafic adapted plant communities. Bordez et al. (2016) found that bacterial and fungal communities in serpentine soils located in New Caledonia inhabited by different plant communities (across different topographies) were not limited compared to non-serpentine systems. Using 16S pyrosequencing, 3477 ± 317 bacterial operational taxonomic units (OTUs) were observed per habitat, with *Proteobacteria, Acidobacteria, Actinobacteria, Planctomycetes, Verrucomicrobia,* and *Chloriflexi* making up the majority of the bacterial community across all plant communities (Bordez et al. 2016). Fungal communities were dominated by *Basidiomycota, Ascomycota, Zygomycota,* and unknown fungi, with an average of 712 ± 43 OTUs observed in each habitat based on ITS pyrosequencing. Additionally, it was found that plant cover was the primary driver of microbial community structure, as opposed to edaphic factors (Bordez et al. 2016).

Pessoa et al. (2015) assessed soil microbiological functioning, diversity, richness, and community structure in Brazilian ultramafic soils populated by tropical savannah vegetation. It was found that soil microbiological functioning was correlated with SOM content more so than soil Ni. Also, Ni did not appear to effect microbial enzyme activities related to the C, P, and S cycles or microbial biomass (Pessoa et al. 2015). While differences in bacterial community structure were observed between serpentine and non-serpentine sites, there was no correlation between community structure and function (based on very limited methods). *Acidobacteria* and *Actinobacteria* were found to be the most abundant phyla in the ultramafic soils. There were no differences in diversity or richness among the sites (Pessoa et al. 2015).

Schipper and Lee (2004) investigated soil microbial biomass, respiration, and diversity under six different plant communities growing on ultramafic soils located in New Zealand. Once again, it was found that vegetation type and total carbon, rather than edaphic factors, were the primary drivers of differences between microbial biomass, respiration, and catabolic evenness (Schipper and Lee 2004). Metal enrichment had no direct effect on microbial diversity, respiration, or biomass, although the authors suggest that the increasing metal concentrations exert an effect on soil microbial communities by altering the plant community structure (Schipper and Lee 2004).

DeGrood et al. (2005) investigated differences in soil microbial communities from non-serpentine soils and disturbed or non-disturbed serpentine soils in Central California, near San Francisco. It was found that microbial biomass was greater in non-serpentine soils when compared to both disturbed and non-disturbed serpentine soils (DeGrood et al. 2005). The most highly disturbed serpentine sight had the lowest microbial biomass and a greater proportion of fungi, and all sites had distinct microbial community structures based on phospholipid fatty acid (PLFA) analysis (DeGrood et al. 2005). It was also found that slow growth and actinomycetes biomarkers were more prominent in the serpentine reference soil than other soils, potentially suggesting an important role for k selected organisms in serpentine systems. Also of note, pH, organic matter, and potassium levels were suggested to play a role in shaping serpentine microbial communities (DeGrood et al. 2005).

Oline (2006) investigated differences among bacterial communities inhabiting serpentine and non-serpentine soils across Northern California and Southern Oregon. It was found that serpentine and non-serpentine communities tended to be different, with geologically separate serpentine communities being more similar to each other than non-serpentine communities, and vice versa (Oline 2006). Serpentine soils were dominated by *Actinobacteria, Acidobacteria, Alphaproteobacteria, Verrucomicrobia*, Greennonsulfur-bacterium related, *Gemmatimonadetes, Planctomycetes*, and *Bacteroidetes* (in that order), while non-serpentine soils were dominated by *Actinobacteria, Alphaproteobacteria, Acidobacteria, Betaproteobacteria, Planctomycetes, Bacteroidetes, and Deltaproteobacteria* (Oline 2006). Additionally, members of the OP8 division were exclusive to serpentine soils, while members of the OP10 division, *Fermicutes*, and *Nitrospirae* were found to be exclusive to non-serpentine systems. Community diversity was not observed to be different by sample, site, or soil type (Oline 2006).

Fitzsimons and Miller (2010) found that the microbial communities associated with the roots of *Avenula sulcata* were not affected by serpentine soil properties, based on PLFA analysis, and found that there were no differences in arbuscular mycorrhizal fungi (AMF) communities (using 18S rDNA) or colonization among serpentine and non-serpentine soils. As noted above, this may indicate that serpentine soils do not pose as much of a challenge for AMF and bacteria as they do for plants. Likewise, Daghino et al. (2012) found that chemical and mineral differences between serpentine soils had no effect on fungal diversity within serpentine systems. On the contrary, Kohout et al.

(2015) found that, in serpentine grasslands of the Czech Republic, serpentine soil characteristics and high Fe concentrations had a negative effect on root colonization by AMF. Additionally, it was observed the high K and Cr concentrations, as well as low pH, had a negative effect on AMF richness. AMF community structure was also found to be correlated with Ni concentration and plant life stage.

In summary, it appears that serpentine syndrome may not pose as much of a challenge to bacteria and fungi as compared to plant life, given that richness and diversity of serpentine microbial communities is often unaffected compared to non-serpentine systems. However, there is some evidence to suggest that while serpentine microbial communities may be just as diverse, there might be some key differences between the community structure of serpentine and non-serpentine systems driven by vegetative cover, pH, and soil carbon. However, other evidence indicates that serpentine soils may pose some unique limitations, as indicated by reduced microbial biomass and respiration in some serpentine soils despite a rich microbial diversity. While these results are interesting, it cannot be ignored that many of these studies utilized methods with course resolution and the number of studies focusing on serpentine microbial communities remains few, indicating a need for more robust studies on ultramafic microbial communities. The fact that serpentine systems are just as diverse as non-serpentine systems and contain similar groups of organisms may be indicative of novel organisms or novel genetic capacity. Additionally, it should be noted that many studies find groups of organisms such as *Actinobacteria*, *Acidobacteria*, *Proteobacteria*, *Firmicutes*, *Cyanoacteria*, *Bacteroidetes*, and *Spirochetes* representing the vast majority of data within genomic databases, presenting the possibility of biased results from molecular ecology studies like those above (Land et al. 2015).

References

Allison, J. D., Brown, D. S., and K. J. Novo-Gradac. 1991. MINTEQA2/PRODEFA2, a geochemical assessment model for environmental systems: Version 3. 0 user's manual. United States: N. p. Web.

Antoniadis, V., E. Levizou, S. M. Shaheen, Y. S. Ok, A. Sebastian, C. Baum, M. N. V. Prasad, W. W. Wenzel, and J. Rinklebe. 2017. Trace elements in the soil–plant interface: Phytoavailability, translocation, and phytoremediation—A review. *Earth-Sci. Rev.* 171:621–645. doi: https://doi.org/10.1016/j.earscirev.2017.06.005.

Baes, C. F. and R. E. Mesmer. 1986. *The Hydrolysis of Cations*. Malabar, FL: Krieger Publishing Company.

Baeyens, B., M. H. Bradbury, and W. Hummel. 2003. Determination of aqueous nickel-carbonate and nickel-oxalate complexation constants. *J. Solution Chem.* 32(4):319–339. doi: 10.1023/a:1023753704426.

Boer, J. L., S. B. Mulrooney, and R. P. Hausinger. 2014. Nickel-dependent metallo-enzymes. *Arch. Biochem. Biophys.* 544:142–152. doi: http://dx.doi.org/10.1016/j.abb.2013.09.002.

Bordez, L., P. Jourand, M. Ducousso, F. Carriconde, Y. Cavaloc, S. Santini, J. M. Claverie, L. Wantiez, A. Leveau, and H. Amir. 2016. Distribution patterns of microbial communities in ultramafic landscape: A metagenetic approach highlights the strong relationships between diversity and environmental traits. *Mol. Ecol.* 25(10):2258–2272. doi: 10.1111/mec.13621.

Boyd, R. S. 2004. Ecology of metal hyperaccumulation. *New Phytol.* 162(3):563–567. doi: 10.1111/j.1469-8137.2004.01079.x.

Boyd, R. S. 2012. Plant defense using toxic inorganic ions: Conceptual models of the defensive enhancement and joint effects hypotheses. *Plant Sci.* 195:88–95. doi: https://doi.org/10.1016/j.plantsci.2012.06.012.

Boyd, R. S., M. A. Davis, and K. Balkwill. 2008. Elemental patterns in Ni hyperaccumulating and non-hyperaccumulating ultramafic soil populations of *Senecio coronatus*. *South Afr. J. Botany* 74(1):158–162. doi: https://doi.org/10.1016/j.sajb.2007.08.013.

Boyd, R. S. and S. N. Martens. 1998. The significance of metal hyperaccumulation for biotic interactions. *CHEMOECOLOGY* 8(1):1–7. doi: 10.1007/s000490050002.

Bradbury, M. H. and B. Baeyens. 2005. Experimental measurements and modeling of sorption competition on montmorillonite. *Geochim. Cosmochim. Acta* 69(17):4187–4197. doi: https://doi.org/10.1016/j.gca.2005.04.014.

Brady, K. U., A. R. Kruckeberg, and H. D. Bradshaw Jr. 2005. Evolutionary ecology of plant adaptation to serpentine soils. *Annu. Rev. Ecol. Evol. System.* 36(1):243–266. doi: 10.1146/annurev.ecolsys.35.021103.105730.

Brown, P. H., R. M. Welch, and E. E. Cary. 1987. Nickel: A micronutrient essential for higher plants. *Plant Physiol.* 85(3):801–803.

Bruins, M. R., S. Kapil, and F. W. Oehme. 2000. Microbial resistance to metals in the environment. *Ecotoxicol. Environ. Safety* 45(3):198–207. doi: http://dx.doi.org/10.1006/eesa.1999.1860.

Bryce, A. L. and S. B. Clark. 1996. Nickel desorption kinetics from hydrous ferric oxide in the presence of EDTA. *Colloids Surf. A Physicochem. Eng. Aspects* 107:123–130. doi: https://doi.org/10.1016/0927-7757(95)03350-5.

Carlini, C. R. and R. Ligabue-Braun. 2016. Ureases as multifunctional toxic proteins: A review. *Toxicon* 110:90–109. doi: 10.1016/j.toxicon.2015.11.020.

Chaney, R., S. Brown, Y. M. Li, J. S. Angle, F. A. Homer, and C. Green. 1995. *Potential Use of Metal Hyperaccumulators.* Vol. 3, *Mining Environmental Management.*

Chen, C., D. Huang, and J. Liu. 2009. Functions and toxicity of nickel in plants: Recent advances and future prospects. *CLEAN—Soil, Air, Water* 37(4–5):304–313. doi: 10.1002/clen.200800199.

Chen, T., G. Hefter, and R. Buchner. 2005. Ion association and hydration in aqueous solutions of nickel(ii) and cobalt(ii) sulfate. *J. Solut. Chem.* 34(9):1045–1066. doi: 10.1007/s10953-005-6993-5.

Coughlin, B. R. and A. T. Stone. 1995. Nonreversible adsorption of divalent metal ions (MnII, CoII, NiII, CuII, and PbII) onto goethite: Effects of acidification, FeII addition, and picolinic acid addition. *Environ. Sci. Technol.* 29(9):2445–2455. doi: 10.1021/es00009a042.

Daghino, S., C. Murat, E. Sizzano, M. Girlanda, and S. Perotto. 2012. Fungal diversity is not determined by mineral and chemical differences in serpentine substrates. *PLoS One* 7(9):e44233. doi: 10.1371/journal.pone.0044233.

Dähn, R., A. M. Scheidegger, A. Manceau, M. L. Schlegel, B. Baeyens, M. H. Bradbury, and D. Chateigner. 2003. Structural evidence for the sorption of Ni(II) atoms on the edges of montmorillonite clay minerals: A polarized X-ray absorption fine structure study. *Geochim. Cosmochim. Acta* 67(1):1–15. doi: https://doi.org/10.1016/S0016-7037(02)01005-0.

Dähn, R., A. M. Scheidegger, A. Manceau, M. L. Schlegel, B. Baeyens, M. H. Bradbury, and M. Morales. 2002. Neoformation of Ni phyllosilicate upon Ni uptake on montmorillonite: A kinetics study by powder and polarized extended X-ray absorption fine structure spectroscopy. *Geochim. Cosmochim. Acta* 66(13):2335–2347. doi: https://doi.org/10.1016/S0016-7037(02)00842-6.

DeGrood, S. H, V. P. Claassen, and K. M. Scow. 2005. Microbial community composition on native and drastically disturbed serpentine soils. *Soil Biol. Biochem.* 37(8):1427–1435.

Deng, T.-H.-B., A. van der Ent, Y.-T. Tang, T. Sterckeman, G. Echevarria, J.-L. Morel, and R.-L. Qiu. 2017. Nickel hyperaccumulation mechanisms: A review on the current state of knowledge. *Plant Soil.* doi: 10.1007/s11104-017-3539-8.

Duke, J. M. 1980. Nickel in rocks and ores. In: *Nickel in the Environment*, edited by J. O. Nriagu, 27–50. New York: John Wiley & Sons.

Dzombak, D. A. and F. M. M. Morel. 1990. *Surface Complexation Modeling: Hydrous Ferric Oxide.* New York: John Wiley & Sons.

Elzinga, E. J. and D. L. Sparks. 2001. Reaction condition effects on nickel sorption mechanisms in illite–water suspensions. *Soil Sci. Soc. Am. J.* 65(1):94–101. doi: 10.2136/sssaj2001.65194x.

Fabiano, C. C., T. Tezotto, J. L. Favarin, J. C. Polacco, and P. Mazzafera. 2015. Essentiality of nickel in plants: A role in plant stresses. *Front. Plant Sci.* 6:754. doi: 10.3389/fpls.2015.00754.

Fitzsimons, M. S. and R. M. Miller. 2010. Serpentine soil has little influence on the root-associated microbial community composition of the serpentine tolerant grass species *Avenula sulcata*. *Plant Soil* 330(1–2):393–405.

Ford, R. G., P. M. Bertsch, and K. J. Farley. 1997. Changes in transition and heavy metal partitioning during hydrous iron oxide aging. *Environ. Sci. Technol.* 31(7):2028–2033. doi: 10.1021/es960824+.

Ford, R. G., K. M. Kemner, and P. M. Bertsch. 1999a. Influence of sorbate–sorbent interactions on the crystallization kinetics of nickel- and lead-ferrihydrite coprecipitates. *Geochim. Cosmochim. Acta* 63(1):39–48. doi: https://doi.org/10.1016/S0016-7037(99)00010-1.

Ford, R. G., A. C. Scheinost, K. G. Scheckel, and D. L. Sparks. 1999b. The link between clay mineral weathering and the stabilization of Ni surface precipitates. *Environ. Sci. Technol.* 33(18):3140–3144. doi: 10.1021/es990271d.

Hoefer, C., J. Santner, S. M. Borisov, W. W. Wenzel, and M. Puschenreiter. 2017. Integrating chemical imaging of cationic trace metal solutes and pH into a single hydrogel layer. *Anal. Chim. Acta* 950:88–97. doi: 10.1016/j.aca.2016.11.004.

Hoffmann, U. and S. L. S. Stipp. 2001. The behavior of Ni^{2+} on calcite surfaces. *Geochim. Cosmochim. Acta* 65(22):4131–4139.

Huerta-Diaz, M. A. and J. W. Morse. 1992. Pyritization of trace metals in anoxic marine sediments. *Geochim. Cosmochim. Acta* 56 (7):2681–2702. doi: https://doi.org/10.1016/0016-7037(92)90353-K.

Hummel, W. and E. Curti. 2003. Nickel aqueous speciation and solubility at ambient conditions: A thermodynamic elegy. *Monatshefte für Chemie/Chemical Monthly* 134(7):941–973. doi: 10.1007/s00706-003-0010-8.

Japan Atomic Energy Agency. 2016. 28-Nickel. http://wwwndc.jaea.go.jp/cgi-bin/nucltab14?28.

Kierczak, J., A. Pędziwiatr, J. Waroszewski, and M. Modelska. 2016. Mobility of Ni, Cr and Co in serpentine soils derived on various ultrabasic bedrocks under temperate climate. *Geoderma* 268:78–91. doi: https://doi.org/10.1016/j.geoderma.2016.01.025.

Kjøller, C., D. Postma, and F. Larsen. 2004. Groundwater acidification and the mobilization of trace metals in a sandy aquifer. *Environ. Sci. Technol.* 38(10):2829–2835. doi: 10.1021/es030133v.

Kohout, P., P. Doubková, M. Bahram, J. Suda, L. Tedersoo, J. Voříšková, and R. Sudová. 2015. Niche partitioning in arbuscular mycorrhizal communities in temperate grasslands: A lesson from adjacent serpentine and nonserpentine habitats. *Mol. Ecol.* 24(8):1831–1843.

Kramer, U. 2010. Metal hyperaccumulation in plants. *Annu. Rev. Plant Biol.* 61:517–534. doi: 10.1146/annurev-arplant-042809-112156.

Land, M., L. Hauser, S.-R. Jun, I. Nookaew, M. R. Leuze, T.-H. Ahn, T. Karpinets, O. Lund, G. Kora, T. Wassenaar, S. Poudel, and D. W. Ussery. 2015. Insights from 20 years of bacterial genome sequencing. *Funct. Integr. Genom.* 15(2):141–161. doi: 10.1007/s10142-015-0433-4.

Larsen, F. and D. Postma. 1997. Nickel mobilization in a groundwater well field: Release by pyrite oxidation and desorption from manganese oxides. *Environ. Sci. Technol.* 31(9):2589–2595. doi: 10.1021/es9610794.

Lindsay, W. L. 1979. *Chemical Equilibria in Soils.* Chichester, Sussex: John Wiley & Sons.

Macomber, L. and R. P. Hausinger. 2011. Mechanisms of nickel toxicity in microorganisms. *Metallomics* 3(11):1153–1162. doi: 10.1039/c1mt00063b.

Manceau, A. 1986. *Nickel-Bearing Clay Minerals: II. Intracrystalline Distribution of Nickel: An X-ray Absorption Study.* Vol. 21.

Manceau, A., G. Calas, and A. Decarreau. 1985. *Nickel-Bearing Clay Minerals: 1. Optical Spectroscopic Study of Nickel Crystal Chemistry.* Vol. 20.

Manceau, A., M. Lanson, and N. Geoffroy. 2007. Natural speciation of Ni, Zn, Ba, and As in ferromanganese coatings on quartz using X-ray fluorescence, absorption, and diffraction. *Geochim. Cosmochim. Acta* 71(1):95–128.

Manceau, A., N. Tamura, M. Marcus, A. Macdowell, R. Celestre, R. E. Sublett, G. Sposito, and H. A. Padmore. 2002. *Deciphering Ni Sequestration in Soil Ferromanganese Nodules by Combining X-ray Fluorescence, Absorption, and Diffraction at Micrometer Scales of Resolution.* Vol. 87.

Mattigod, S. V., D. Rai, A. R. Felmy, and L. Rao. 1997. Solubility and solubility product of crystalline $Ni(OH)_2$. *J. Solut. Chem.* 26(4):391–403. doi: 10.1007/bf02767678.

McNear, D. H. Jr., R. L. Chaney, and D. L. Sparks. 2010. The hyperaccumulator *Alyssum murale* uses complexation with nitrogen and oxygen donor ligands for Ni transport and storage. *Phytochemistry* 71(2–3):188–200. doi: 10.1016/j.phytochem.2009.10.023.

McNear, D. H. Jr., E. Peltier, J. Everhart, R. L. Chaney, S. Sutton, M. Newville, M. Rivers, and D. L. Sparks. 2005a. Application of quantitative fluorescence and absorption-edge computed microtomography to image metal compartmentalization in *Alyssum murale*. *Environ. Sci. Technol.* 39(7):2210–2218. doi: 10.1021/es0492034.

McNear, D. H. Jr., R. Tappero, and D. L. Sparks. 2005b. Shining light on metals in the environment. *Elements* 1(4):211–216. doi: 10.2113/gselements.1.4.211.

Merlen, E., P. Gueroult, J. B. d'Espinose de la Caillerie, B. Rebours, C. Bobin, and O. Clause. 1995. Hydrotalcite formation at the alumina/water interface during impregnation with Ni (II) aqueous solutions at neutral pH. *Appl. Clay Sci.* 10(1):45–56. doi: https://doi.org/10.1016/0169-1317(95)00015-V.

Nachtegaal, M. and D. L. Sparks. 2003. Nickel sequestration in a kaolinite–humic acid complex. *Environ. Sci. Technol.* 37(3):529–534. doi: 10.1021/es025803w.

Nies, D. H. 1999. Microbial heavy-metal resistance. *Appl. Microbiol. Biotechnol.* 51(6):730–750.

Nishida, S., C. Tsuzuki, A. Kato, A. Aisu, J. Yoshida, and T. Mizuno. 2011. AtIRT1, the primary iron uptake transporter in the root, mediates excess nickel accumulation in *Arabidopsis thaliana*. *Plant Cell Physiol.* 52(8):1433–1442. doi: 10.1093/pcp/pcr089.

Nishida, S., A. Aisu, and T. Mizuno. 2012. Induction of IRT1 by the nickel-induced iron-deficient response in *Arabidopsis*. *Plant Signal. Behav.* 7(3):329–331. doi: 10.4161/psb.19263.

Nowack, B., H. Xue, and L. Sigg. 1997. Influence of natural and anthropogenic ligands on metal transport during infiltration of river water to groundwater. *Environ. Sci. Technol.* 31(3):866–872. doi: 10.1021/es960556f.

Oline, D. K. 2006. Phylogenetic comparisons of bacterial communities from serpentine and nonserpentine soils. *Appl. Environ. Microbiol.* 72(11):6965–6971.

Pędziwiatr, A., J. Kierczak, J. Waroszewski, G. Ratié, C. Quantin, and E. Ponzevera. 2017. Rock-type control of Ni, Cr, and Co phytoavailability in ultramafic soils. *Plant Soil.* doi: 10.1007/s11104-017-3523-3.

Peltier, E., R. Allada, A. Navrotsky, and D. Sparks. 2006. *Nickel Solubility and Precipitation in Soils: A Thermodynamic Study*. Vol. 54.

Pessoa, M., C. C. Barreto, F. B. dos Reis, R. R. Fragoso, F. S. Costa, I. D. Mendes, and L. R. M. de Andrade. 2015. Microbiological functioning, diversity, and structure of bacterial communities in ultramafic soils from a tropical savanna. *Antonie Van Leeuwenhoek Int. J. Gen. Mol. Microbiol.* 107(4):935–949. doi: 10.1007/s10482-015-0386-6.

Polacco, J. C., P. Mazzafera, and T. Tezotto. 2013. Opinion: Nickel and urease in plants: Still many knowledge gaps. *Plant Sci.* 199–200:79–90. doi: 10.1016/j.plantsci.2012.10.010.

Pollard, A. J., R. D. Reeves, and A. J. Baker. 2014. Facultative hyperaccumulation of heavy metals and metalloids. *Plant Sci.* 217–218:8–17. doi: 10.1016/j.plantsci.2013.11.011.

Rascio, N. and F. Navari-Izzo. 2011. Heavy metal hyperaccumulating plants: How and why do they do it? And what makes them so interesting? *Plant Sci.* 180(2):169–181. doi: 10.1016/j.plantsci.2010.08.016.

Registry, Agency for Toxic Substances and Disease. 2005. Toxicological Profile for Nickel (Update). ed. Public U.S. Department of Public Health and Human Services and Health Service. Atlanta, GA.

Richter, R. O. and T. L. Theis. 1980. Nickel speciation in a soil/water system. In: *Nickel in the Environment*, ed. J. O. Nriagu, pp. 189–202, New York: John Wiley & Sons.

Santner, J., M. Larsen, A. Kreuzeder, and R. N. Glud. 2015. Two decades of chemical imaging of solutes in sediments and soils—A review. *Anal. Chim. Acta* 878:9–42. doi: 10.1016/j.aca.2015.02.006.

Schaaf, G., U. Ludewig, B. E. Erenoglu, S. Mori, T. Kitahara, and N. von Wirén. 2004. ZmYS1 functions as a proton-coupled symporter for phytosiderophore- and nicotianamine-chelated metals. *J. Biol. Chem.* 279(10):9091–9096. doi: 10.1074/jbc.M311799200.

Scheckel, K. G., A. C. Scheinost, R. G. Ford, and D. L. Sparks. 2000. Stability of layered Ni hydroxide surface precipitates—A dissolution kinetics study. *Geochim. Cosmochim. Acta* 64(16):2727–2735. doi: https://doi.org/10.1016/S0016-7037(00)00385-9.

Scheckel, K. G. and D. L. Sparks. 2000. Kinetics of the formation and dissolution of Ni precipitates in a gibbsite/amorphous silica mixture. *J. Colloid Interface Sci.* 229(1):222–229. doi: 10.1006/jcis.2000.7001.

Scheckel, K. G. and D. L. Sparks. 2001. Temperature effects on nickel sorption kinetics at the mineral–water interface. *Soil Sci. Soc. Am. J.* 65(3):719–728.

Scheidegger, A. M., D. G. Strawn, G. M. Lamble, and D. L. Sparks. 1998. The kinetics of mixed Ni-Al hydroxide formation on clay and aluminum oxide minerals: A time-resolved XAFS study. *Geochim. Cosmochim. Acta* 62(13):2233–2245. doi: https://doi.org/10.1016/S0016-7037(98)00136-7.

Scheinost, A. C., R. G. Ford, and D. L. Sparks. 1999. The role of Al in the formation of secondary Ni precipitates on pyrophyllite, gibbsite, talc, and amorphous silica: A DRS study. *Geochim. Cosmochim. Acta* 63(19):3193–3203. doi: https://doi.org/10.1016/S0016-7037(99)00244-6.

Scheinost, A. C. and D. L. Sparks. 2000. Formation of layered single- and double-metal hydroxide precipitates at the mineral/water interface: A multiple-scattering XAFS analysis. *J. Colloid Interface Sci.* 223(2):167–178. doi: 10.1006/jcis.1999.6638.

Schipper, L. A. and W. G. Lee. 2004. Microbial biomass, respiration and diversity in ultramafic soils of West Dome, New Zealand. *Plant Soil* 262(1–2):151–158.

Schultz, M. F., M. M. Benjamin, and J. F. Ferguson. 1987. Adsorption and desorption of metals on ferrihydrite: Reversibility of the reaction and sorption properties of the regenerated solid. *Environ. Sci. Technol.* 21(9):863–869. doi: 10.1021/es00163a003.

Schulze, D. G. 2002. An introduction to soil minerology. In: *Soil Mineralogy with Environmental Applications*, eds. J. B. Dixon, D. G. Schulze, pp. 1–35, SSSA Book Ser. 7. Madison, WI: SSSA. doi:10.2136/sssabookser7.c1.

Seregin, I. and A. Kozhevnikova. 2006. *Physiological Role of Nickel and Its Toxic Effects on Higher Plants.* Vol. 53.

Strathmann, T. J. and S. C. B. Myneni. 2004. Speciation of aqueous Ni(II)-carboxylate and Ni(II)-fulvic acid solutions: Combined ATR-FTIR and XAFS analysis. *Geochim. Cosmochim. Acta* 68(17):3441–3458. doi: https://doi.org/10.1016/j.gca.2004.01.012.

Tashakor, M., W. Z. W. Yaacob, and H. Mohamad. 2013. Serpentine soil, adverse habitat for plants. *Am. J. Environ. Sci.* 9(1):82–87. doi: 10.3844/ajessp.2013.82.87.

Thoenen, T. 1999. Pitfalls in the use of solubility limits for radioactive waste disposal: The case of nickel in sulfidic groundwaters. *Nucl. Technol.* 126(1):75–87. doi: 10.13182/NT99-A2959.

Tonkin, J. W., L. S. Balistrieri, and J. W. Murray. 2004. Modeling sorption of divalent metal cations on hydrous manganese oxide using the diffuse double layer model. *Appl. Geochem.* 19:29–53.

United States Environmental Protection Agency. 2000. Nickel Compounds. ed. Office of Research and Development National Center for Environmental Assessment. Washington, D.C.

United States Environmental Protection Agency. 2017. Toxic Release Inventory. https://iaspub.epa.gov/triexplorer/tri_release.chemical.

United States Geological Survey. 1997–2017. Mineral commodities survey.

United States Geological Survey. 2016. 2013 Minerals Yearbook.

United States Geological Survey. 2017. Mineral commodity summaries 2017. https://doi.org/10.3133/70180197: United States Geological Survey.

van der Ent, A., A. J. M. Baker, R. D. Reeves, A. J. Pollard, and H. Schat. 2013. Hyperaccumulators of metal and metalloid trace elements: Facts and fiction. *Plant Soil* 362(1):319–334. doi: 10.1007/s11104-012-1287-3.

Vaughan, A. P. M. and J. H. Scarrow. 2003. Ophiolite obduction pulses as a proxy indicator of superplume events? *Earth Planet. Sci. Lett.* 213(3–4):407–416. doi: http://dx.doi.org/10.1016/S0012-821X(03)00330-3.

Visioli, G., S. D'Egidio, and A. M. Sanangelantoni. 2014. The bacterial rhizobiome of hyperaccumulators: Future perspectives based on omics analysis and advanced microscopy. *Front. Plant Sci.* 5:752. doi: 10.3389/fpls.2014.00752.

Zachara, J. M., J. K. Fredrickson, S. C. Smith, and P. L. Gassman. 2001. Solubilization of Fe(III) oxide-bound trace metals by a dissimilatory Fe(III) reducing bacterium. *Geochim. Cosmochim. Acta* 65(1):75–93.

4

Nickel Resources and Sources

Paweł Harasim

CONTENTS

4.1 Introduction

Nickel is a naturally occurring metallic element. The first known unintentional use of nickel is dated 4000–3500 BC when civilizations of Sumeria, Syria, and the Indus Valley made some bronzes containing over 1% of Ni (Cheng and Schwitter 1957). In 1751, Baron Axel Fredrik Cronstedt identified a new metal in the Kupfernickel (niccolite or nickeline) mineral from Freiberg in Germany. After isolation in an impure sample and further studies, he proposed the name nickel for the new element. The word "nickel" is a shortened translation of the German word "Kupfernickel" (copper-nickel), so called because although it looks similar to copper, it does not contain copper. In 1804, the German scientist Richter achieved pure nickel, which allowed the new metal's physical and chemical properties to be explored.

Two-thirds of all elements found in nature are metals. Metals are contained in the Earth's crust and in parent rocks, by whose weathering soils are formed, so their presence differs in different geographic regions (Zovko and Romić 2011). Nickel's abundance in the Earth's crust is 80–90 mg kg^{-1}, but the largest deposits of this metal are concentrated in the core, which makes it the fifth most common element on the planet (Moss et al. 2011). Nickel occurs in nature as a trace constituent in a wide variety of minerals (Table 4.1), but globally, more than 90% is obtained from pentlandite (Eisler 2007).

Nickel is a silvery white metal that belongs to the transition metal series. For that reason, it is resistant to corrosion (by air, water, and alkali) but dissolves

TABLE 4.1

The Most Common Nickel-Bearing Minerals Found in Economic Deposits

Name	Group	Most Common Mode of Occurrence	Example Deposits
Pentlandite	Sulfide	In mafic intrusions or remobilized phase after metamorphism	Norilsk, Russia Bushveld, South Africa Voisey's Bay, Canada Kambalda, Western Australia
Ni replacement in pyrrhotite	Sulfide	In mafic intrusions	Norilsk, Russia Bushveld, South Africa Voisey's Bay, Canada Kambalda, Western Australia
Garnierite	Hydrous nickel silicate (serpentine)	In laterites related to ultramafic rocks	New Caledonia Sulawesi, Indonesia Cerro Matoso, Colombia
Nickeliferous limonite	Hydroxide	In laterites related to ultramafic rocks	New Caledonia Sulawesi, Indonesia Euboea, Greece
Millerite	Sulfide	Mafic intrusions where metasomatism has remobilized Ni and S from pentlandite; also from metamorphism of olivine	Silver Swan, Western Australia Sudbury, Canada
Niccolite	Nickel arsenide	Hydrothermal replacement of pentlandite in mafic intrusions or metasomatism of low-sulfur mafic rocks. Seafloor manganese nodules	Cobalt, Ontario Widgiemooltha Dome and Kambalda, Western Australia
Nickeliferous goethite	Hydrated oxide	In laterites related to ultramafic rocks	Koniambo Massif, New Caledonia
Siegenite	Sulfide	Hydrothermal veins	Siegen, Germany Jachymov, Czech Republic

Source: Bide, T., Hetherington, L., Gunn, G., and A. Minks. 2008. Nickel. British geological survey (https://www.bgs.ac.uk/downloads/start.cfm?id=1411).

in dilute oxidizing acids. Since the beginning of the nineteenth century, nickel-containing alloys have found an ever-increasing use in modern technologies (Sreekanth et al. 2013). Owing mainly to the corrosive resistance of this metal, its global annual demand reached roughly 1800 kilotons (kt) in 2013.

4.2 Nickel Resources and Sources

In the middle of the nineteenth century, commercial nickel exploitation and electroplating began, which coincided with the Second Industrial Revolution. In those days, the majority of nickel mining was from laterites of

New Caledonia. In 1873, polymetallic nodules containing nickel on the ocean floor were discovered (Sharma 2010). The twentieth century was marked by a significant increase in Ni mining. In 1920, most of the global nickel supply (80%) came from sulfide ores mined in the Sudbury region of Ontario, Canada (Lightfoot 2016), and Canada was the largest producer of Ni in the world for over 50 years of the last century. In Canada, nickel is obtained commercially from pentlandite and pyrrhotite minerals of the Sudbury region. The main source of Ni in that region is from outer space. Originally thought of as a meteor, it is now believed that a comet struck the Earth about 1.8 billion years ago, making the second largest crater in the world—the Sudbury basin. This region is currently responsible for about 10% of the world's nickel production.

Ni-bearing ore deposits are important for economic exploitation. Nickel resources are found in either sulfide- or laterite-type ores (Figure 4.1). Sulfide ore grades range from 0.15% to 7.77% Ni, but most of them are in the 0.2% to 2% Ni range (Hoatson et al. 2006). Globally, the bulk of historic Ni production has been derived from sulfide ores, while the majority of known Ni resources are contained in laterite ores. This unusual difference is due mainly to the challenges of processing laterite compared to sulfide ores—leading to a historical preference for sulfide ores. To meet future demands for Ni, however, there is an increasing amount of Ni being mined from laterite ores (Mudd 2010). Nowadays, about 60% of production comes from sulfide and 40% comes from laterite deposits.

Production of nickel in recent years has continued to increase; in 2014, global production reached 2450 kt while top 10 nickel-producing countries reached 1935 kt (this is a sum of all 10 countries from Figure 4.2).

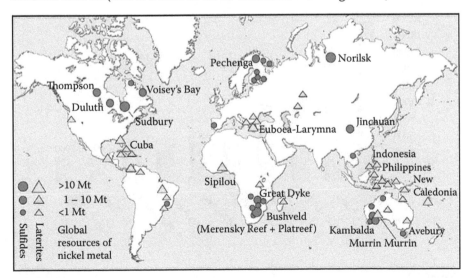

FIGURE 4.1
Global resources of nickel. (From Bide, T., L. Hetherington, G. Gunn, and A. Minks. 2008. Nickel. British geological survey [accessed via https://www.bgs.ac.uk/downloads/start.cfm?id=1411].)

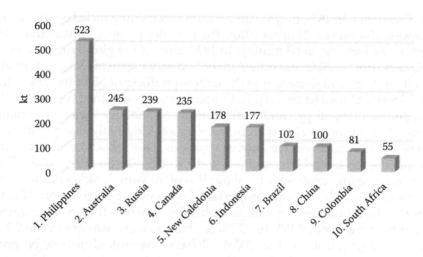

FIGURE 4.2
Nickel mines production in 2014. (Adapted from Kuck, P. H. 2016. Nickel: U.S. Geological Survey Mineral Commodity Summaries 2016, pp. 114–115 [http://minerals.usgs.gov/minerals /pubs/commodity/nickel/mcs-2016-nicke.pdf].)

More than 20% of the world's nickel supply is being mined in the Philippines (523 kt); however, as far as Ni reserves are concerned, the highest estimated (19,000 kt) lie in Australia (Figure 4.3). Out of five countries having the biggest Ni reserves (Figure 4.3), Australia has nickel resources with both sulfide- and laterite-type ore deposits. Brazil, New Caledonia, and Cuba

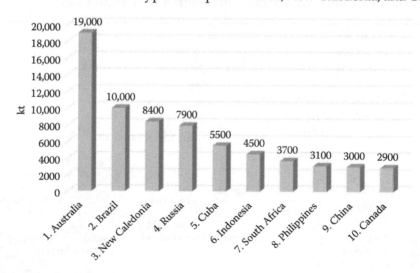

FIGURE 4.3
Nickel reserves. (Adapted from Kuck, P. H. 2016. Nickel: U.S. Geological Survey Mineral Commodity Summaries 2016, pp. 114–115 [http://minerals.usgs.gov/minerals/pubs/commodity /nickel/mcs-2016-nicke.pdf].)

have laterite-type (or oxide-type from weathered rock) deposits, and Russia mainly has sulfide-type deposits (Arndt et al. 2015) (Figure 4.1).

The global nickel demand is constantly growing and is heavily dependent on China, whose nickel consumption is increasing rapidly, driven by strong stainless steel demand (Moss et al. 2011). In 2005, the global nickel sector was dominated by alloy production: 68% was used to manufacture stainless steel (which consists of 8% nickel, 18% chromium, and 74% iron); 11%, nickel and copper alloys; 7%, alloy steels; 7%, plating (to slow down corrosion); 4%, batteries, catalyst, and other; 3%, foundry (Reck and Rotter 2012). Apart from metallic nickel, commercially important Ni compounds include the following: acetate, carbonate, carbonyl, chloride, nitrate, oxide, sulfate, and sulfide.

Due to rising nickel prices in the global market, in 2005, China could not fulfill its growing demand and found a solution in the form of nickel pig iron (NPI), which is a low-grade ferronickel used for the production of stainless steel. NPI contains up to 7% Ni and is a cheaper alternative to pure nickel sold in the world market. NPI was at first imported from Indonesia to China, but after the ban incorporated in 2014, the Philippines became the main supplier of NPI while New Caledonia is gaining importance. As a result of massive stockpiles built by Chinese NPI makers, nickel price has dropped until now. In 2016, over a dozen nickel mines in the Philippines have been suspended because of environmental crackdown. These mines fail to meet environmental and welfare standards, and they account for over 50% of nickel ore output in the Philippines based on last year's production. As a result, the global production of nickel in 2017 is expected to be lower than the demand and the nickel price will presumably rise.

Global nickel mineral resources are nearly 300 million tons (Mudd and Jowitt 2014). Identified land-based resources averaging 1% nickel or greater contain at least 130 million tons of nickel, with about 60% in laterites and 40% in sulfide deposits. Extensive nickel resources are also found in manganese crusts and nodules on the ocean floor (Kuck 2016). It is estimated that 290 million tons of nickel is contained in deep sea nodules (Bide et al. 2008), which are also called polymetallic nodules because they contain nickel, cobalt, iron, and manganese. Though nodules occur in all oceans, those of the highest economical value containing Ni are, on average, at depths of 4000–6000 m, mainly in the Pacific Ocean and the Indian Ocean (Cronan 1980; Halbach and Fellerer 1980). Nearly 150 years have passed since the discovery of polymetallic nodules, but the exploitation of this nickel source has not yet begun, but this will most likely change soon. The first commercial extraction is in the planning stages. Seabed mining is at the very beginning, facing the problem of ownership of deposits, which lie mainly in international waters, and raising the question of what impact it will have on the aquatic environment especially in the long term. If it becomes profitable, the great ocean floor resources will extend available Ni reserves, providing ample time to search for Ni alternatives. Kuck (2016) reported some alternatives, which are used nowadays, including low-nickel, duplex, or ultrahigh-chromium stainless steels that are being substituted for austenitic grades in construction. Nickel-free specialty steels are sometimes used in place of stainless steel in the power-generating and petrochemical industries. Titanium alloys can substitute for nickel metal or nickel-base alloys in corrosive chemical environments. Lithium-ion batteries instead of nickel-metal hydride may be used in certain applications.

Nickel is an important metal with wide applications in our day-to-day life. The demand for nickel has been significantly increased in recent decade for infrastructural development and industrial applications to achieve inclusive growth in world economy. However, the mineral resources for nickel production are limited. The sulfidic minerals, which have been the prime sources for nickel, are depleting in alarming rate due to extensive mining for the metal. In this context the abandoned low-grade laterites are gaining importance as an alternative resource for nickel. The nickel present in nickel laterites is in very complex form as well as the content of the metal is very less, hence the laterite ores have been neglected for nickel extraction. Because of such reasons the conventional mineral processing (pyrometallurgical and hydrometallurgical) techniques are not economically suitable for nickel extraction. In this context, microbial mineral processing (biohydrometallurgy) routes are leading aspects for extraction of metal values from such complex lateritic ores. The microbial processing of low-grade ores offers many advantages over conventional methods due to its relative simplicity, low energy input and being environmentally friendly (Sukla et al. 2014).

4.3 Environmental Occurrence of Nickel

In the last few decades, the importance of nickel in the environment is an issue that is gaining broader recognition owing to the interest of scientists as a result of its increasing industrial and commercial significance as well as the improvement of in situ and remote analytical methods of Ni detection (Cempel and Nikel 2006; Harasim and Filipek 2015). Nickel is an omnipresent trace metal released into the environment from natural and anthropogenic sources; hence, it is found in soil, water, air, and the biosphere (Barrie 1981; Berkovitz et al. 2014). Nickel has oxidation states of −1, 0, +1, +2, +3, and +4; however, the most common valence state in the environment is Ni^{2+} (Cotton and Wilkinson 1988; Nieboer et al. 1988). In the Earth's crust, which is derived from igneous rocks, nickel occurs solely in one natural valence state (Ni^{2+}). The contents of Ni in igneous rocks are highly elevated and range from 1400 to 2000 mg kg^{-1} in ultramafic rocks, 130 mg kg^{-1} in basaltic rocks, and 4.5–15 mg kg^{-1} in granitic rocks; a crustal abundance is 99 mg kg^{-1} (Kabata-Pendias and Szteke 2015; Mielke 1979). The nickel content of igneous rocks occurring in continental regions is 10 times as high as that from oceanic regions (Shoji and Kaneda 1980). Besides anthropogenic sources, trace metals can also be found in the parent material from which the soils developed (Zovko and Romić 2011). Serpentine soils are derived from ultramafic rocks. The way that ultramafic rocks alter into serpentine is central to a variety of processes at the interface between the solid earth, the hydrosphere, the atmosphere, and the biosphere (Müntener 2010) (Figure 4.4). Such soils are widespread in the Balkans and Sabah (Malaysia), which has one of the largest surface expressions of ultramafic rocks on Earth (Bani et al. 2010; van der Ent et al. 2015). Serpentine soils are characterized by elevated concentrations of heavy metals, including nickel, cobalt, and chromium, while simultaneously having cation imbalances (low Ca:Mg ratio) and general deficiencies of essential macronutrients (Brooks 1987; Doubková et al. 2013; Proctor 2003).

Nickel is constantly cycling between land, water, and air owing to physical and chemical processes including erosion, leaching, and precipitation (Yadav and Sharma 2016). From the atmosphere where nickel can be present as a component or suspended particulate matter (aerosols), it is removed in the form of wet or dry deposition. As a suspended particulate matter, Ni can remain in the air for a long time. Duration depends on particle size; in the case of larger particles, residence time is up to 8 days; however, in the case of submicron particles, it can take much longer. At this time, particles are being transported and eventually accumulated onto a surface (land or water) during the dry periods of no precipitation.

FIGURE 4.4
(See color insert.) Ultramafic rock in the process of altering to serpentine. (From Hayes, G. [http://geotripper.blogspot.com/2015/12/a-netherworld-incompatible-with.html].)

4.4 Natural Sources of Nickel

Natural sources of airborne nickel include soil dust from erosion, sea salt spray, volcano activity, forest fires, meteoric dust, and vegetation exudates (Godish 2004; Nriagu 1989; Richardson et al. 2001). The particulate size ranges from 2 to 10 μm (Yu and Tsunoda 2004). On a global scale, windblown soil particles from eroded areas and volcanic activity (a source of Earth's mantle, dust, and gas emissions) account for at least 70% of natural Ni emissions (Nriagu 1990). Estimates for the emission of nickel into the atmosphere from natural sources range from 8.5 kt/year in the 1980s to 30 kt/year in the early 1990s (ATSDR 2005). Barbante et al. 2002 (based on data from Nriagu 1989 and Pacyna et al. 1995) show medians from ranges in estimates that resulted in the following values: volcanoes, 14 kt; dust from rock and soil, 11 kt; wild forest fires, 2.3 kt; spray from sea salt, 1.3 kt; continental particulates, 0.51 kt; marine, 0.12 kt; continental volatiles, 0.10 kt, which have a total of 30 kt. This is in agreement with the latest calculations made by Pacyna et al. (2016). However, according to Richardson et al. (2001), global emission of Ni based on natural sources (wind erosion of soil particulate matter, sea salt spray, volcanic emissions, forest and brush fires, and meteoric dust) is much higher, and it is estimated to be 1800 kt, whereas in North America and Canada, the global emission values are 38 kt and 1 kt, respectively. The authors concluded that, for nickel, soil dust particles are a prevalent source of natural emissions to the atmosphere.

Nickel is redistributed by geological and biologic cycles. Nickel enters surface waters from three natural sources: as particulate matter in rainwater, through the dissolution of primary bedrock materials, and from secondary soil phases (EU Risk Assessment 2008). Nickel-rich serpentinite formations around the San Francisco Bay estuary are eroded, transported, and accumulated in estuarine sediment, providing a natural source of nickel (Topping and Kuwabara 2003). Weathering of serpentinite sediments produces effluent-enriched in nickel (Rytuba et al. 2000). Natural sources of aqueous nickel derive from biological cycles and solubilization of nickel compounds from soils (Sunderman 2004).

4.5 Anthropogenic Sources of Nickel

According to EEA (2014), there are several main anthropogenic sources of nickel emissions into the air: combustion of oil for the purposes of heating, shipping, or power generation; Ni mining and primary production; incineration of waste and sewage sludge; steel manufacture; electroplating; and coal combustion. Anthropogenic nickel exposure of the environment occurs locally from emissions of metal mining, smelting, and refining operations; from combustion of fossil fuels; from industrial activities, such as nickel plating and alloy manufacturing; from land disposal of sludges, solids, and slags; and from disposal as effluents. Other more diffuse sources may arise from combustion of fossil fuels, waste incineration, wood combustion, and so on (EU Risk Assessment 2008). Nickel may be released to the environment from the stacks of large furnaces used to make alloys or from power plants and trash incinerators. The nickel that is released from the stacks of power plants attaches to small particles of dust that settle to the ground or are removed from the air in rain or snow (ATSDR 2005). From anthropogenic sources, nickel is emitted as oxides, sulfides, silicates, and soluble compounds, and as a metallic nickel (Aydin et al. 2013), in the form of particulate matter with particle sizes ranging from 0.1 to 2 μm (Yu and Tsunoda 2004).

According to Nriagu and Pacyna (1988), the calculated annual emission of Ni from anthropogenic sources (fossil fuels, mainly oil) is 55.6 kt, but current estimation of Ni released to the atmosphere is 95.3 kt (Pacyna et al. 2016). In this new assessment, total global emission of Ni is 533.3 kt per year. Besides the aforementioned atmosphere, 113.0 kt of Ni enters aquatic ecosystems and 325.0 kt goes into soil (Pacyna et al. 2016).

Approximately 40% of the world electricity production uses coal (Gaffney and Marley 2009). In coal, fly ash and crude oil nickel amounts highly vary. Nickel content in coal depends on the region and can reach up to 300 mg kg^{-1}, although most samples contain well below 100 mg kg^{-1}. In fly ash, nickel content is lower, up to 90 mg kg^{-1}, whereas in crude oil, it is up to 150 mg kg^{-1} (Kabata-Pendias and Szteke 2015; Shang et al. 2016; Swaine 1980). Fossil fuel combustion is responsible for the release of 70 kt year^{-1} of nickel (Sharma 2012).

Ross (1994) classified Ni (next to As, Co, Cd, Cr, Cu, Zn, and Hg) to an industry group of anthropogenic sources of metal contamination. High anthropogenic pressure, which took place in the twentieth century, led to many local pollutions with toxic substances; thus, Ni pollution has become a serious concern. It is estimated that concentrations of this metal in polluted soils, surface waters, and air have reached up to 26,000 mg kg^{-1}, 0.2 mg L^{-1}, and 2000 ng m^{-3}, respectively, which repeatedly exceeded those found in unpolluted areas (Kabata-Pendias and Mukherjee 2007). Bencko (1983) reported average levels of airborne nickel of 3–30 ng m^{-3} in urban areas and 70–770 ng m^{-3} in nickel processing areas with metallurgical industry. A substantial part of atmospheric releases come from nickel production (nickel mining, smelting, and refining), which is a part of the anthropogenic cycle of Ni (Figure 4.5). The most important Ni source comes from milling through waste tailings and from smelting as a result of refining processes. Groundwater or surface water pollution may result from leaching of bottom ash, which comes from coal-fired plants. This ash is usually disposed of in landfills or used as a fill material in road construction (Gaffney and Marley 2009). The EU target

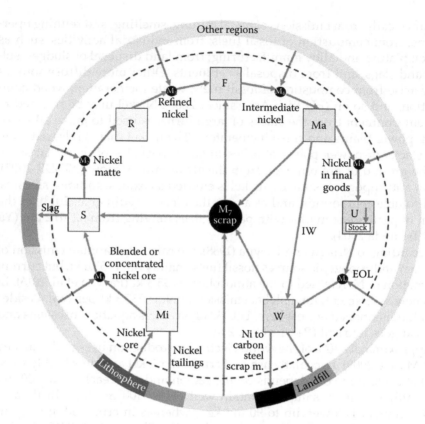

FIGURE 4.5
Circular diagram for nickel, with the main processes mining/milling (Mi), smelting (S), refining (R), fabrication (F), manufacturing (Ma), use (U), and waste management and recycling (W). The processes are connected through seven markets, each related to other regions through net import flows. "Carbon steel scrap market" stands for "carbon steel and copper scrap markets"; EOL, for end-of-life; IW, for industrial wastes. (Reprinted with permission from Reck, B. K., D. B. Müller, K. Rostkowski, and T. E. Graedel. 2008. Anthropogenic nickel cycle: Insights into use, trade, and recycling. *Environ. Sci. Technol.* 42(9): 3394–3400. Copyright 2008, American Chemical Society.)

value for nickel in the air is 20 ng m^{-3}. Concentrations of nickel, arsenic, cadmium, and lead in air are generally low in Europe, with few exceedances of limit or target values. Air pollution episodes happen when emissions suddenly increase from their baseline levels, when weather conditions favor the buildup of pollution in the air masses, or as a combination of both (EEA 2014). There has been a continuous reduction of heavy metal emissions in Europe during the last 40 years. Better knowledge of heavy metal sources, emissions, pathways, and the fate in the environment and progress in developing efficient emissions control equipment has resulted in more efficient regulatory efforts to curb heavy metal emissions from anthropogenic sources substantially (Pacyna et al. 2007). Ni emissions decreased in the EU-28 and EEA-33 countries by 44% between 2003 and 2012 (EEA 2014) and by 73% (from 2258 t to 621 t) for EU-28 between 1990 and 2014. The industry and energy sectors have equivalent contributions of Ni emissions, at 36% for the EU-28 and the EEA-33 in 2014 (EEA 2016). Pacyna et al. (2016) point out that, in the last two decades, emission of Ni decreased in North America as well. On the contrary, developing countries are confronted with the great challenge of controlling atmospheric pollution, especially in the rapidly growing megacities. Living in Kinshasa is associated with elevated levels of Ni in ambient air, which is contaminated with toxic heavy metals (Kabamba et al. 2016). Tian et al. (2012) have estimated that the total anthropogenic atmospheric nickel emissions from all the sources in China have increased from 1096.07 t in 1980 to 3933.71 t in 2009, at an average annual growth rate of 4.5%. Therein, coal combustion is the leading source, contributing 63.4% of the national total nickel emissions in 2009; liquid fuels consumption ranks the second, contributing 12.4% of the totals; biofuels burning accounts for 8.4% and the remaining sources together contribute 15.8% of the total. Significant spatial variations are demonstrated among provincial emissions, and the most concentrated regions are the highly industrialized and densely populated areas.

Though distribution of nickel in the soil profile is rather uniform, industrial deposition and agricultural activities are responsible for Ni accumulation at the surface (Bencko 1983). Cement plant surroundings may be exposed to emissions of polluting gases but also to dust from alkaline reactions, containing Cd, Pb, Cr, Cu, Zn, and Ni (Dąbkowska-Naskręt et al. 2014). Another source of Ni and Ni compounds being released in to the environment is through waste streams, that is, wastewater treatment systems and landfills (SCHER 2009). Wastewater containing heavy metals (Zn, Cd, Cu, Ni, Hg, Pb, Cr, Ag, Ti, and V) directly or indirectly is released into the environment. The problem concerns developing countries in particular. Contaminated aqueous wastes originate from many sources including electroplating, mining operations, tanneries, and smelting (Barakat 2011; Fu and Wang 2011; Kadirvelu et al. 2001). In order to prevent nickel from entering the environment via factory effluents, strict environmental regulations have been put on the electroplating industry (Di Bari 2010).

Soils are the major sink for metal contaminants released into the environment by anthropogenic activities (Kirpichtchikova et al. 2006; Pacyna et al. 2016). Of particular concern is the increasing concentration of Ni deposited in agricultural soils by airborne Ni particles (Sreekanth et al. 2013). Generally, the metal plating industry, combustion of fossil fuels, and nickel mining and refining are important sources of Ni in the soil (Khodadoust et al. 2004; Osman 2014). Podlesakova et al. (2001) reported that plant uptake of nickel from anthropogenic sources was higher than that from natural occurring sources.

Nickel may also be introduced into the soil directly with inorganic and organic fertilizers as well as indirectly by foliar feeding of plants. Phosphate fertilizers and multicomponent fertilizers are the source of heavy metals

into environment. The basic sources of metals polluting fertilizers are phos-
phorites used as a source of phosphorus. These fertilizers have variable levels of impurities such as Ni and Cd and can contain significant amounts of
metals depending on phosphorite sources. An increased amount of nickel
is also found in the fertilizers that have magnesium among their components. Magnesium derives from ground dolomite supplement in the fertilizers (Gambuś and Wieczorek 2012; Kabata-Pendias and Pendias 1992;
Sharma and Agarwal 2005; Zhang and Shan 2008). Most countries do not
have regulations on Ni content in fertilizers, but in Canada, the maximum
acceptable Ni concentration is 180 mg kg^{-1} d.w. (Benson et al. 2014). Gambuś
and Wieczorek (2012) reported Ni content in phosphate and multicomponent
fertilizers in the range of 7.6–396.0 mg kg^{-1} d.m. Long-term and extensive
use of agricultural land with frequent application of growing practices and
use of pesticides may cause heavy metals such as copper, nickel, zinc, and
cadmium to be strongly accumulated in the topsoil (Zovko and Romić 2011).
According to Ociepa et al. (2008), to use fertilizers safely, one should respect
existing legislation on fertilization use and carry out the monitoring of soils,
plants, and residues. Nickel is the latest (17th) element added to the list of
micronutrients based on its essentiality for plant growth and development
(Chen et al. 2009; Liu 2001). According to Wood et al. (2004), deficiencies of
this micronutrient to pecan plants manifest as a growth abnormality called
mouse ear of pecan, and to correct this deficiency, nickel can be supplied to
plants by foliar feeding. That way, Ni can also unintentionally get to the soil.

In 2005, globally, from the extracted nickel, 18% was lost as tailings and
slag. The remaining 82% went into fabrication and manufacturing; 52% of Ni
was recycled and another 17% and 13% were lost through landfill and non-
functional recycling, respectively (Reck and Graedel 2012). Nickel is recycled
in different ways depending on its original application. Nickel alloys are
often recycled as the same alloys, for example, the nickel in stainless steel,
where about 40% of the nickel used in the production of stainless steel originates from post-consumer stainless steel scrap. Other secondary nickel aris-
ings tend to be recycled by primary nickel smelters (BIR 2008).

4.6 Conclusions

Over 150 years have passed since commercial nickel mining began.
Nowadays, more nickel is extracted from sulfides than from laterites, while
about 60% of nickel-bearing ores can be found in laterites and the remaining 40% can be found in sulfides. Owing to advancements in the extraction
process, for example, microbial mineral processing (biohydrometallurgy),
laterites are gaining importance. Nickel is omnipresent in the environment.
Apart from natural sources, nickel is introduced into the environment also

through anthropogenic activity. Windblown soil particles are the predominant natural source of Ni, while main anthropogenic sources are coal combustion (e.g., in China) and combustion of liquid fuels (e.g., in Europe).

Industrial use of nickel is mainly related to production of stainless steel, the biggest importer of which is the fast-developing nation of China. The growing nickel demand and its decreasing resource force governments to seek new sources of Ni and to exploit resources having small amounts of this metal. In the near future, a new method of Ni acquisition called seabed mining will most likely begin. In seabed mining, Ni sources are polymetallic nodules and manganese crust located on the ocean floor. Extraction of Ni from asteroids is likewise planned. The main goal is to retrieve water and metals; however, even if this proves to be successful, Ni will probably be used in situ for construction and not to be brought to Earth.

Based on actual demand, global resources will suffice for several decades, but potentially, this period can be extended if seabed mining proves to be successful and economically justified. The price of nickel is unstable on the global market. Few years back, growing nickel prices caused increased interest in NPI. Apart from products with lower nickel content, substitute materials that do not contain Ni are rather limited and involve chromium, titanium, and lithium; however, nickel's advantage is its possibility to recover and recycle. As a fully recyclable metal from stainless steel or other nickel–iron alloys, it can be recovered from scrap metal independently of its grade (high or low). Currently, more than 50% of nickel is recycled, and in recent years, more attention is being paid to collecting and recycling used batteries.

Nickel extraction from its beginning was important for humanity. A consequence of industrialization was an increasing demand for nickel, which, due to mining, smelting, and refining, inevitably caused environmental changes. Local or diffuse nickel pollution of air, soil, and water occurred. Owing to regulations and actions aimed at environmental protection, and despite increasing production and consumption of nickel, there has been a decrease in air pollution in Europe and North America in recent years. Globally, however, developing countries are having difficulties keeping their air clean.

Because of the wide range of nickel applications and the limited alternatives, as well as knowledge about Ni gained recently, it is expected that Ni will remain a highly valued metal, the resources of which may deplete relatively quickly as a result of high demand.

References

Arndt, N., Kesler, S., and C. Ganino. 2015. Classification, distribution and uses of ores and ore deposits. In: *Metals and Society. An Introduction to Economic Geology.* Springer, Switzerland. pp. 15–40.

ATSDR (Agency for Toxic Substances and Disease Registry). 2005. Toxicological Profile for Nickel. U.S. Department of Health and Human Services. Atlanta, U.S. [http://www.atsdr.cdc.gov/ToxProfiles/tp15.pdf].

Aydin, I., Aydin, F., Kilinc, E., Duz, M. Z., and C. Hamamci. 2013. Chemical fractionation of nickel in asphaltite based bottom ash. *Chem. Spec. Bioavail.*, 25(2): 113–118.

Bani, A., Pavlova, D., Echevarria, G., Mullaj, A., Reeves, R. D., Morel, J. L., and S. Sulçe. 2010. Nickel hyperaccumulation by the species of Alyssum and Thlaspi (*Brassicaceae*) from the ultramafic soils of the Balkans. *Botanica Serbica*, 34(1): 3–14.

Barakat, M. A. 2011. New trends in removing heavy metals from industrial wastewater. *Arab. J. Chem.*, 4(4): 361–377.

Barbante, C., Boutron, C., Moreau, A.-L., Ferrari, C., Van de Velde, K., Cozzi, G., Turetta, C. and P. Cescon. 2002. Seasonal variations in nickel and vanadium in Mont Blanc snow and ice dated from the 1960s and 1990s. *J. Environ. Monit.*, 4: 960–966.

Barrie, L. A. 1981. Atmospheric nickel in Canada. In: *Effects of nickel in the Canadian environment*. National Research Council of Canada No. 18568. 3, Ottawa, pp. 55–76.

Bencko, V. 1983. Nickel: A review of its occupational and environmental toxicology. *J. Hyg. Epidem. Micro. Immun.*, 27(2): 237–247.

Benson, N. U., Anake, W. U., and U. M. Etesin. 2014. Trace metals levels in inorganic fertilizers commercially available in Nigeria. *J. Sci. Res. Rep.*, 3(4): 610–620.

Berkovitz, B., Dror, I., and B. Yaron. 2014. *Contaminant Geochemistry. Interactions and Transport in the Subsurface Environment*. 2nd edition. Springer-Verlag, Berlin, Heidelberg.

Bide, T., Hetherington, L., Gunn, G., and A. Minks. 2008. Nickel. British geological survey [https://www.bgs.ac.uk/downloads/start.cfm?id=1411].

BIR (Bureau of International Recycling). 2008. Report on the Environmental Benefits of Recycling.

Brooks, R. R. 1987. *Serpentine and Its Vegetation: A Multidisciplinary Approach*. Dioscorides Press, Portland.

Cempel, M. and G. Nikel. 2006. Nickel: A review of its sources and environmental toxicology. *Pol. J. Environ. Stud.*, 15: 375–382.

Chen, C., Huang, D., and J. Liu. 2009. Functions and toxicity of nickel in plants: Recent advances and future prospects. *Clean Soil Air Water*, 37(4–5): 304–313.

Cheng, C. F. and C. M. Schwitter. 1957. Nickel in ancient bronzes. *Am. J. Archaeol.* 61(4): 351–365.

Cotton, F. A. and G. Wilkinson. 1988. *Advanced Inorganic Chemistry: A Comprehensive Text*. 5th edition. Wiley-Interscience, New York, NY.

Cronan, D. S. 1980. *Underwater Minerals*. Academic Press, London.

Dąbkowska-Naskręt, H., Jaworska, H. and J. Długosz. 2014. Assessment of the total nickel content and its available forms in the soils around cement plant Lafarge Poland. *Int. J. Environ. Res.*, 8(1): 231–236.

Di Bari, G. A. 2010. *Electrodeposition of Nickel, in Modern Electroplating*. 5th edition (eds. M. Schlesinger and M. Paunovic). John Wiley & Sons, Hoboken, NJ.

Doubková, P., Vlasáková, E. and R. Sudová. 2013. Arbuscular mycorrhizal symbiosis alleviates drought stress imposed on *Knautia arvensis* plants in serpentine soil. *Plant Soil*, 370(1): 149–161.

Eisler, R. 2007. *Encyclopedia of Environmentally Hazardous Priority Chemicals.* Elsevier Science, Amsterdam.

EEA (European Environment Agency). 2014. Air quality in Europe—2014 report [http://www.eea.europa.eu/publications/air-quality-in-europe-2014/at_down load/file].

EEA (European Environment Agency). 2016. European Union emission inventory report 1990–2014 under the UNECE Convention on Long-range Transboundary Air Pollution (LRTAP) [http://www.eea.europa.eu/publications/lrtap-emission -inventory-report-2016/at_download/file].

EU Risk Assessment. 2008. European Union Risk Assessment Report. Nickel and nickel compounds [https://echa.europa.eu/documents/10162/cefda8bc-2952 -4c11-885f-342aacf769b3].

Fu, F. and Q. Wang. 2011. Removal of heavy metal ions from wastewaters: A review. *J. Environ. Manage.,* 92(3): 407–418.

Gaffney, J. S., and N. A. Marley. 2009. The impacts of combustion emissions on air quality and climate—From coal to biofuels and beyond. *Atmos. Environ.,* 43: 23–36.

Gambuś, F. and J. Wieczorek. 2012. Pollution of fertilizers with heavy metals. *Ecol. Chem. Eng. A,* 19(4–5): 353–360.

Godish, T. 2004. *Air Quality.* 4th edition. Lewis Publishers, CRC Press, Boca Raton, FL.

Halbach, P. and R. Fellerer. 1980. The metallic minerals of the Pacific Seafloor. *Geojournal,* 4: 407–421.

Harasim, P. and T. Filipek. 2015. Nickel in the environment. *J. Elem.,* 20(2): 525–534.

Hayes, G. [http://geotripper.blogspot.com/2015/12/a-netherworld-incompatible-with .html].

Hoatson, D. M., Jaireth, S., and A. L. Jaques. 2006. Nickel sulfide deposits in Australia: Characteristics, resources, and potential. *Ore Geol. Rev.,* 29: 177–241.

Kabamba, M., Basosila, N., Mulaji, C., Mata, H., and J. Tuakuila. 2016. Toxic heavy metals in ambient air of Kinshasa, Democratic Republic Congo. *J. Environ. Anal. Chem.* 3: 178.

Kabata-Pendias, A. and A. B. Mukherjee. 2007. *Trace Elements from Soil to Human.* Springer, New York, Berlin, Heidelberg, 550 pp.

Kabata-Pendias, A. and H. Pendias. 1992. *Trace Elements in Soils and Plants.* CRC Press, London.

Kabata-Pendias, A. and B. Szteke. 2015. *Trace Elements in Abiotic and Biotic Environments.* CRC Press, Boca Raton, FL.

Kadirvelu, K., Thamaraiselvi, K., and C. Namasivayam. 2001. Removal of heavy metals from industrial wastewaters by adsorption onto activated carbon prepared from an agricultural solid waste. *Bioresource Technol.,* 76: 63–65.

Khodadoust, A. P., Reddy, K. R., and K. Maturi. 2004. Removal of nickel and phenanthrene from kaolin soil using different extractants. *Environ. Eng. Sci.,* 21(6): 691–704.

Kirpichtchikova, T. A., Manceau, A., Spadini, L., Panfili, F., Marcus, M. A., and T. Jacquet. 2006. Speciation and solubility of heavy metals in contaminated soil using X-ray microfluorescence, EXAFS spectroscopy, chemical extraction, and thermodynamic modeling. *Geochim. Cosmochim. Acta,* 70(9): 2163–2190.

Kuck, P. H. 2016. Nickel: U.S. Geological Survey Mineral Commodity Summaries 2016, pp. 114–115 [http://minerals.usgs.gov/minerals/pubs/commodity/nickel/mcs -2016-nicke.pdf].

Lightfoot, P. C. 2016. *Nickel Sulfide Ores and Impact Melts: Origin of the Sudbury Igneous Complex*. Elsevier, Amsterdam.

Liu, G. D. 2001. A new essential mineral element—Nickel. *Plant Nutr. Fertil. Sci.*, 7(1): 101–103.

Mielke, J. E. 1979. Composition of the Earth's crust and distribution of the elements. In: *Review of Research on Modern Problems in Geochemistry* (ed. F. R. Siegel). UNESCO Report, Paris, pp. 13–37.

Moss, R. L., Tzimas, E., Kara, H., Willis, P., and J. Kooroshy. 2011. Critical Metals in Strategic Energy Technologies. JRC. Scientific and Technical Reports [https://setis.ec.europa.eu/system/files/CriticalMetalsinStrategicEnergyTechnologies-def.pdf].

Mudd, G. M. 2010. Global trends and environmental issues in nickel mining: Sulfides versus laterites. *Ore Geol. Rev.*, 38: 9–26.

Mudd, G. M. and S. M. Jowitt. 2014. A detailed assessment of global nickel resource trends and endowments. *Econ. Geol.*, 109(7): 1813–1841.

Müntener, O. 2010. Serpentine and serpentinization: A link between planet formation and life. *Geology*, 38(10): 959–960.

Nieboer, E., Tom, R. T., and W. E. Sanford. 1988. Nickel metabolism in man and animals. In: *Metal Ions in Biological Systems* (eds. H. Sigel and A. Sigel). Marcel Dekker, New York, NY. vol. 23, pp. 91–121.

Nriagu, J. O. 1989. A global assessment of natural sources of atmospheric trace metals. *Nature*, 338: 47–49.

Nriagu, J. O. 1990. Global metal pollution poisoning the biosphere? *Environment*, 32: 7–33.

Nriagu, J. O. and J. M. Pacyna. 1988. Quantitative assessment of worldwide contamination of air, water and soils by trace metals. *Nature*, 333: 134–139.

Ociepa, A., Pruszek, K. Lach, J. and E. Ociepa. 2008. Influence of long-term cultivation of soils by means of manure and sludge on the increase of heavy metals content in soils [in Polish]. *Ecol. Chem. Eng.*, 15(1): 103–109.

Osman, K. T. 2014. *Soil Degradation, Conservation and Remediation*. Springer, Dordrecht, Netherlands.

Pacyna, E. G., Pacyna, J. M., Fudała, J., Strzelecka-Jastrząb, E., Hławiczka, S., Panasiuk, D., Nitter, S., Pregger, T., Pfeiffer, H., and R. Friedrich. 2007. Current and future emissions of selected heavy metals to the atmosphere from anthropogenic sources in Europe. *Atmos. Environ.*, 41: 8557–8566.

Pacyna, J. M., Scholtz, M. T., and Y.-F. A. Li. 1995. Global budget of trace metal sources. *Environ. Rev.*, 3: 145–159.

Pacyna, J. M, Sundseth, K., and E. G. Pacyna. 2016. Sources and fluxes of harmful metals. In: *Environmental Determinants of Human Health* (eds. Pacyna and Pacyna). Springer, Switzerland. pp. 1–26.

Podlesakova, E., Nemecek, J., and R. Vacha. 2001. The transfer of less hazardous trace elements with a high mobility from soils into plants. *Rostlinna Vyroba*, 47(10): 433–439.

Proctor, J. 2003. Vegetation and soil and plant chemistry on ultramafic rocks in the tropical Far East. *Perspect. Plant Ecol.*, 6(1–2): 105–124.

Reck, B. K. and T. R. Graedel. 2012. Challenges in metal recycling. *Science*, 337: 690–695.

Reck, B. K., D. B. Müller, K. Rostkowski, and T. E. Graedel. 2008. Anthropogenic nickel cycle: Insights into use, trade, and recycling. *Environ. Sci. Technol.*, 42(9): 3394–3400.

Reck, B. K. and V. S. Rotter. 2012. Comparing growth rates of nickel and stainless steel use in the early 2000s. *J. Indust. Ecol.*, 16(4): 518–528.

Richardson, G. M., Garrett, R., Mitchell, I., Mah-Paulson, M., and T. Hackbarth. 2001. Critical review on natural global and regional emissions of six trace metals to the atmosphere. Risklogic Scientific Services, Ottawa [https://www.echa .europa.eu/documents/10162/13630/vrar_appendix_p2_en.pdf].

Ross, S. 1994. *Toxic Metals in Soil–Plant Systems*. John Wiley & Sons, Chichester, UK.

Rytuba, J. J., Enderlin, D., Ashley, R., Seal R., and M. P. Hunerlach. 2000. Evolution of the McLaughlin Gold Mine Pit Lakes, California. In: *Proceedings of the Fifth International Conference on Acid Rock Drainage*. Society of Mining, Metallurgy, and Exploration (SME), Littleton.

SCHER (Scientific Committee on Health and Environmental Risks). 2009. Risk Assessment Report on nickel and its compounds. Environmental Part, 13 January 2009 [http://ec.europa.eu/health/archive/ph_risk/committees/04_scher /docs/scher_o_112.pdf].

Shang, H., Liu, Y., Shi, J.-C., Shi, Q., and W.-H. Zhang. 2016. Microwave-assisted nickel and vanadium removal from crude oil. *Fuel Process. Technol.*, 142: 250–257.

Sharma, R. 2010. First nodule to first mine-site: Development of deep-sea mineral resources from the Indian Ocean. *Curr. Sci.*, 99(6).

Sharma, R. K. and M. Agarwal. 2005. Biological effects of heavy metals: An overview, *J. Environ. Biol.*, 26(2): 301–313.

Sharma, Y. C. 2012. *A Guide to Economic Removal of Metals from Aqueous Solutions*. Scrivener, Salem, MA; Wiley, Hoboken, NJ.

Shoji, T. and H. Kaneda. 1980. Classification of igneous rocks, based on the relationship among nickel, cobalt, and silica contents. *Mining Geol.*, 30(5): 289–297.

Sreekanth, T. V. M., Nagajyothi, P. C., Lee, K. D., and T. N. V. K. V. Prasad. 2013. Occurrence, physiological responses and toxicity of nickel in plants. *Int. J. Environ. Sci. Technol.*, 10: 1129–1140.

Sukla, L. B., Behera, S. K., and N. Pradhan. 2014. Microbial recovery of nickel from lateritic (oxidic) nickel ore: A review. In: *Geomicrobiology and Biogeochemistry* (eds. N. Parmar and A. Singh). Volume 39 of the series *Soil Biology*, pp. 137–151.

Sunderman, F. 2004. Nickel. In: *Elements and their Compounds in the Environment: Occurrence, Analysis and Biological Relevance*. 2nd edition (eds. E. Merian, M. Anke, M. Ihnat, and M. Stoeppler). WILEY-VCH Verlag GmbH&Co. KGaA, Weinheim.

Swaine, D. J. 1980. Nickel in coal and fly ash. In: *Nickel in the Environment* (ed. J. O. Nriagu). Wiley, New York, pp. 67–92.

Tian, H. Z., Lu, L., Cheng, K., Hao, J. M., Zhao, D., Wang, Y., Jia, W. X., and P. P. Qiu. 2012. Anthropogenic atmospheric nickel emissions and its distribution characteristics in China. *Sci. Total Environ.*, 417–418.

Topping, B. R. and J. S. Kuwabara. 2003. Dissolved nickel and benthic flux in South San Francisco Bay: A potential for natural sources to dominate. *Bull. Environ. Contam. Toxicol.*, 71: 46–51.

van der Ent, A., Erskine, P., and S. Sumail. 2015. Ecology of nickel hyperaccumulator plants from ultramafic soils in Sabah (Malaysia). *Chemoecology*, 25: 243–259.

Wood, B. W., Reilly, C. C., and A. P. Nyczepir. 2004. Mouse-ear of pecan: A nickel deficiency. *HortScience*, 39(6): 1238–1242.

Yadav, N. and S. Sharma. 2016. An account of nickel requirement, toxicity and oxidative stress in plants. *Biol. Forum*, 8(1): 414–419.

Yu, M.-H. and H. Tsunoda. 2004. *Environmental Toxicology: Biological and Health Effects of Pollutants*. 2nd edition. CRC Press, Boca Raton, FL.

Zhang, H. and B. Shan. 2008. Historical distribution and partitioning of phosphorus in sediments rich in an agricultural watershed in the Yangtze–Huaihe Region, China. *Environ. Sci. Technol.*, 42: 2328–2333.

Zovko, M. and M. Romic. 2011. *Soil Contamination by Trace Metals: Geochemical Behaviour as an Element of Risk Assessment, Earth and Environmental Sciences* (ed. I. Ahmad Dar). InTech [http://www.intechopen.com/books/earth-and -environmental-sciences/soil-contamination-by-trace-metals-geochemical -behaviour-as-an-element-of-risk-assessment].

5

The Origin of Nickel in Soils

Dionisios Gasparatos and Nikolaos Barbayiannis

CONTENTS

5.1 Introduction

Nickel (Ni) is the 24th most abundant element in the Earth's crust with an average concentration of 75–80 mg kg^{-1} (Adriano 2001; McGrath 1995). Nickel is a transition metal with five stable isotopes in nature (^{58}Ni, ^{60}Ni, ^{61}Ni, ^{62}Ni, and ^{64}Ni) and ^{58}Ni (68.077%) is the most common. Although Ni can occur in a number of oxidation states, only Ni(II) oxidation state is stable in the soil environment.

The most important use of Ni is in manufacturing of stainless steel (65%) (>50% of the produced steel used in construction, food processing, and transportation sectors) and alloy production (21%); therefore, in recent years, Ni has become a strategic metal. The world Ni mining production increased from 0.9×10^6 Mg in 1994 to 2.4×10^6 Mg in 2014, and the primary producers of Ni are Canada, Russia, Philippines, Australia, Indonesia, and New Caledonia (Table 5.1).

There are two principal types of ore deposits that can be economically used for Ni mining:

1. The magmatic sulfide deposits consisting mainly of pyrrhotite Fe_{1-x} S ($x = 0.0$ to $x = 0.2$), associated pentlandite $(Ni,Fe)_9S_8$, and chalcopyrite $CuFeS_2$. Pentlandite is the most commercially important Ni ore, which also contains minor but recoverable amounts of cobalt. According to Hoatson et al. (2006), 93% of known sulfide deposits are in the range 0.2–2% Ni. These deposits contain about 40% of the world's Ni resources.

2. The nickeliferous laterites formed by the prolonged weathering of peridotite and serpentinite. The main ore minerals are garnierite $(Ni,Mg)_3Si_2O_5(OH)$ (the nickel silicate variety) and limonite (Fe,Ni) O(OH) (the nickeliferous iron variety). Laterite deposits contain approximately 60% of the world's Ni resources.

TABLE 5.1

World Ni Mine Production 2014 (Metric Tons)

Country	Mine Production
Philippines	440,000
Russia	260,000
Indonesia	240,000
Canada	233,000
Australia	220,000
New Caledonia	165,000
Brazil	126,000
China	100,000
Colombia	75,000
Cuba	66,000
South Africa	54,700
Madagascar	37,800
United States	3600
Other countries	410,000

Source: U.S. Geological Survey. 2015. Mineral commodity summaries 2015: U.S. Geological Survey, 196 pp., http://minerals.usgs.gov /minerals/pubs/mcs/2015/mcs2015.pdf (last accessed February 1, 2017).

5.2 Nickel in Rocks and Minerals

Nickel, is a siderophile element with great affinity for sulfur (S) and forms more than 100 mineral species. Most widespread minerals containing Ni include gersdorffite NiAsS, millerite NiS, nikeline NiAs, and ullmanite NiSbS (Kabata-Pendias and Mukherjee 2007). Nickel concentrations of the common rocks vary widely between different rock types. With the exception of carbonates, rocks low in SiO_2 (<45 wt.% of SiO_2), like peridodite, are high in Ni, and those high in SiO_2, like granite, are low in Ni. Because Ni has the same oxidation state as and a similar ionic size to iron (Fe) and magnesium (Mg), it can substitute these metals and therefore partition into ferromagnesian minerals (olivine, pyroxene, etc.). For this reason, Ni is strongly enriched in mafic and ultramafic igneous rock with concentrations of up to 2000 mg kg^{-1}. Felsic igneous rocks have much less Ni and sedimentary rocks contain particularly low Ni (Table 5.2).

TABLE 5.2

Concentrations of Ni in Various Rocks

Rock Type	Ni (mg kg^{-1})
Igneous Rocks	
Ultramafic (peridotites, pyroxenites)	1400–2000
Mafic (basalts, gabbros)	130–160
Intermediate (diorites)	5–55
Felsic/granitic (granites)	5–15
Felsic/volcanic (rhyolites, dacites)	20
Sedimentary Rocks	
Black shales	10–500
Shales	50–70
Argillaceous sediments	40–90
Sandstones	5–20
Limestones	7–20
Metamorphic Rocks	
Gneisses	5–15

Sources: Kabata-Pendias, A. and A. B. Mukherjee. 2007. *Trace Elements from Soil to Human.* Berlin, New York: Springer-Verlag; McGrath, S. P. 1995. Chromium and nickel. In: *Heavy Metals in Soils*, ed. B. J. Alloway, 152–178. New York: Blackie Academic and Professional.

108

5.3 Origin of Ni in Soils

Nickel levels in the soil environment may be attributed to the natural weathering processes of parent materials and pedogenesis (lithogenic/pedogenic origin) and to various anthropogenic sources (anthropogenic origin). Understanding the origin of Ni in soils is essential to assess human impact on the environment and is a strategic aim of soil protection policies that can ensure water and food security.

5.3.1 Lithogenic/Pedogenic Sources

A literature review shows that the mean content of Ni in world soils is estimated to be 20 mg kg^{-1} with a very broad range of concentrations from 0.2 to 450 mg kg^{-1} (Adriano 2001; Ma and Hooda 2010; McGrath 1995). The content of natural Ni in soils depends predominantly on the nature of the parent material in the "young" soils, while its content in the "mature" soils reflects the additional impact of pedogenic processes (Gonnelli and Renella 2013).

Data from McGrath (1995) demonstrate that sandy and coarse loamy soils contain less than the median values of Ni, while clay and loamy soils contain above median concentrations. According to Kabata-Pendias and Pendias (2001), the mean Ni content in sandy soils and in clay loamy soils of Poland is 5 and 22 mg kg^{-1}, respectively. Values (in mg kg^{-1}) of Ni for soils from various countries have been reported by Adriano (2001) and Kabata-Pendias and Pendias (2001): England and Wales (mean, 26; range, 4.4–228), Canada (mean, 20; range, 5–50), Austria (mean, 21; range, 6–38), Italy (mean, 28; range, 4.0–97.5), Japan (mean, 28; range 2–660), and the United States (mean, 16.5; range, 0.7–269).

High concentrations exceeding 10,000 mg kg^{-1} of Ni can be found in "serpentine" soils (Antibachi et al. 2012; Brooks 1987; Hseu et al. 2016; Kanellopoulos et al. 2015; Quantin et al. 2008). The term "serpentine soils" is used to describe any soil that is formed from ultramafic rocks, for example, peridotites (igneous rocks) and serpentinites (metamorphic rock derived from hydrothermal transformations of peridotite). Although these rocks cover minimal area (less than 1% of the Earth's surface), mostly within the Circum-Pacific margin and the Mediterranean area, ultramafic rocks have been found at many places of the world, such as Greece, South Africa, the Italian Alps, France, Portugal, Spain, Cameroon, Morocco, Poland, Zimbabwe, India, Sri Lanka, Brazil, Costa Rica, Taiwan, Turkey, New Caledonia, New Zealand, Indonesia, Japan, Canada, and the United States.

The chemical weathering of ultramafic rocks produces soils with high levels of Ni, which in most cases commonly range in several thousands of mg kg^{-1} (Table 5.3). The distribution of Ni in serpentine soils and the formation

TABLE 5.3

Total Ni Concentrations in Serpentine Soils from Various
Countries

Country	Total Ni (mg kg^{-1})
Italy	2600
Portugal	4000
New Zealand	3300
Greece	1390
Australia (western)	700
New Caledonia	5400
Japan	326 520
Albania	3180
Poland	2229
United States	2020
Spain	940

Source: Adriano, D. C. 2001. *Trace Elements in Terrestrial Environments: Biogeochemistry, Bioavailability and Risks of Metals.* 2nd ed. New York: Springer-Verlag.

of Ni-bearing minerals are highly variable and are strongly influenced by the geochemistry and mineralogy of the parent material, as well as the other soil-forming factors including climatic conditions, topography, biota, and time (Bonifacio et al. 2010; Caillaud et al. 2009; Kierczak et al. 2016; Oze et al. 2008). For example, Massoura et al. (2006) found that soils with moderate development formed in temperate and Mediterranean ultramafic environments have total Ni concentrations from 863 to 2600 mg kg^{-1}.

The primary sources of Ni in ultramafic rocks are pyroxene, olivine, spinel, amphibole, and serpentine (chrysotile, antigorite, and lizardite), which are unstable in the soil environment (Massoura et al. 2006; Quantin et al. 2008; Ratie et al. 2015). The chemical weathering of these Ni-bearing minerals leads to the release of Ni and subsequent association with secondary minerals, that is, clay minerals and Fe–Mn oxides (Gasparatos 2013; Massoura et al. 2006). In surface horizons, Fe (oxy)hydroxides of different crystallization stages are generally considered as important hosts for Ni, whereas in the lower parts of soil profile, Ni is mainly associated with clay minerals such as serpentine, talc, chlorite, and smectite (Ratie et al. 2015). Progressively, the pedogenic processes promote the incorporation of Ni into the lattice of secondary minerals; hence, geogenic Ni tend to be found in mineral structures (residual fraction according to Tessier's sequential extraction scheme) rather than the surface charge of soil particles (Hseu and Iizuka 2013). The Ni-bearing phases that formed during the course of weathering and pedogenesis finally control the mobility and availability of Ni in the soil profile.

Nickel in Soils and Plants

5.3.1.1 Focus on Europe and Greece

Although anthropogenic activities may also be responsible for Ni accumulation in parts of Europe, according to recent European surveys (LUCAS, FOREGS, and GEMAS), soil Ni distribution can be attributed to natural factors to a great extent (Reimann et al. 2014; Salminen et al. 2005; Tóth et al. 2016a). According to FOREGS Geochemical Atlas of Europe, samples with

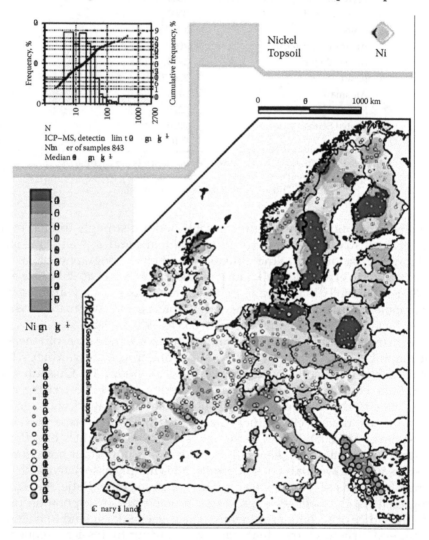

FIGURE 5.1
(See color insert.) Geochemical map of Europe showing the distribution of total Ni in topsoils. The gray shaded areas represent major parts of the ophiolite complexes and mafic and ultramafic rocks across Europe. (Modified after the EuroGeoSurveys—FOREGS Geochemical Baseline Database; Salminen R. et al. 2005. Geochemical atlas of Europe. Part 1: background information, methodology and maps. Espoo Geological Survey of Finland.)

high concentrations of geogenic Ni can be found in soils that are mainly located in the Mediterranean region where major parts of the ophiolitic complexes of Europe occur (Figures 5.1 and 5.2). In Greece, there are various areas with ophiolitic appearances located on Pindos, Vourino, Othris, Euboea, Kastoria, Koziakas, Vermio, East Thessaly, Oiti, Argolida, Thessaloniki (Triadi), central Chalkidiki, Oraiokastro, Gevgeli, Evros (Soufli and Dadia),

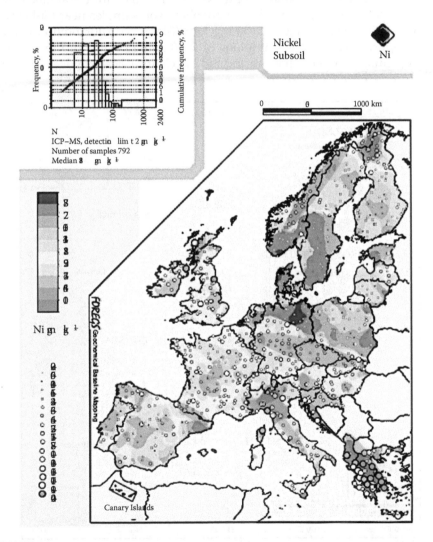

FIGURE 5.2
(See color insert.) Geochemical map of Europe showing the distribution of total Ni in subsoils. The gray shaded spots represent major parts of the ophiolite complexes and mafic and ultramafic rocks across Europe. (Modified after the EuroGeoSurveys—FOREGS Geochemical Baseline Database; Salminen R. et al. 2005. Geochemical atlas of Europe. Part 1: background information, methodology and maps. Espoo Geological Survey of Finland.)

Rodopi (Organi and Murtiskos), Samothrace, Lesvos, Rhodes, and Crete (Figure 5.3) (Kaprara et al. 2015; Kelepertzis 2014; Megremi 2010).

Several studies in Greek areas with ultramafic rocks identified high Ni concentrations in soils. Antibachi et al. (2012) and Kelepertzis et al. (2013) reported Ni levels up to 2640 mg kg^{-1} in agricultural soils of the Thiva valley (Central Greece) due to the weathering of ultramafic rocks. In the Asopos basin, Lilli et al. (2015) found 940 mg Ni kg^{-1} in soils with ultramafic components. In Central Euboea, Ni in rhizosphere soil samples near Ni-laterite deposits had a mean value of 2800 mg kg^{-1} (Megremi 2010). Bompoti et al. (2015) reported soil Ni concentrations of 1200 and 2357 mg kg^{-1} in Thermi and Vergina, two sites that are located in Central Macedonia and their

FIGURE 5.3
The distribution of ophiolites and ultramafic rocks in Greece (IGME 1983). (Reprinted from *Journal of Hazardous Material*, Vol. 281, Kaprara et al., Occurrence of Cr(VI) in drinking water of Greece and relation to the geological background, 2–11. Copyright (2015), with permission from Elsevier.)

stratigraphic column consists partially of strongly serpentinized and weathered ultrabasic and basic rocks. In a study of ultramafic-affected soils in the area of Atalanti (Central Greece), Kanellopoulos et al. (2015) found a mean Ni content of 533 mg kg^{-1} (range, 44–2730 mg kg^{-1}). These results support the hypothesis that the parent material influences predominantly the distribution of Ni in Greek soils, which is totally in line with the Ni maps from the Geochemical Atlas of Europe.

5.3.2 Anthropogenic Sources

The major anthropogenic sources of Ni to soils are the atmospheric deposition from various emissions, the mining and smelting activities, and the application of sewage sludge and phosphate fertilizers, which may contain a wide variety of heavy metals as impurities.

5.3.2.1 Atmospheric Deposition

It is estimated that 26,300–28,100 tons of Ni each year are emitted into the atmosphere from natural sources such as windblown dust, volcanic activity, forest fires, meteoritic dust, sea salt spray, and small biogenic emissions (Eisler 2007). However, more than 80% of Ni total emissions are of anthropogenic origin, for example, combustion of oil, Ni metal refining, steel production, waste incineration, Ni-Cd battery production, combustion of coal, and Ni mining–smelting. According to Eisler (2007), human activities release 47,200–99,800 tons of Ni into the atmosphere and 50,800 tons of Ni of the total emissions deposited on land annually. Airborne Ni remains in the atmosphere normally for 5–8 days, a time period that depends on climatic conditions, the Ni phase (solid, liquid, condensed vapor, or gas), the chemical form of Ni (Ni-oxides, Ni-salts, Ni-sulfates), and the size of the particles. The largest anthropogenic source of Ni is the combustion of oil (accounts for 62% of anthropogenic emissions) with increasing emissions from 40,833 tons of Ni in 1983 to 86,100 tons of Ni in 1995. In China, because of the rise in industrialization, the total atmospheric Ni emissions have increased from 1096.07 tons in 1980 to 3933.71 tons in 2009, at an average annual growth rate of 4.5% (Tian et al. 2012).

Measured total deposition of Ni varies greatly from 54.6 µg m^{-2} at Irafoss, Iceland, to 7732.7 µg m^{-2} in Svanvik, Norway, a site under the effect of Ni–Cu smelters in the Kola Peninsula (McGrath 1995). Settimo and Viniano (2015) reported that the bulk deposition of Ni is 0.03–4.3 µg m^{-2} day^{-1} in rural, 5–11 µg m^{-2} day^{-1} in urban, and 2.3–22 µg m^{-2} day^{-1} in industrial areas.

5.3.2.2 Mining and Smelting

The mining and smelting of metal deposits are one of the important sources of soil contamination by metals (Nriagu 1996). Because of the worldwide increasing demand for metals, Ni production has increased from 1,400,000 tons to 2,400,000 tons over the last 10 years. Therefore, Ni mining and smelting release large amounts of anthropogenic Ni. There have been several studies of airborne Ni contamination of landscapes around smelters. Norseth (1994) reported that Ni smelters in the Kola Peninsula (Russia) released approximately 504 tons of Ni annually into the atmosphere that spread according to the wind direction and particle size. Barcan and Kovnatsky (1998) studied the metal contamination of soils around the Cu–Ni Severonickel Smelter Complex (Kola Peninsula). They reported that Ni concentrations were elevated above background levels almost 450 times and ranged between 26 and 9288 mg kg^{-1}, depending on the soil type and the distance from the smelter. A soil-based geochemical survey in an area of about 350 km^2 around the Zvečan Pb–Zn smelter (northern Kosovo) have shown that Ni concentrations were in the range 12.3–2842 mg kg^{-1}, with values above 1000 mg kg^{-1} in about 3% (Borgna et al. 2009). In the Sudbury area in Ontario, "the nickel capital of the world," Hutchinson and Whitby (1974) reported elevated levels of Ni (3000–5000 mg kg^{-1}) for surface soils close to the smelters. Twenty years later, Dudka et al. (1995) showed a marked reduction of total Ni concentrations due to the reduction of atmospheric emissions, leaching, and erosion processes.

5.3.2.3 Sewage Sludge

Information on the effect of sewage sludge on the Ni content of soils is inconsistent given the highly variable characteristics of sewage sludges and the great variety of application rates. As sewage sludges can contain a wide range of Ni concentrations (6–5300 mg kg^{-1}), their application on soils can cause increase of soil total Ni especially at high application rates. Nriagu and Pacyna (1988) estimated 4.04%–4.72% of global Ni inputs to soils, as a result of disposal of sewage sludge. However, over the last 25 years, heavy metal loadings in sewage sludge have been reduced as a result of industry adaptation to cleaner technologies (Antoniadis et al. 2006). At the same time, many countries regulated the amounts of metals that can be added to soils with sewage sludge applications (Haynes et al. 2009). Total Ni concentrations (various countries) and permissible limits in sewage sludge for land applications are shown in Table 5.4. Recent studies reported that low rates of sewage sludge applied to land do not increase substantially total concentrations of Ni in soils. Antoniadis et al. (2010) reported that there was no increase in Ni concentrations in a slight alkaline calcareous clay loam soil (Typic Xerochrept) after receiving various sewage sludge treatments (10, 30, and 50 tons ha^{-1}).

In an experiment in NW Spain, two different types of sludges (W1 and W2) were applied at two rates (a: 40 tons ha^{-1} and b: 400 tons ha^{-1}) in two different soils (loamy sand Typic Rhodoxeralf—S1 and sandy loam Typic Xeropsamment—S2). The data in Table 5.5 show that although total Ni content increased after the addition of sludges, especially at the highest rate, the concentrations were quite lower than the permissible limits in the EU

TABLE 5.4

Total Ni Concentrations in Sewage Sludge from Various Countries and Permissible Limits for Agricultural Use

Sewage Sludge	Ni (mg kg^{-1})
Canada	23–410
United States	36–562
China	15.8–233
Australia	60
United Kingdom	20
New Zealand	25
France	26.4–44.0
Spain	9.8–36.5
Egypt	39.0–271
Poland	21.7–155
Denmark	8–141
Greece	300 ± 76
Slovenia	372–995
European Union	9–90
Permissible limits in the United States	420
Permissible limits in the European Union (pH > 7)	400
Permissible limits in the European Union (pH < 7)	300

Sources: Cheng, M., L. Wu, Y. Huang, Y. Luo, and P. Christie. 2014. Total concentrations of heavy metals and occurrence of antibiotics in sewage sludges from cities throughout China. *Journal of Soils and Sediments* 14:1123–1135.; Antoniadis, V., C. Tsadilas, V. Samaras, and J. Sgouras. 2006. Availability of heavy metals applied to soil through sewage sludge. In: *Trace Elements in the Environment*, eds. M. Narasimha, V. Prasad, K. Sajawan, and R. Naidu, 39–57. New York: Taylor & Francis.; McGrath, S. P. 1995. Chromium and nickel. In: *Heavy Metals in Soils*, ed. B. J. Alloway, 152–178. New York: Blackie Academic and Professional.

(Sanchez-Martin et al. 2007). In a clayey Oxisol (Rhodic Hapludox), although total Cu and Zn concentrations increased linearly with the increasing rates of sewage sludge (maximum rate, 80 tons ha^{-1}), the total Ni was not affected (Martins et al. 2003).

Although there is a debate in the literature about the downward movement of metals and its depth distribution in the soil profile, a plethora of studies support that Ni movement after the application of sewage sludge is not significant, and in most cases, Ni accumulates only in the upper soil layer (Antoniadis et al. 2006).

5.3.2.4 Fertilizers

Agricultural inputs such as fertilizers can cause potential accumulation of trace elements in soils (Jiao et al. 2012). Among mineral fertilizers, phosphate

TABLE 5.5

Total Ni Content in Unamended Soils (S1 and S2), in Sewage
Sludge (W1 and W2), and in Sludge Amended Soils at Two
Rates of Application (a and b)

Samples	Ni (mg kg^{-1})
S1	13.3 ± 2.8
S2	12.0 ± 1.2
W1	45.5 ± 7.9
W2	37.2 ± 7.8
S1–W1a	17.7 ± 4.6
S1–W1b	22.5 ± 3.6
S1–W2a	16.8 ± 2.8
S1–W2b	22.5 ± 4.9
S2–W1a	20.3 ± 3.9
S2–W1b	24.6 ± 2.9
S2–W2a	16.5 ± 3.7
S2–W2b	24.9 ± 2.8

Source: Adapted from Sanchez-Martin, M. J., M. Garcia-Delgado,
L. F. Lorenzo, M. S. Rodriguez-Cruz, and M. Arienzo. 2007.
Heavy metals in sewage sludge amended soils determined
by sequential extractions as a function of incubation time of
soils. *Geoderma* 142: 262–273.

fertilizers contain significant amounts of Ni, depending on the source of
phosphate rocks. de Lopez Camelo et al. (1997) reported that the continuous
application of phosphate fertilizers in Argentina led to soil enrichment with
Ni but without any risk of soil pollution. Luo et al. (2009) indicate a content of
Ni in different fertilizers ranging from 9.6 to 11.4 mg kg^{-1}. According to Molina
et al. (2009), the concentrations of Ni showed differences and high coefficient
variation when analyzing triple superphosphates (TSP), monoammonium
phosphates (MAP), and diammonium phosphates (DAP). Nickel concen-
trations in the samples on P basis were in the following ranges: 12–89 mg
Ni kg^{-1} (TSP), 8–51 mg Ni kg^{-1} (MAP), and 12–58 mg Ni kg^{-1} (DAP). The
AROMIS project (Assessment and reduction of heavy metal input into agroecosystems) using a large set of data collected from 21 European countries
(Austria, Belgium, Switzerland, Czech Republic, Germany, Denmark, Spain,
Finland, France, Greece, Hungary, Ireland, Italy, The Netherlands, Norway,
Poland, Portugal, Romania, Sweden, Slovenia, and United Kingdom) showed
that the mean input of Ni from fertilizers was 6.7 g ha^{-1} year^{-1} (Alloway 2013).

5.3.3 Distinguishing between Natural and Anthropogenic
Sources of Ni in Soils

The knowledge of natural versus anthropogenic sources of heavy metals
in soils is imperative for assessing human impact on the environment and

developing appropriate strategies for protecting ecosystem quality (Hu and Cheng 2016; Micó et al. 2006). Especially for elements like Ni, the identification of sources is challenging because of the great spatial heterogeneity of soil parent materials and the high variability of anthropogenic inputs affecting both agricultural and urban soils.

Although, on a global basis, the geological formations and parent materials control the heavy metal concentrations in soils, human activities can cause either localized contamination on a regional scale or diffuse contamination of larger land surfaces (Tóth et al. 2016b). For example, Freedman and Hutchison (1980) reported extremely high Ni concentrations, up to 26,000 mg kg^{-1} in top soils near the Ni–Cu smelter at Sudbury, Canada. However, separating out the lithogenic and anthropogenic origin of heavy metals in soils is often not a very easy task, especially if the point source cannot be identified (Lin et al. 2017).

There are a number of approaches for assessing the sources of heavy metal in soils, such as chemical speciation analysis, isotope tracer analysis, multivariate statistics, and geostatistical analysis (Chai et al. 2015; Schneider et al. 2016).

Chemical speciation of metals is important to determine their bioavailability, mobility, and toxicity in soils. Sequential extraction procedures that have been developed to extract the chemical forms of metals provide critical information on metal origin, bioavailability, and mobility under different pedoenvironmental conditions. Heavy metals originating from anthropogenic sources often occur in forms different from native metal content in the soil, for example, in easily labile forms, and tend to be more mobile and bioavailable than lithogenic ones (Chai et al. 2015). Therefore, metals with anthropogenic origin are mainly extracted in the first step of sequential extraction procedures (soluble and exchangeable fractions), while lithogenic metals are found mostly in the last step corresponding to the residual fraction (metals retained within the crystal structure of soil minerals). According to Massas et al. (2013) and Borgese et al. (2013), metals of anthropogenic origin are usually more weakly bound to the soil components and therefore more easily released to soil solution.

Several researchers have used the isotopic composition of soils as a tool for differentiating natural from anthropogenic sources. Currently, lead (Pb) isotope analyses are widely used in these studies as Pb is ubiquitous in many industrial processes and the Pb isotope composition does not change during industrial and environmental processes but always reflects the source origin (Cicchella et al. 2016; Grezzi et al. 2011; Tarzia et al. 2002).

Multivariate statistical analysis (principal component analysis and cluster analysis) provides a useful technique that has been widely applied to identify natural or anthropogenic sources of heavy metals in agricultural (Facchinelli et al. 2001; Kelepertzis 2014; Micó et al. 2006) and urban soils (Abollino et al. 2002; Acosta et al. 2010).

Finally, geostatistical tools that manage the spatial distribution of heavy metals can be a useful tool to determine their sources in a complex soil environment (Sun et al. 2013; Zhang et al. 2016).

5.3.3.1 Agricultural Soils

Normal practices such as the application of fertilizers and sewage sludges can generally cause enrichment of Ni in agricultural soils (Cai et al. 2012; Nicholson et al. 2003). Recently, Nziguheba and Smolders (2008) estimated an average annual input of Ni to agricultural soils from phosphate fertilizers in Europe to 3.6 g Ni ha^{-1}, based on statistics on fertilizer use of 1999/2000. In addition, some agricultural areas near industrial plants or the combustion of fossil fuels can be affected by atmospheric deposition (Zhou et al. 2014). Pandey and Pandey (2009) reported that the annual atmospheric deposition of Ni in soils under an organic farming system in India was 0.073 kg Ni ha^{-1} due to proximity of industries, including a zinc smelter, pesticide, and phosphate fertilizer factories. These anthropogenic sources of Ni may contribute to the great spatial variation of Ni in soils that mainly depends on the heterogeneity of parent material (Li et al. 2015).

Many studies on the heavy metal status of agricultural soils in the world have been conducted (Table 5.6). The majority of these studies have pointed out that Cr and Ni usually come from natural sources, whereas As, Pb, Cd, Cu, Zn, and Hg are mainly of anthropogenic origin (Bai et al. 2015; Facchinelli et al. 2001; Hani and Pazira 2011; Li et al. 2015; Nanos and Rodríguez Martín 2012; Tian et al. 2017; Yang et al. 2009).

Recently, the Geochemical Mapping of Agricultural and Grazing Land Soil (GEMAS) project focused on the productive soils of Europe has identified

TABLE 5.6

Nickel Mean (or Median) Concentrations in Agricultural Soils from Various Countries

Location	Ni (mg kg^{-1})	References
Almería (Spain)	38.6	Gil et al. (2004)
Alicante (Spain)	20.9	Micó et al. (2006)
Bandirma (Turkey)	97	Dartan et al. (2015)
Murcia (Spain)	13.5	Acosta et al. (2011)
Castellón (Spain)	19.3	Peris et al. (2008)
Piemonte (Italy)	83.2	Facchinelli et al. (2001)
Beijing (China)	29.2	Lin et al. (2017)
Duero basin (Spain)	15.08	Nanos and Rodríguez Martín (2012)
Zagreb (Croatia)	49.5	Romic and Romic (2003)
Dehui (China)	20.8	Sun et al. (2013)
Huizhou (China)	14.89	Cai et al. (2012)
Jiangsu (China)	38.5	Huang et al. (2007)
Argolida basin (Greece)	146.8	Kelepertzis (2014)
Çanakkale (Turkey)	205.3	Sungur (2016)
Larissa (Greece)	165.3	Skordas et al. (2013)
Tehran (Iran)	36.9	Hani and Pazira (2011)

central European and Mediterranean areas with high Ni values, equal to or above the corresponding medians (Albanese et al. 2015). In this project, although most soils of Europe are characterized by low Ni concentrations <48 mg kg^{-1} (agricultural soils) and <51 mg kg^{-1} (grazing land samples), the highest Ni concentrations are found in Greece (2475 mg kg^{-1}) for agricultural soil samples and in Serbia (2466 mg kg^{-1}) for grazing land soil samples. These concentrations obviously reflect the impact of lithogenic sources (ophiolite complex and mafic–ultramafic rocks) on Ni content in agricultural soils in these areas of Europe.

In the Argolida basin, a region in Greece with intensive agricultural activities, Kelepertzis (2014) reported that the continuous application of large amounts of agrochemicals has resulted in Cu, Zn, Cd, Pb, and As accumulation in soils, whereas Ni, Cr, Co, and Fe amounts are controlled primarily by parent material (ultramafic materials composed of small serpentinitic bodies). In greenhouse soils of Wuwei District, China, Bai et al. (2015) found that greenhouse cultivation had little impact on the accumulation of Cr and Ni, indicating that Ni status in these agricultural soils was highly dependent on the Ni content in parent materials.

5.3.3.2 Urban Soils

Soils in urban and industrial areas are characterized by unpredictable layering, poor structure, and high concentrations of heavy metals (Ajmone-Marsan and Biasioli 2010; Massas et al. 2010). Because urban soils are greatly influenced by anthropogenic activities (industrial and residential wastes, vehicle emissions, coal burning waste, etc.), numerous studies have been conducted on urban soil contamination with heavy metals, including Ni (Table 5.7). The reported concentrations (with some exceptions) are roughly comparable with the global range of 20 to 50 mg kg^{-1} (Gonnelli and Renella 2013) or the average total concentration of Ni in soils of the world (34 mg kg^{-1}) (McGrath 1995), demonstrating the limited effect of anthropogenic sources on Ni content in urban soils.

In China, a country where urban contamination has been extensively investigated, numerous studies report that Ni and Cr were the least polluted elements in urban soils, supporting the limited contribution of anthropogenic sources to Ni enrichment (Lv et al. 2015; Wei and Yang 2010). The elevated Ni concentrations in cities like Athens (Greece) are consistent with the presence of serpentinized ophiolithic rocks in local geology (Argyraki and Kelepertzis 2014). Recently, Gasparatos et al. (2015) reported that the vast majority of Ni content in the urban soils of the Thriassio plain, a residential, agricultural, and industrial area near Athens (Greece), was associated with the residual fraction, indicating that the presence of Ni in the studied soils is mainly controlled by the soil's parent material.

TABLE 5.7

Nickel Mean or Median Concentrations in Urban Soils from Various Cities around the World

City	Ni (mg kg^{-1})	References
Baltimore (USA)	18.4	Yesilonis et al. (2008)
Ajka (Hungary)	19.6	Zacháry et al. (2015)
Beijing (China)	23.8	Wang et al. (2012)
Berlin (Germany)	7.7	Birke and Rauch (2000)
Bristol (UK)	21	Giusti (2011)
Talcahuano (Chile)	30	Tume et al. (2014)
Chicago (USA)	31	Cannon and Horton (2009)
Damascus (Syria)	35	Möller et al. (2005)
Galway (Ireland)	22	Zhang (2006)
Greater Athens and Piraeus (Greece)	102	Argyraki and Kelepertzis (2014)
Hong Kong (China)	11.2	Li et al. (2004)
Athens (Greece)	81.6	Massas et al. (2013)
Ibadan (Nigeria)	16.5	Odewande and Abimbola (2008)
Lisbon (Portugal)	20	Cachada et al. (2013)
Mexico City (Mexico)	39	Morton-Bermea et al. (2009)
Napoli (Italy)	8.9	Cicchella et al. (2008)
New Orleans (USA)	8	Ajmone-Marsan and Biasoli (2010)
Oslo (Norway)	24.1	Tijhuis et al. (2002)
Seville (Spain)	23.1	Madrid et al. (2004)
Sicily (Italy)	17.8	Manta et al. (2002)
Stockholm (Sweden)	13	Ajmone-Marsan and Biasoli (2010)
Turku (Finland)	24.1	Salonen and Korkka-Niemi (2007)
Uppsala (Sweden)	19	Ajmone-Marsan and Biasoli (2010)
Zagreb (Croatia)	48.7	Romic and Romic (2003)

5.4 Conclusions

Nickel is ubiquitous in soils with a very broad range from trace amounts to relatively high concentrations, as compared to other heavy metals. Soil parent material, and especially the presence of ultramafic rocks, is the most important factor that determines the distribution of Ni on a global area basis. In this chapter, the origin of Ni in soils has been reviewed extensively and the predominance of the natural (lithogenic–pedogenic origin) over anthropogenic sources has been clearly demonstrated in many countries (e.g., China, Iran, Italy, Greece, Spain). Anthropogenic sources including fertilizers and sewage sludges for agricultural soils and industrial-residential wastes, vehicle emissions, and coal burning waste for urban soils were also significant, especially at the regional scale in areas with low Ni geochemical background. The identification of natural and anthropogenic sources of Ni in soils is important in order to reduce inputs and to establish reliable strategies for ecosystem quality protection.

References

Abollino, O., M. Aceto, M. Malandrino, E. Mentasti, C. Sarzanini, and F. Petrella. 2002. Heavy metals in agricultural soils from Piedmont, Italy, distribution, speciation and chemometric data treatment. *Chemosphere* 49:545–557.

Acosta, J. A., A. Faz, and S. M. Martinez. 2010. Identification of heavy metal sources by multivariable analysis in a typical Mediterranean city (SE Spain). *Environmental Monitoring and Assessment* 169:519–530.

Acosta, J.A., A. Faz, S. Martínez-Martínez, and J. M. Arocena. 2011. Enrichment of metals in soils subjected to different land uses in a typical Mediterranean environment (Murcia City, southeast Spain). *Applied Geochemistry* 26:405–414.

Adriano, D. C. 2001. *Trace Elements in Terrestrial Environments: Biogeochemistry, Bioavailability and Risks of Metals.* 2nd ed. New York: Springer-Verlag.

Ajmone-Marsan, F. and M. Biasioli. 2010. Trace elements in soils of urban areas. *Water, Air, & Soil Pollution* 213:121–143.

Albanese, S., M. Sadeghi, A. Lima, D. Cicchella, E. Dinelli, P. Valera, M. Falconi, A. Demetriades, and B. Vivo. 2015. GEMAS: Cobalt, Cr, Cu and Ni distribution in agricultural and grazing land soil of Europe. *Journal of Geochemical Exploration* 154:81–93.

Alloway, B. 2013. Sources of heavy metals and metalloids in soils. In: *Heavy Metals in Soils—Trace Metals and Metalloids in Soils and Their Bioavailability,* ed. B. Alloway, 11–50. Dordrecht: Springer.

Antibachi, D., E. Kelepertzis, and A. Kelepertis. 2012. Heavy metals in agricultural soils of the Mouriki-Thiva area and environmental impact implications. *Soil and Sediment Contamination* 21:434–450.

Antoniadis, V., C. Tsadilas, V. Samaras, and J. Sgouras. 2006. Availability of heavy metals applied to soil through sewage sludge. In: *Trace Elements in the Environment,* eds. M. Narasimha, V. Prasad, K. Sajawan, and R. Naidu, 39–57. New York: Taylor & Francis.

Antoniadis, V., C. D. Tsadilas, and V. Samaras. 2010. Trace element availability in a sewage sludge-amended cotton grown Mediterranean soil. *Chemosphere* 80:1308–1313.

Argyraki, A. and E. Kelepertzis. 2014. Urban soil geochemistry in Athens, Greece: The importance of local geology in controlling the distribution of potentially harmful trace elements. *Science of the Total Environment* 482–483:366–377.

Bai, L.Y., X. B. Zeng, S. M. Su, R. Duan, Y. N. Wang, and X. Gao. 2015. Heavy metal accumulation and source analysis in greenhouse soils of Wuwei District, Gansu Province, China. *Environmental Science and Pollution Research* 22:5359–5369.

Barcan, V. and E. Kovnatsky. 1998. Soil surface geochemical anomaly around the copper–nickel metallurgical smelter. *Water, Air, & Soil Pollution* 103:197–218.

Birke, M. and U. Rauch. 2000. Urban geochemistry: Investigations in the Berlin metropolitan area. *Environmental Geochemistry and Health* 22:233–248.

Bompoti, N., M. Chrysochoou, and D. Dermatas. 2015. Geochemical characterization of Greek ophiolitic environments using statistical analysis. *Environmental Processes* 2:5–21.

Bonifacio, E., G. Falsone, and S. Piazza. 2010. Linking Ni and Cr concentrations to soil mineralogy: Does it help to assess metal contamination when the natural background is high? *Journal of Soils and Sediments* 10:1475–1486.

Borgese, L., S. Federici, A. Zacco, A. Gianoncelli, L. Rizzo, R. D. Smith, F. Donna, R. Lucchini, L. E. Depero, and E. Bontempi. 2013. Metal fractionation in soils and assessment of environmental contamination in Vallecamonica Italy. *Environmental Science and Pollution Research* 20:5067–5075.

Borgna, L., L. A. Di Lella, F. Nannonni, A. Pisani, E. Pizzatti, G. Protano, F. Riccobono, and S. Rossi. 2009. The high contents of lead in soils of northern Kosovo. *Journal of Geochemical Exploration* 101:137–146.

Brooks, R. R. 1987. *Serpentine and Its Vegetation: A Multidisciplinary Approach*. Portland, Oregon: Dioscorides Press.

Cachada, A., A. C. Dias, P. Pato, C. Mieiro, T. Rocha-Santos, M. E. Pereira, E. Ferreira da Silva, and A. Duarte. 2013. Major inputs and mobility of potentially toxic elements contamination in urban areas. *Environmental Monitoring and Assessment* 185:279–294.

Cai, L., Z. Xu, M. Ren, Q. Guo, X. Hu, G. Hu, H. Wan, and P. Peng. 2012. Source identification of eight hazardous heavy metals in agricultural soils of Huizhou, Guangdong Province, China. *Ecotoxicology and Environmental Safety* 78:2–8.

Caillaud, J., D. Proust, S. Philippe, C. Fontaine, and M. Fialin. 2009. Trace metals distribution from a serpentinie weathering at the scales of the weathering profile and its related weathering Microsystems and clay minerals. *Geoderma* 149:199–208.

Cannon, W. F. and J. D. Horton. 2009. Soil geochemical signature of urbanization and industrialization—Chicago, Illinois, USA. *Applied Geochemistry* 24:1590–1601.

Chai, Y., J. Guo, S. Chai, J. Cai, L. Xue, and Q. Zhang. 2015. Source identification of eight heavy metals in grassland soils by multivariate analysis from the Baicheng–Songyuan area, Jilin Province, Northeast China. *Chemosphere* 134:67–75.

Cheng, M., L. Wu, Y. Huang, Y. Luo, and P. Christie. 2014. Total concentrations of heavy metals and occurrence of antibiotics in sewage sludges from cities throughout China. *Journal of Soils and Sediments* 14:1123–1135.

Cicchella, D., B. De Vivo, A. Lima, S. Albanese, R. A. R. McGill, and R. R. Parrish. 2008. Heavy metal pollution and Pb isotopes in urban soils of Napoli, Italy. *Geochemistry: Exploration, Environment, Analysis* 8:103–112.

Cicchella, D., J. Hoogewerff, S. Albanese, P. Adamo, A. Lima, M. V. E. Taiani, and B. De Vivo. 2016. Distribution of toxic elements and transfer from the environment to humans traced by using lead isotopes. A case of study in the Sarno river basin, south Italy. *Environmental Geochemistry and Health* 38:619–637.

Dartan, G., F. Taspinar and İ. Toröz. 2015. Assessment of heavy metals in agricultural soils and their source apportionment: A Turkish district survey. *Environmental Monitoring and Assessment* 187:99.

de Lopez Camelo, L. G., S. R. de Miguez, and L. Marban. 1997. Heavy metals input with phosphate fertilizers used in Argentina. *Science of the Total Environment* 204:245–250.

Dudka, S., R. Ponce-Hernandez, and T. C. Hutchinson. 1995. Current level of total element concentrations in the surface layer of Sudbury's soils. *Science of the Total Environment* 162:161–171.

Eisler, R. 2007. *Eisler's Encyclopedia of Environmentally Hazardous Priority Chemicals*, 1st ed. Amsterdam, the Netherlands: Elsevier.

Facchinelli, A., E. Sacchi, and L. Mallen. 2001. Multivariate statistical and GIS-based approach to identify heavy metal sources in soils. *Environmental Pollution* 114:313–324.

Freedman, B. and T. C. Hutchinson. 1980. Pollutants inputs from the atmosphere and accumulations in soils and vegetation near a nickel-copper smelter at Sudbury, Ontario, Canada. *Canadian Journal of Botany* 58:108–131.

Gasparatos, D. 2013. Sequestration of heavy metals from soil with Fe–Mn concretions and nodules. *Environmental Chemistry Letters* 11:1–9.

Gasparatos, D, G. Mavromati, P. Kotsovilis, and I. Massas. 2015. Fractionation of heavy metals and evaluation of the environmental risk for the alkaline soils of the Thriassio plain: A residential, agricultural, and industrial area in Greece. *Environmental Earth Sciences* 74:1099–1108.

Gil, C., R. Boluda, and J. Ramos. 2004. Determination and evaluation of cadmium lead and nickel in greenhouse soils of Almería (Spain). *Chemosphere* 55:1027–1034.

Giusti, L. 2011. Heavy metals in urban soils of Bristol (UK). Initial screening for contaminated land. *Journal of Soils and Sediments* 11:1385–1398.

Gonnelli, C. and G. Renella. 2013. Nickel and chromium. In *Heavy Metals in Soils: Trace Metals and Metalloids in Soils and Their Availability*, ed. B. J. Alloway, 335–366. Dordrecht: Springer.

Grezzi, G., R. A. Ayuso, B. De Vivo, A. Lima, and S. Albanese. 2011. Lead isotopes in soils and groundwaters as tracers of the impact of human activities on the surface environment: The Domizio-Flegreo Littoral (Italy) case study. *Journal of Geochemical Exploration* 109:51–58.

Hani, A. and E. Pazira. 2011. Heavy metals assessment and identification of their sources in agricultural soils of Southern Tehran, Iran. *Environmental Monitoring and Assessment* 176:677–691.

Haynes, R. J., G. Murtaza, and R. Naidu. 2009. Inorganic and organic constituents and contaminants of biosolids: Implications for land application. *Advances in Agronomy* 104:165–267.

Hoatson, D. M., S. Jaireth, and A. L. Jaques. 2006. Nickel sulfide deposits in Australia: Characteristics, resources, and potential. *Ore Geology Reviews* 29:177–241.

Hseu, Z. Y. and Y. Iizuka. 2013. Pedogeochemical characteristics of chromite in a paddy soil derived from serpentinites. *Geoderma* 202–203:126–133.

Hseu, Z. Y., T. Watanabe, A. Nakao, and S. Funakawa. 2016. Partition of geogenic nickel in paddy soils derived from serpentinites. *Paddy and Water Environment* 14:417–426.

Hu, Y. and H. Cheng. 2016. A method for apportionment of natural and anthropogenic contributions to heavy metal loadings in the surface soils across large-scale regions. *Environmental Pollution* 214:400–409.

Huang, S. S., Q. L. Liao, M. Hua, X. M. Wu, K. S. Bi, C. Y. Yan, B. Chen, and X. Y. Zhang. 2007. Survey of heavy metal pollution and assessment of agricultural soil in Yangzhong district, Jiangsu province, China. *Chemosphere* 67:2148–2155.

Hutchinson, T. C. and L. M. Whitby. 1974. Heavy-metal pollution in the Sudbury mining and smelting region of Canada, I. Soil and vegetation contamination by nickel, copper, and other metals. *Environmental Conservation* 1:123–132.

Jiao, W. T., W. P. Chen, A. C. Chang, and A. L. Page. 2012. Environmental risks of trace elements associated with long-term phosphate fertilizers applications: A review. *Environmental Pollution* 168:44–53.

Kabata-Pendias, A. and A. B. Mukherjee. 2007. *Trace Elements from Soil to Human.* Berlin, New York: Springer-Verlag.

Kabata-Pendias, A. and H. Pendias. 2001. *Trace Elements in Soils and Plants,* 3rd ed. Boca Raton: CRC Press.

Kanellopoulos, C., A. Argyraki, and P. Mitropoulos. 2015. Geochemistry of serpentine agricultural soil and associated groundwater chemistry and vegetation in the area of Atalanti, Greece. *Journal of Geochemical Exploration* 158:22–33.

Kaprara, E., N. Kazakis, K. Simeonidis, S. Coles, A. I. Zoumboulis, P. Samaras, and M. Mitrakas. 2015. Occurrence of Cr(VI) in drinking water of Greece and relation to the geological background. *Journal of Hazardous Materials* 281:2–11.

Kelepertzis, E. 2014. Accumulation of heavy metals in agricultural soils of Mediterranean: Insights from Argolida basin, Peloponnese, Greece. *Geoderma* 221–222:82–90.

Kelepertzis, E., E. Galanos, and I. Mitsis. 2013. Origin, mineral speciation and geochemical baseline mapping of Ni and Cr in agricultural top soils of Thiva Valley (Central Greece). *Journal of Geochemical Exploration* 125:56–68.

Kierczak, J., A. Pędziwiatr, J. Waroszewski, and M. Modelska. 2016. Mobility of Ni. Cr and Co in serpentine soils derived on various ultrabasic bedrocks under temperate climate. *Geoderma* 268:78–91.

Li, C., F. Li, Z. Wu, and J. Cheng. 2015. Effects of landscape heterogeneity on the elevated trace metal concentrations in agricultural soils at multiple scales in the Pearl River Delta, South China. *Environmental Pollution* 206:264–274.

Li, X., S. L. Lee, S. C. Wong, W. Shi, and I. Thornton. 2004. The study of metal contamination in urban soils of Hong Kong using a GIS-based approach. *Environmental Pollution* 129:113–24.

Lilli, M. A., D. Moraetis, N. P. Nikolaidis, G. P. Karatzas, and N. Kalogerakis. 2015. Characterization and mobility of geogenic chromium in soils and river bed sediments of Asopos basin. *Journal of Hazardous Materials* 281:12–19.

Lin, Y., P. Han, Y. Huang, G. L. Yuan, L. X. Guo, and J. Li. 2017. Source identification of potentially hazardous elements and their relationships with soil properties in agricultural soil of the Pinggu district of Beijing, China: Multivariate statistical analysis and redundancy analysis. *Journal of Geochemical Exploration* 173:110–118.

Luo, L., Y. Ma, S. Zhang, D. Wei, and Y. Zhu. 2009. An inventory of trace element inputs to agricultural soils in China. *Journal of Environmental Management* 90: 2524–2530.

Lv, J., Y. Liu, Z. Zhang, R. Zhou, and Y. Zhu. 2015. Distinguishing anthropogenic and natural sources of trace elements in soils undergoing recent 10-year rapid urbanization: A case of Donggang, Eastern China. *Environmental Science and Pollution Research* 22:10539–10550.

Ma, Y. and P. S. Hooda. 2010. Chromium, nickel and cobalt, In *Trace Elements in Soils,* ed. P. Hooda, 461–480, United Kingdom: John Wiley & Sons.

Madrid, L., E. Díaz-Barrientos, R. Reinoso, and F. Madrid. 2004. Metals in urban soils of Sevilla: Seasonal changes and relations with other soil components and plant contents. *European Journal of Soil Science* 55:209–17.

Manta, D. S., M. Angelone, A. Bellanca, R. Neri, and M. Sprovieri. 2002. Heavy metals in urban soils: A case study from the city of Palermo (Sicily), Italy. *Science of Total Environment* 300:229–243.

Martins, A. L. C., O. C. Bataglia, and O. A. Camargo. 2003. Copper, nickel and zinc phytoavailability in an Oxisol amended with sewage sludge and liming. *Scientia Agricola* 60:747–754.

Massas, I., C. Ehaliotis, D. Kalivas, and G. Panagopoulou. 2010. Concentrations and availability indicators of soil heavy metals; the case of children's playgrounds in the city of Athens (Greece). *Water, Air, & Soil Pollution* 212:51–63.

Massas, I., D. Kalivas, C. Ehaliotis, and D. Gasparatos. 2013. Total and available heavy metal concentrations in soils of the Thriassio plain (Greece) and assessment of soil pollution indexes. *Environmental Monitoring and Assessment* 185:6751–6766.

Massoura, S. T., G. Echevarria, T. Becquer, J. Ghanbaja, E. Leclere-Cessac, and J. L. Morel. 2006. Control of nickel availability by nickel bearing minerals in natural and anthropogenic soils. *Geoderma* 136:28–37.

McGrath, S. P. 1995. Chromium and nickel. In: *Heavy Metals in Soils*, ed. B. J. Alloway, 152–178. New York: Blackie Academic and Professional.

Megremi, I. 2010. Distribution and bioavailability of Cr in central Euboea, Greece. *Central European Journal of Geosciences* 2:103–123.

Micó, C., L. Recatalá, M. Peris, and J. Sánchez. 2006. Assessing heavy metal sources in agricultural soils of a European Mediterranean area by multivariate analysis. *Chemosphere* 65:863–872.

Molina, M., F. Aburto, R. Calderóón, M. Cazanga, and M. Escudey. 2009. Trace element composition of selected fertilizers used in Chile: Phosphorus fertilizers as a source of long-term soil contamination. *Soil and Sediment Contamination* 18: 497–511.

Möller, A., H. W. Müller, A. Abdullah, G. Abdelgawad, and J. Utermann. 2005. Urban soil pollution in Damascus, Syria: Concentrations and patterns of heavy metals in the soils of the Damascus Ghouta. *Geoderma* 124:63–71.

Morton-Bermea, O., E. Hernández-Álvarez, G. González-Hernández, F. Romero, R. Lozano, and L. E. Beramendi-Orosco. 2009. Assessment of heavy metal pollution in urban topsoils from the metropolitan area of Mexico City. *Journal of Geochemical Exploration* 101:218–224.

Nanos, N. and J. A. Rodríguez Martín. 2012. Multiscale analysis of heavy metal contents in soils: Spatial variability in the Duero river basin (Spain). *Geoderma* 189–190:554–562.

Nicholson, F. A., S. R. Smith, B. J. Alloway, C. Carlton-Smith, and B. J. Chambers. 2003. An inventory of heavy metals inputs to agricultural soils in England and Wales. *Science of the Total Environment* 311:205–219.

Norseth, T. 1994. Environmental pollution around nickel smelters in the Kola Peninsula (Russia). *Science of the Total Environment* 148:103–108.

Nriagu, J. O. 1996. A history of global metal pollution. *Science* 272:223–224.

Nriagu, J. O. and J. Pacyna. 1988. Quantitative assessment of worldwide contamination of air, water and soil by trace metals. *Nature* 333:134–139.

Nziguheba, G. and E. Smolders. 2008. Inputs of trace elements in agricultural soils via phosphate fertilizers in European countries. *Science of Total Environment* 390:53–57.

Odewande, A. A. and A. F. Abimbola. 2008. Contamination indices and heavy metal concentrations in urban soil of Ibadan metropolis, southwestern Nigeria. *Environmental Geochemistry and Health* 30:243–254.

Oze, C., C. Skinner, and A. Schroth. 2008. Growing up green on serpentine soils: Biogeochemistry of serpentine vegetation in the Central Coast Range of California. *Applied Geochemistry* 23:3391–3403.

Pandey, J. and U. Pandey. 2009. Atmospheric deposition and heavy metal accumulation in organic farming system in seasonally dry tropical region of India. *Journal of Sustainable Agriculture* 33:361–378.

Peris, M., L. Recatalá, C. Micó, R. Sánchez, and J. Sánchez. 2008. Increasing the knowledge of heavy metal contents and sources in agricultural soils of the European Mediterranean region. *Water, Air, & Soil Pollution* 192:25–37.

Quantin, C., V. Ettler, J. Garnier, and O. Šebek. 2008. Sources and extractability of chromium and nickel in soil profiles developed on Czech serpentinites. *Comptes Rendus Geosciences* 340:872–882.

Ratie, G., D. Jouvin, J. Garnier, O. Rouxel, S. Miska, E. Guimarães, L. Cruz Vieira, Y. Sivry, I. Zelano, E. Montarges-Pelletier, F. Thil, and C. Quantin. 2015. Nickel isotope fractionation during tropical weathering of ultramafic rocks. *Chemical Geology* 402:68–76.

Reimann, C., M. Birke, A. Demetriades, P. Filzmoser, P., and O'Connor (Eds.). 2014. *Chemistry of Europe's Agricultural Soils—Part A: Methodology and Interpretation of the GEMAS Data.* Stuttgart: Schweizerbart Science Publishers.

Romic, M. and D. Romic. 2003. Heavy metals distribution in agricultural topsoils in urban area. *Environmental Geology* 43:795–805.

Salminen R., M. J. Batista, M. Bidovec, A. Demetriades, B. De Vivo, W. De Vos, M. Duris, A. Gilucis, V. Gregorauskiene, J. Halamic, P. Heitzmann, A. Lima, G. Jordan, G. Klaver, P. Klein, J. Lis, J. Locutura, K. Marsina, A. Mazreku, P. J. O'Connor, S. A. Olsson, R. T. Ottesen, V. Petersell, J. A. Plant, S. Reeder, I. Salpeteur, H. Sandstrom, U. Siewers, A. Steenfelt, and T. Tarvainen. 2005. Geochemical atlas of Europe. Part 1: Background information, methodology and maps. Espoo Geological Survey of Finland.

Salonen, V. and K. Korkka-Niemi. 2007. Influence of parent sediments on the concentration of heavy metals in urban and suburban soils in Turku, Finland. *Applied Geochemistry* 22:906–918.

Sanchez-Martin, M. J., M. Garcia-Delgado, L. F. Lorenzo, M. S. Rodriguez-Cruz, and M. Arienzo. 2007. Heavy metals in sewage sludge amended soils determined by sequential extractions as a function of incubation time of soils. *Geoderma* 142:262–273.

Schneider, A. R., X. Morvan, N. P. A. Saby, B. Cancès, M. Ponthieu, M. Gommeaux, and B. Marin. 2016. Multivariate spatial analyses of the distribution and origin of trace and major elements in soils surrounding a secondary lead smelter. *Environmental Science and Pollution Research* 23:15164–15174.

Settimo, S. and G. Viviano. 2015. Atmospheric depositions of persistent pollutants: Methodological aspects and values from case studies. *Annali dell'Istituto Superiore di Sanità* 51:298–304.

Skordas, K., G. Papastergios, and A. Filippidis. 2013. Major and trace element contents in apples from a cultivated area of Central Greece. *Environmental Monitoring and Assessment* 185:8465–8471.

Sun, C., J. Liu, Y. Wang, L. Sun, and H. Yu. 2013. Multivariate and geostatistical analyses of the spatial distribution and sources of heavy metals in agricultural soil in Dehui, Northeast China. *Chemosphere* 92:517–523.

Sungur, A. 2016. Heavy metals mobility sources, and risk assessment in soils and uptake by apple (*Malus domestica* Borkh.) leaves in urban apple orchards. *Archives of Agronomy and Soil Science* 62:1051–1065.

Tarzia, M., B. De Vivo, R. Somma, R. A. Ayuso, R. A. R. Mc Gill, and R. R. Parrish. 2002. Anthropogenic vs. natural pollution: An environmental study of an industrial site under remediation (Naples, Italy). *Geochemistry: Exploration, Environmental Analysis* 2:45–56.

Tian, H. Z., L. Lu, K. Cheng, J. M. Hao, D. Zhao, Y. Wang, W. X. Jia, and P. P. Qiu. 2012. Anthropogenic atmospheric nickel emissions and its distribution characteristics in China. *Science of Total Environment* 417:148–157.

Tian, K., B. Huang, Z. Xing, and W. Hu. 2017. Geochemical baseline establishment and ecological risk evaluation of heavy metals in greenhouse soils from Dongtai, China. *Ecological Indicators* 72:510–520.

Tijhuis, L., B. Brattli, and O. M. Saether. 2002. A geochemical survey of topsoil in the city of Oslo, Norway. *Environmental Geochemistry and Health* 24:67–94.

Tóth, G., T. Hermann, M. R. Da Silva, and L. Montanarella. 2016a. Heavy metals in agricultural soils of the European Union with implications for food safety. *Environment International* 88:299–309.

Tóth, G., T. Hermann, G. Szatmári, and L. Pásztor. 2016b. Maps of heavy metals in the soils of the European Union and proposed priority areas for detailed assessment. *Science of Total Environment* 565:1054–1062.

Tume, P., R. King, E. González, G. Bustamante, F. Reverter, N. Roca, and J. Bech. 2014. Trace element concentrations in schoolyard soils from the port city of Talcahuano, Chile. *Journal of Geochemical Exploration* 147 (Part B):229–236.

U.S. Geological Survey. 2015. Mineral commodity summaries 2015: U.S. Geological Survey, 196 pp., http://minerals.usgs.gov/minerals/pubs/mcs/2015/mcs2015 .pdf (last accessed February 1, 2017).

Wang, M., B. Markert, W. Chen, C. Peng, and Z. Ouyang. 2012. Identification of heavy metal pollutants using multivariate analysis and effects of land uses on their accumulation in urban soils in Beijing, China. *Environmental Monitoring and Assessment* 184:5889–5897.

Wei, B. G. and L. S. Yang. 2010. A review of heavy metal contaminations in urban soils, urban road dusts and agricultural soils from China. *Microchemical Journal* 94:99–107.

Yang, P., R. Mao, H. Shao, and Y. Gao. 2009. The spatial variability of heavy metal distribution in the suburban farmland of Taihang Piedmont Plain, China. *Comptes Rendus Biologies* 332:558–566.

Yesilonis, I. D., R. V. Pouyat, and N. K. Neerchal. 2008. Spatial distribution of metals in soils in Baltimore: Role of native parent material, proximity to major roads, housing age and screening guidelines. *Environmental Pollution* 156:723–731.

Zacháry, D., G. Jordan, P. Völgyesi, A. Bartha, and C. Szabó. 2015. Urban geochemical mapping for spatial risk assessment of multisource potentially toxic elements—A case study in the city of Ajka, Hungary. *Journal of Geochemical Exploration* 158:186–200.

Zhang, C. 2006. Using multivariate analyses and GIS to identify pollutants and their spatial patterns in urban soils in Galway, Ireland. *Environmental Pollution* 142:501–511.

Zhang, J., Y. Wang, J. Liu, Q. Liu, and Q. Zhou. 2016. Multivariate and geostatistical analyses of the sources and spatial distribution of heavy metals in agricultural soil in Gongzhuling, Northeast China. *Journal of Soils and Sediments* 16:634–644.

Zhou, L., B. Yang, N. Xue, F. Li, H. M. Seip, X. Cong, Y. Yan, B. Liu, B. Han, and H. Li. 2014. Ecological risks and potential sources of heavy metals in agricultural soils from Huanghuai Plain, China. *Environmental Science and Pollution Research* 21:1360–1369.

6

Nickel Forms in Soils

Christos D. Tsadilas

CONTENTS

6.1 Introduction

Nickel (Ni) is an element the use of which is traced to the beginning of the Bronze Age in 3500 BC. The bronzes produced in these ancient times contained 2% Ni. Since then, Ni has been an important element for industry. The main use of Ni is in the manufacture of stainless steel, consisting of 8% Ni. Today, there are more than 3000 known Ni-containing alloys. Batteries also contain Ni, and various Ni compounds are used as pigments (Nieminen et al. 2007). It is estimated that in 2008, 1.61 Mt of Ni were mined all over the world (Kabata-Pendias 2011).

After the discovery of Ni essentiality to plant growth (Brown et al. 1987a), and its well-established toxic effects on human health (Denkhaus and Salnikow 2002), the importance of Ni in the environment gained significant recognition worldwide. Nickel is now generally accepted as an essential nutrient for plant growth, although the knowledge of the whole biological significance for plant productivity is still limited. Besides the experimental evidence of Ni essentiality to plant growth for barley, wheat, and oats as well as for other cops, field experiments demonstrated that crops like potato, wheat, and bean respond to Ni fertilizers application and soybean (Brown 2007). When Ni is at inadequate levels, plants show relevant symptoms. In legumes and other dicotyledonous plants, Ni deficiency results in suppression of plant growth, followed by development of leaf tip necrosis and pale green leaves, symptoms attributed to the accumulation of toxic levels of urea in the leaf tissues (Eskew et al. 1983).

When Ni is in excessive levels above the tolerable limit, an oxidative stress to plants is generated, although the operating mechanisms are not well known yet (Yusuf et al. 2011). In excessive levels, Ni causes toxicity symptoms similar to those of iron deficiency (i.e., interveinal chlorosis in monocotyledons or mottling in dicotyledons) or zinc deficiency (i.e., chlorosis and restricted expansion of leaves). At severe cases of Ni toxicity, necrosis leads to plant death (Brown et al. 1987b; Chapman 1966; Wood et al. 2004). Through relevant researches, critical concentrations of Ni in plants have been established. For example, for barley, wheat, oats, and soybean, deficient levels for Ni are considered to be 0.2, 0.037, 0.017, and 0.002–0.004 mg Ni kg^{-1}, respectively, while concentration values of 63–113 and 10–83 mg Ni kg^{-1} are considered toxic for wheat and beans, respectively. Concentration values of 0.084 and 0.10 mg Ni kg^{-1} are considered adequate for wheat and oats, respectively (Brown et al. 1987b; Eskew et al. 1984; Macnicol and Beckett 1985; Singh et al. 1990).

Although the essentiality of Ni in animal organisms is questionable, it is well known that Ni binds, complexes, and chelates with a number of substances that are of biological interest, and it is a functional constituent of several enzymes occurring in bacterial and animal systems (Brown 2007). Since Ni is in abundance in nature, Ni deficiency in higher organisms does not occur. The daily dietary intake of Ni has been estimated to be from 25 to 300 µg of Ni, which is more than triple the daily requirement (Denkhaus and Salnikow 2002). However, at high doses and in certain forms, Ni is toxic to human and animals, causing teratogenesis, and has carcinogenic potential in animals. In humans, it was found that Ni in excess may cause abortions in females (Chashschin et al. 1994) and systemic allergy (Peltonen 1979). Nickel is taken up by humans mainly through foods. Via suitable experiments, it was found that from the Ni contained in foods, only 10%–25% is absorbed by humans and the rest is excreted (Horak and Sunderman 1972; Myron et al. 1978). It was also revealed that women retain quite less Ni than men. Although Ni intake via inhalation is negligible to humans, it was found to be carcinogenic in rats (Oller et al. 1997).

Nickel is an abundant element in the Earth, amounting to about 3% of its composition. On average, Ni is contained in soils in about 50 mg Ni kg^{-1}, but its range is very wide, that is, between 5 and 5300 mg Ni kg^{-1}. The higher values are met near metal refineries or in places where sewage sludge was deposited. Agricultural soils contain Ni in values between 3 and 1000 mg Ni kg^{-1}. Soils derived from basic igneous rocks contain high Ni amounts, ranging from 2000 to 6000 mg Ni kg^{-1}. Chemical weathering of ultramafic rocks produces soils with high amounts of Ni and Cr, compared to soils derived from non-ultramafic rocks. Under temperate Mediterranean conditions, where mineral weathering is slight, secondary Mg-rich or Fe-rich clay minerals and Ni- and Cr-bearing smectites, vermiculites, and chromite-silicate mixtures are formed. In addition, in the amorphous Fe oxides and goethite formed by weathering of clay minerals, Ni and Cr are accumulated, since such soils tend to act as hosts of Ni and Cr. In soils like them, which have ultramafic origin, soil Ni is high and its bioavailability may cause potential toxicity to plants, depending on Ni speciation (Cheng et al. 2011; Kierczak et al. 2008; Mizuno and Kirihata 2015).

Similar to the majority of the heavy metals, Ni sources in the environment are natural and anthropogenic. Nickel-bearing rocks are among the main Ni natural sources. Basaltic igneous rocks contain 45–240 mg Ni kg^{-1}, shales and clays contain 20–250 mg Ni kg^{-1}, and black shales contain 10–500 mg Ni kg^{-1}, while sandstones are free of Ni. Another process that is referred to as a source of airborne emission of Ni is the so-called "bubble bursting" as well as prairie fires (Nagajyoti et al. 2010). It is estimated that on a global scale, soil particles from eroded areas transferred by winder account for about 30%–50% of natural Ni emissions, while volcano-ejected material may account for about 40%–50% of Ni in the air (Nriagu 1990). Nriagu and Pacyna (1988) estimated that, in 1983, about 24,000–87,000 tons of Ni were emitted to the atmosphere, and 106,000 to 544,000 tons of Ni per year are released to the soils from atmospheric fallout, wastes, fertilizers, sewage sludge, and so on. Nickel concentration in river waters is 0.35–5.06 µg L^{-1} and the fluxes in riverines are estimated to be 30 kilotons per year (Kabata-Pendias 2011). Important anthropogenic sources of Ni are agrochemicals (pesticides and fungicides), liming, and sewage sludge. Industrial sources are also among the most important Ni sources. They include mining, refinement and tailings, ore transport, smelting and metal finishing, and recycling. Effluents from industries such as metallurgy and electroplanting, ink manufacturing, glass, textiles, fertilizers, and petroleum refining are common examples of industrial sources of Ni (Nagajyoti et al. 2010). Other anthropogenic sources of Ni are coal burning, oil burning, and emissions from transportation of vehicles (Gimero-Garcia et al. 1996). The concentration of Ni in fertilizers, sewage sludge, and fly ash is 11, 16–5300, and 87.9 mg kg^{-1}, respectively (Kabata-Pendias 2011).

6.2 Nickel Forms in Soils

The average world Ni background content of noncontaminated surface soils is 29 mg Ni kg^{-1} (Kabata-Pendias 2011). Between the soil orders with higher Ni content are Calcisols (2–450 mg Ni kg^{-1}; mean value, 34), followed by Cambisols (3–110 mg Ni kg^{-1}; mean value, 26) and Chernozems (6–61 mg Ni kg^{-1}; mean value, 25). Nickel ion forms in soil solution are both cationic $\left[Ni^{2+}; NiOH^+, NiHCO_3^+\right]$ and anionic $\left[HNiO^{2-} \text{ and } Ni(OH)_3^-\right]$, depending mainly on soil solution pH (Kabata-Pendias and Sadurski 2004; Speir et al.

2003; Tsadilas 2001). What is important, from an environmental point of view, is to know the processes or mechanisms that govern the mobility of trace metals in soils and their movement. These processes determine the distribution of metals between liquid and solid phases and therefore their availability to plants, as well as their leachability. The basic processes involved are dissolution/sorption, hydrolysis, oxidation/reduction, and carbonization. Most of these are significantly affected by pH and Eh values; the granulometric fraction, organic matter, oxides, and hydroxides of Fe, Mn, and Al content; and microorganisms. Thus, a great effort has been made so far to study the constituents and the mechanisms associated with metals retention in the solid phase of soil. There are many analytical procedures that may separate trace metal forms in soils.

6.3 Methods for Assessing Soil Ni Forms

Total soil Ni, as it stands for the rest of the heavy metals, reflects both geological and anthropogenic inputs and is not a good indicator of its mobility. Therefore, it cannot be used for estimating its availability to plants or its leachability to groundwater. Total Ni concentrations express the whole amount of Ni existing in soils in several forms, which are associated to soil compounds in different ways. When we are talking about total Ni or heavy metal content, we have to distinguish the extraction procedures usually used, which involve digestion with $HF + HClO_4$ ("total" metal content) and strong acids or aqua regia digests giving the so-called "pseuodototal" heavy metal content. This heavy metal content gives the maximum potentially soluble or mobile contents of metals that are not bound to silicates, which gives the maximum potential hazard that could happen in the long term or under extreme environmental conditions.

However, in environmental studies, it is necessary to have a more reliable estimation on the Ni quantities that may be taken up by plants or that might move down the profile causing groundwater pollution. In these cases, a number of selective extraction methods have been used that may determine target element species in soils, bound or associated with particular soil phases or compounds. By these methods, a better estimation of the potential effects of Ni to plants, animals, or the environment can be obtained. In addition, by these methods, elucidation of soil chemistry issues such as the structure and composition of soil components and processes in soils that control the retention of heavy metals and transport mechanisms may be facilitated.

There are different approaches to determine heavy metal forms in soils, and they are summarized as follows:

- Column leaching techniques using water or very weak solutions by which the very easily mobile metal forms are determined. By these techniques, usually a highest percentage of about 5% are extracted.
- Another technique usually employed for determining the plant-available heavy metals is the one that uses single extractants (Table 6.1), which extracts about 10% of the total heavy metal content. They are considered as models simulating raining and flooding events.
- The third technique is the sequential extraction procedure that uses several extractants thus sequentially separating the metal forms in various "fractions," which are usually called "exchangeable," "reducible," "oxidizable," "residual," and "silicate bound."
- The other two approaches have already been mentioned, and they are the "pseudototal" and "total" by using strong acids or aqua regia and HF + $HClO_4$ as extractants.

Menzies et al. (2007) evaluated several extractants for their suitability to determine heavy metals forms "available" to plants, including Ni. They selected an extensive data set from the literature of a wide range of levels

TABLE 6.1

Single Extractants Used for Plant-Available Ni Determination

Extractant	References
1 M NH_4NO_3	Merkel (1996); Rekasi and Filep (2006); Symeonides and McRae (1977)
0.1 M $NaNO_3$	Gupta and Aten (1993); Hani and Gupta (1986); Sanka and Dolezal (1992)
0.005 DTPA	Gupta and Sinha (2006); Kukier and Chaney (2001); Lindsay and Norvell (1978)
1 M EDTA	Clayton and Tiller (1979); Quevauviller et al. (1997)
0.01 M $CaCl_2$	Houba et al. (1996); Novozamsky et al. (1993)
0.1 M HCl	Baker and Amacher (1982)
1 M NH_4OAc	Sanka and Dolezal (1992)
Melich 1	Korcak and Fanning (1978)
0.01 M $Sr(NO_3)_2$ 1:4, 2 h shaking	Kukier and Chaney (2001)

Source: Compiled from Rao et al. (2008). A review of the different methods applied in environmental geochemistry for single and sequential extraction of trace elements in soils and related materials. *Water Air Soil Pollut.* 189: 291–333.

of contamination and different substrates from about 500 studies and 2000 individual data points. After the appropriate treatment of the data, they concluded that, for Ni (as well as for Cd, Zn, Cu, and Pb), the total concentration should be used only to establish threshold values for further investigations, since there were no significant correlations with plant Ni concentration. Similarly, the complexing agents DTPA and EDTA, as well as acid extractants, were generally poorly correlated to plant uptake. Neutral salts such as 0.01 M $CaCl_2$, 0.1 M $NaNO_3$, and 1.0 M NH_4OAc, however, provide the most useful indication of Ni phytoavailability. Further research was suggested by the authors for investigation of the effectiveness of these neutral salt extractants.

Meers et al. (2007) assessed the phytoavailability of heavy metals, including Ni extracted with several single extractants methods, for phaseolous vulgaris in a big number of soils. They found that among the single extractants tested, $CaCl_2$, NH_4NO_3, $MgCl_2$, NH_4OAc, DTPA, and NH_4OAc-EDTA values of Ni extracted were significantly correlated with the Ni values in plant tissues.

Kukier and Chaney (2001) in a pot experiment with a highly polluted soil with Ni and wheat and oats among others have tested the suitability of DTPA (Lindsay and Norvell 1978) and 0.01 M $Sr(NO_3)_2$ to assess Ni availability to plants and found very good correlation between both the extractants and the plant tissues of wheat and oats tested. The correlation was better for DTPA when the bulk density of soils was included in the calculations.

6.4 Sequential Extraction

In this procedure, different reagents are used with increasing dissolution power, in a sequence where the earlier reagent is less strong and most specific while the subsequent extractants are progressively more destructive and less specific. Fractionation is the process of classification of any analyte or group of analytes from a certain sample according to physical (e.g., size, solubility) or chemical (bonding, reactivity) properties (Sahuquillo and Rauret 2003). This procedure is not the same described as chemical speciation in the sense of determination of chemical species, such as Cr(III) and Cr(VI), although in the literature it is referred so.

There are several different schemes using various extraction agents that have been used for heavy metal partitioning. In general, they use a series of reagents that are, in summary, presented in Table 6.2 (Sahuquillo and Rauret 2003).

However, two of them are the most well known and the most broadly used. The first one is the five-step Tessier procedure and the four-step Community Bureau of Reference (BCR) procedure described below.

The Tessier extraction procedure (Tessier et al. 1979) includes four steps shown in Table 6.3 (also see Table 6.4).

TABLE 6.2

Reagents Used for Extraction of Heavy Metal Fractions (Including Ni) and Fractions Targeted

Chemical Form of Extractants	Type of Extractant	Metal Fraction
Water	Distilled or deionized water	Water-soluble fraction
KNO_3 1 M (pH 7) $Mg(NO_3)_2$ 0.005 M (pH 7) $CaCl_2$ 0.01–0.05 M $MgCl_2$ or $BaCl_2$ 1 M (pH 7) $NaCH_3COO$ or NH_4CH_3COO 0.1–1.0 M (pH 7 or 8.2)	Salts of strong acids and bases or salts of weak acids	Exchangeable and weak adsorbed fraction
CHCOOH 25% or 1 M $NaCH_3COO$ 1 M/ CH_3COOH (pH 5) HCl or CH_3COOH (unbuffered solution) EDTA 0.2 M (pH 10–12)	Acid or buffer solutions	Carbonate-bound fraction
0.2% hydroquinone in NH_4CH_3COO 1 M (pH 7) $NH_2OH \cdot HCl$ 0.02–1 M in CH_3COOH or HNO_3 $(NH_4)_2C_2O_4$ 0.2 M (pH 3) $(NH_4)_2C_2O_4$ 0.2 M/$H_2C_2O_4$ in ascorbic acid 0.1 M $Na_2S_2O_4$/Na-citrate/citric acid $Na_2S_2O_4$/Na-citrate/$NaHCO_3$ (pH 7.3) $Na_2S_2O_4$/$K_4P_2O_7$ H_2O_2 10% in 0.0001 N HNO_3 HCl 20% EDTA 0.02–0.1 M (pH 8–10.5) Na_2EDTA (buffered with NH_4CH_3COO 1 M) Hydrazine chloride (pH 4.5)	Reducing solutions, other agents	Fractions bound to oxides of Fe, Mn, and Al
H_2O_2 10% in HNO_3 + extraction with NH_4CH_3COO or $MgCl_2$ NaClO (pH 9.5) Alkalipyrophosphate ($Na_4P_2O_7$ or $K_4P_2O_7$ 0.1 M) H_2O_2/ascorbic acid HNO_3/tartaric acid $KClO_3$/HCl $Na_2B_4O_7$ (addition of surfactant)	Oxidizing reagents	Organically bound fractions or sulfidic phase
Alkaline fusion HF/$HClO_4$/HNO_3 Aqua regia HNO_3/H_2O_2 HCl/HF/HNO_3	Strong acids	Residual fractions bound to aluminosilicates

TABLE 6.3

Stages of Tessier's Sequential Extraction Procedure

Step	Reagent	Experimental Conditions	Fraction
1	8 ml of 1 mol L^{-1} MgC1$_2$, pH 7.0	1 h at 25°C	Exchangeable
2	8 ml of 1 mol L^{-1} NaOAc, pH 5 with acetic acid	5 h at 25°C	Associated with carbonates
3	20 ml of NH$_2$OHHCL 0.04mol L^{-1} in 25%w/vHOAc, pH2	6 h at 96°C	Associated with Fe-Mn oxides
4	3 ml of 0.02 mol L^{-1} HNO$_3$/5 ml of m/vH$_2$O$_2$ +3 ml of 30% m/v H$_2$O$_2$ +5 ml of 3.2 mol L^{-1} NH$_4$OAc	2 h at 85°C 3 h at 85°C 30 min at 25°C	Associated with organic matter

Source: Tessier et al. (1979). Sequential extraction procedure for the speciation of particulate trace metals. *Anal. Chem.* 51: 844–851.

TABLE 6.4

The Sequential Extraction Procedure of BCR Improved by Rauret et al. (1999)

Step	Reagent	Experimental Conditions	Fraction
1	40 ml of 0.11 mol L^{-1} acetic acid	16 h, at room temperature	Exchangeable, acid and water soluble, soil solution, exchangeable cations, carbonates
2	40 ml of 0.5 mol L^{-1} hydroxylamine hydrochloride solution at pH 1.5 30% w/v H$_2$O$_2$	16 h shaking at room temperature	Reducible, iron and manganese oxyhydroxides
3	10 ml of 30% w/v H$_2$O$_2$ 10 ml of 30% w/v H$_2$O$_2$ 50 ml of 1 mol L^{-1} then 1 M ammonium acetate at pH 2	1 h at room temperature, agitation at 85°C 1 h at 85°C to reduce the volume to a few milliliters 16 h at room temperature	Oxidizable, associated with organic matter and sulfides
4	Aqua regia	ISO 11466 (1995)	Residual, associated with non-silicate minerals

Source: Rauret et al. (1999). Improvement of the BCR three step sequential extraction procedure prior to the certification of new sediment and soil reference materials. *J. Environ. Monit.* 1(1): 57–61.

6.5 Bioavailability of Ni Fractions

Bioavailable fraction is considered the fraction of the total amount of a heavy metal existing in the soil that is available to receptor organisms (Vig et al. 2003).

Jordao et al. (2016) in a pot experiment with lettuce and soils amended with vermicompost enriched with Cu, Zn, and Ni found that lettuce plants had uptaken Ni above the critical limit for Ni toxicity but the plants did not show toxicity symptoms. In addition, they found that all the fractions determined (i.e., exchangeable, reducible, oxidizable, and residual) were significantly correlated with Ni concentration in lettuce tissues, concluding that all these fractions are easily available to plants. Regarding the residual fraction, they suggested that the high level of Ni (and Cu and Zn) in this fraction reflects the tendency of contaminant metals to be associated with this fraction. Only complete destruction of the soil matrix with strong treatment (HF/HNO$_3$) would release metal amounts from the residual fraction for determination.

In a study with soils irrigated with wastewater and cultivated with spinach and bitter gourd, Bashir et al. (2014) reported that, from the fractions of Ni determined by the Tessier method, the exchangeable and extractable and the acid-soluble fraction were the most available to the vegetables studied, while the residual fraction was negatively correlated with Ni uptake, showing that this fraction is not available to plants. In a study with representative agricultural soils from Poland, Rozanski (2013) investigated the distribution of heavy metal fractions including Ni by using the BCR procedure. He found that most of the Ni was in the residual fraction, ranging from about 49% to 85% of the total Ni. Total Ni was positively correlated with the fractions associated with iron and manganese compounds and the free and amorphous iron oxides. The exchangeable fraction and organic matter associated accounted for a small percentage of total Ni content.

In a study with soils from the Thriasio plain (i.e., one of the most polluted areas in Greece), Gasparatos et al. (2015) fractionated the metals Cu, Zn, Pb, Ni, and Cr of 50 surface agricultural soils by using the Tessier procedure. They found that mean total Ni content was 103 mg Ni kg^{-1} with a range of 41 to 175 mg Ni kg^{-1}. The dominant fraction of Ni was the residual (about 44%), indicating that the main source of the metal was the parent material. The second fraction was the reducible, suggesting that Ni was adsorbed in Mn oxides through specific adsorption. The oxidizable and the acid-soluble fractions covered around 12%, while the exchangeable fraction covered the lowest percentage amounting to 5%.

Jalali and Afrania (2011), working with a sandy calcareous soil amended with municipal sewage sludge at different rates, found that Ni, along with Cd and Pb, was associated in a major proportion with organic matter followed by exchangeable fraction and residual, but Cu and Zn were associated mainly with the residual fraction. Ni was also found to leach more easily compared to Cu and Zn, a behavior with potential environmental significance.

Alfaro et al. (2015) studied a big number of Cuban soils representing all the soil orders existing in the country, with regard to their heavy metal content and their distribution in soil fractions. They found that the average total Ni content was very high around 294 mg Ni kg^{-1} and in any case higher than the values reported for several countries such as China (23 mg Ni kg^{-1}), the United States (3 mg Ni kg^{-1}), Brazil (1.3–45 mg Ni kg^{-1}), and the whole world (2–750 mg Ni kg^{-1}). Fractionation of Ni by the USEPA method 3015 (USEPA 1998) showed that very little amount was in the soluble form (only 0.12 mg

Ni kg^{-1}), exchangeable form (0.91 mg Ni kg^{-1}), and organic matter associated (0.02 mg Ni kg^{-1}). A high amount of Ni was associated with crystalline iron oxides (37.53 mg Ni kg^{-1}) and the rest was in the residual form. They concluded that since most of the Ni amount was associated with the residual fraction, most of the Ni was originated by parent rock, and for this reason, it was not available to plants.

In a similar pot experiment study, Tsadilas et al. (1995), working with an acid unproductive soil amended with municipal sewage sludge, found that available Ni, determined by the DTPA extraction method, increased significantly with the increase of sewage sludge rate. DTPA extractable Ni was significantly correlated with total Ni concentration and organic matter content. Heavy metal fractionation made according to Emmerich et al. (1982) revealed that most of the Ni was found in the residual fraction, although a significant percentage was also found in organic matter and carbonate fractions.

References

Alfaro, M.R., A. Montero, O.M. Ugarte, C.W. Araujo do Nascimento, A.M. de Aguiar Accioli, C.M. Biondi, and Y.J.A.B. da Silva. 2015. Background concentrations and reference values for heavy metals in soils of Cuba. *Environ. Monit. Assess.* 187: 4198–4208.

Baker, D.E. and M.C. Amacher. 1982. Nickel, copper, zinc, and cadmium. In: Page, A.L., Miller, R.H., Keeney, D.R. (Eds.), *Methods of Soil Analysis. Part 2. Chemical and Microbiological Methods.* American Society of Agronomy/Soil Science Society of America, Madison, WI, pp. 323–336.

Bashir, F., M. Tariq, M.H. Khan, R.A. Khan, and S. Aslam. 2014. Fractionation of heavy metals and their uptake by vegetables growing in soils irrigated with sewage effluent. *Turk. J. Eng. Environ. Sci.* 38: 1–10.

Brown, P.H. 2007. Nickel. In: Barker, A.V. and Pilbeam, D.J. (Eds.), *Handbook of Plant Nutrition*, pp. 395–410.

Brown, P.H., R.M. Welch, and E.E. Cary. 1987a. Nickel: A micronutrient essential for higher plants. *Plant Physiol.* 85: 801–803.

Brown, P.H., R.M. Welch, E.E. Cary, and R.T. Checkai. 1987b. Beneficial effects of nickel on plant growth. *J. Plant Nutr.* 10: 2125–2135.

Chapman, H.D. 1966. *Diagnostic Criteria for Plants and Soils.* Riverside: Division of Agricultural Science, University of California.

Chashschin, V.P., G.P. Artunina, and T. Norseth. 1994. Congenital defects, abortion and other health effects in nickel refinery workers. *Sci. Total Environ.* 148(2–3): 287–291.

Cheng, C.-H., S.-H., Jien, Y. Lizuka, H. Tsai, and Z.-H, Hseu. 2011. Pedogenic chromium and nickel partitioning is sermpentine soils along a toposequence. *Soil Sci. Soc. Am. J.* 75: 659–668.

Clayton, P.M. and K.G. Ttiller. 1979. A chemical method for determination of the heavy metal content of soils in environmental studies. Division of Technical paper (Australia, Commonwealth Scientific and Industrial Research Organization 41), 17 pp.

Denkhaus, E. and K. Salnikow. 2002. Nickel essentiality, toxicity, and carcinogenicity. *Crit. Rev. Oncol. Hematol.* 42: 35–36.

Emmerich, W.E., L.J. Lund, A.L. Page, and A.C. Chang. 1982. Solid phase forms of heavy metal in sewage sludge treated soils. *J. Environ. Qual.* 11(2): 178–181.

Eskew, D.L., R.M. Welch, and W.A. Norvell. 1983. Nickel: An essential micronutrient for legumes and possibly all higher plants. *Science* 222: 621–623.

Eskew, D.L., R.M. Welch, and W.A. Norvell. 1984. Nickel in higher plants. Further evidence for an essential role. *Plant Physiol.* 76: 691–693.

Gasparatos, D., G. Mavromati, P. Kotsovolis, and I. Massas. 2015. Fractionation of heavy metals and evaluation of the environmental risk for the alkaline soils of the Thriassio plain: A residential, agricultural, and industrial area in Greece. *Environ. Earth Sci.* 74: 1099–1108.

Gimero-Garcia, E. and A.V. Boluda. 1996. Heavy metals incidence in the application of inorganic fertilizers and pesticides to rice farming soils. *Environ. Pollut.* 92: 19–25.

Gupta, S.K. and C. Aten. 1993. Comparison and evaluation of extraction media and their suitability in a simple model to predict the biological relevance of heavy metal concentrations in contaminated soil. *Int. J. Anal. Chemist.* 51: 25–46.

Gupta, A.K. and S. Sinha. 2006. Role of *Brassica juncea* (L.) Czem. (var. Vaibhav) in the phytoextraction of Ni from soil amended with fly ash: Selection of extractant for metal bioavailability. *J. Hazard. Mater.* 136(2): 371–378.

Hani, H. and S. Gupta. 1986. Chemical methods for the biological characterization of metal in sludge and soil. Commission of the European Communities, Report 10361, 157–167.

Horak, S.E. and F.W. Sunderman. 1972. Fecal nickel excretion by healthy adults. *Clin. Chem.* 19: 429–430.

Houba, V.J.G., T.M. Lexmond, I. Novozamsky, and J.J. van der Lee. 1996. State of the art and future developments in soil analysis for bioavailability assessment. *Sci. Total Environ.* 178(1–3): 21–28.

Jalali, M. and H. Afrania. 2011. Distribution and fractionation of cadmium, copper, lead, nickel, and zinc in a calcareous sand soil receiving municipal sewage sludge. *Environ. Monit. Assess.* 173: 241–250.

Jordao, C.P., R.P. de Andrade, A.J.B. Cotta, P.R. Cecon, J.C.L. Neves, M.P.F. Fontes, and R.B.A. Fernandes. 2016. Copper, nickel and zinc accumulations in lettuce grown in soil amended with contaminated cattle manure vermicompost after sequential cultivations. *Environ. Technol.* 34(6): 765–777.

Kabata-Pendias, A. 2011. *Trace Elements in Soils and Plants*, 4th edition. CRC Press, 534 pp.

Kabata-Pendias, A. and W. Sadurski. 2004. Trace elements and compounds in soil. In: Merian, E., Anke, M., Inhant, M., and Stoeppler, M. (Eds.), *Elements and Their Compounds in the Environment*. Wiley-VCA Verlag GmbH & Co, KGaA, pp. 79–100.

Kierczak, J., C. Neel, U. Aleksander-Kwaterczak, E., Helios-Rybicka, E. Brill, and J. Puziewicz. 2008. Solid speciation and mobility of potentially toxic elements from natural and contaminated soils: A combined approach. *Chemosphere* 73: 776–784.

Korcak, R.F. and D.S. Fanning. 1978. Extractability of cadmium, copper, nickel, and zinc by double acid versus DTPA and plant content at excessive soil levels. *J. Environ. Qual.* 7: 506–512.

Kukier, U. and R.L. Chaney. 2001. Amelioration of nickel phytotoxicity in muck and mineral soils. *J. Environ. Qual.* 30: 1949–1960.

Lindsay, W.L. and W.A. Norvell. 1978. Development of a DTPA test for zinc, iron, manganese, and copper. *Soil Sci. Soc. Am. J.* 42: 421–428.

Macnicol, R.D. and P.H.T. Beckett. 1985. Critical tissue concentrations of potentially toxic elements. *Plant Soil* 85: 107–129.

Meers, E., R. Samson, F.M.G. Tack, A. Ruttens, M. Vandehuchte, J. Vangronsveld, and M.G. Verloo. 2007. Phytoavailability assessment of heavy metals in soils by single extractions and accumulation by *Phaseolus vulgaris*. *Environ. Exp. Botany* 60: 385–396.

Menzies, N.W., M. J. Donn, and P.M. Kopittke. 2007. Evaluation of extractants for estimation of the phytoavailable trace metals in soils. *Environ. Pollut.* 145(1): 121–130.

Merkel, D. 1996. Cd-, Cu-, Ni-, Pb- and Zn-contents of wheat grain and soils, extracted with $CaCl_2$/DTPA (CAD), $CaCl_2$, and NH_4NO_3, respectively. *Agribiol. Res.* 49(1): 30–37.

Mizuno, T. and Y. Kirihata. 2015. Elemental composition of plants from the serpentine soil of Sugashima Island, Japan. *Austral. J. Botany*. http://dx.doi.org/10.1071/BT14226.

Myron, D.R., T.J. Zimmermann, and T.R. Shuler. 1978. Intake of nickel and vanadium by humans. *Am. J. Clin. Nutr.* 31: 527–531.

Nagajyoti, P.C., K.D. Lee, and T.V.M. Sreekanth. 2010. Heavy metals and toxicity for plants: A review. *Environ. Chem. Lett.* 8: 199–216.

Nieminen, T.M., L. Ukonmaanaho, N. Rausch, and W. Shotyk. 2007. Biogeochemistry of nickel and its release into the environment. In: Sigel, A., Nriagu, H., and Pacyna, J.M. 1985. *Nickel and Its Surprising Impact in Nature*. John Wiley & Sons, pp. 134–139.

Novozamsky, I., T.T. Lexmond, and V.J.G. Houba. 1993. A single extraction procedure of soil for evaluation of uptake of some heavy metals by plants. *Int. J. Environ. Anal. Chem.* 51: 47–58.

Nriagu, J.O. 1990. Global metal pollution poisoning the biosphere? *Environment* 32(7): 7–33

Nriagu, J.O. and J.M. Pacyna. 1988. Quantitative assessment of worldwide contamination of air, water and soils by trace metals. *Nature* 333: 134–139.

Oller, A.R., Costa, M., and Oberdorster, G. 1997. Carcinogenicity assessment of selected nickel compounds. *Toxicol. Appl. Pharmacol.* 143: 152–166.

Peltonen, L. 1979. Nickel sensitivity in the general population. *Contact Dermatitis* 5: 27–32.

Quevauviller, P., R. Rauret, G. Rubio, J.F. Lopezsanchez, A.M. Ure, J.R. Bacon, and H. Muntau. 1997. Certified reference materials for the quality control of EDTA- and acetic acid-extractable contents of trace elements in sewage sludge amended soils (CRMs 483 and 484). *Fres. J. Anal. Chem.* 357: 611e618.

Rao, C. R. M., A. Sahuquillo, and J. F. Lopez Sanchez. 2008. A review of the different methods applied in environmental geochemistry for single and sequential extraction of trace elements in soils and related materials. *Water Air Soil Pollut.* 189: 291–333.

Rauret, G., J. F. Lopez-Sanchez, A. Sahuquillo, R. Rubio, C. Davidson, and A. Ure. 1999. Improvement of the BCR three step sequential extraction procedure prior to the certification of new sediment and soil reference materials. *J. Environ. Monit.* 1(1): 57–61.

Rekasi, M. and T. Filep. 2006. Effect of microelement loads on the element fractions of soil and plant uptake. *Agrokemia Talajtan* 55(1): 213–222.

Rozanski, S. 2013. Fractionation of selected heavy metals in agricultural soils. *Ecol. Chem. Eng. S.* 20: 117–125.

Sahuquillo, A. and G. Rauret. 2003. Sequential extraction. In: Mester, Z. and Sturgeon, R. (Eds.), *Comprehensive Analytical Chemistry XLI.* Elsevier B.V., pp. 1–24.

Sanka, M. and M. Dolezal. 1992. Prediction of plant contamination by cadmium and zinc based on soil extraction method and contents in seedlings. *Int. J. Environ. Anal. Chem.* 46: 87–96.

Singh, B., Y.P. Dang, and S.C. Mehta. 1990. Influence of nitrogen on the behavior of nickel in wheat. *Plant Soil* 127: 213–218.

Speir, T.W., A.P. van Schaik, H.J. Percival, M.E. Close, and L. Pang. 2003. Heavy metals in soil, plants and groundwater following high-rate sewage sludge application land. *Water Air Soil Pollut.* 150: 319–358.

Symeonides, C. and S.G. McRae. 1977. The assessment of plant-available cadmium in soils. *J. Environ. Qual.* 6: 120–123.

Tessier, A., P.G.C. Campbell, and M. Bissom. 1979. Sequential extraction procedure for the speciation of particulate trace metals. *Anal. Chem.* 51: 844–851.

Tsadilas, C.D. 2001. Soil pH effect on the distribution of heavy metals among soil fractions. In Iskandar, I.K. (Ed.), *Environmental Restoration of Metals—Contaminated Soils*, pp. 107–119.

Tsadilas, C.D., T. Matsi, N. Barbayaiannis, and D. Dimoyiannis. 1995. Influence of sewage sludge application on soil properties and on the distribution and availability of heavy metal fractions. *Commun. Soil Sci. Plant Anal.* 26(1–16): 2603–2619.

Vig, K., M. Megharaj, N. Sethunathan, and R. Naidu. 2003. Bioavailability and toxicity of cadmium to microorganisms and their activities in soil: A review. *Adv. Environ. Res.* 8: 121–135.

Wood, B.W., C.C. Reilly, and A.P. Nyczepir. 2004. Mouse-ear of pecan: A nickel deficiency. *HortScience* 39(6): 1238–1242.

United States Environmental Protection Agency (USEPA). 1998. Method 3051a—microwave assisted acid digestion of sediments, sludges, soils, and oils. http://www.epa.gov/SW-846/pdfs/3051a.pdf.

Yusuf, M., Fariduddin, Q., Hayat, S., and Ahmad, A. 2011. Nickel: An overview of uptake, essentiality and toxicity in plants. *Bull. Environ. Contam. Toxicol.* 86: 1–17.

7

A Review of Nickel in Sediments

Anna Sophia Knox, Michael H. Paller, and Dien Li

CONTENTS

7.1 Introduction

The two main types of aquatic ecosystems are freshwater and marine. There are three basic types of freshwater ecosystems: lentic, lotic, and wetland. Lentic ecosystems include ponds, lakes, and any other water bodies with still water. Lotic ecosystems have flowing water, for example, creeks, streams, and rivers. Wetlands are characterized by saturated or inundated soils for at least part of the year. Marine ecosystems cover approximately 71% of the Earth's surface and contain about 97% of the planet's water. They are distinguished from freshwater ecosystems by the presence of dissolved compounds, especially salts, in the water. Approximately 85% of the dissolved materials in seawater are sodium and chlorine. Seawater has an average salinity of 35 parts per thousand (ppt), although actual salinity varies among different marine ecosystems. Marine ecosystems can be divided into many zones depending on water depth, shoreline characteristics, and other factors. The oceanic zone is the vast open part of the ocean; the benthic zone consists of substrates below water where many invertebrates live; the intertidal zone is the area between high and low tides; and nearshore zones can include estuaries, salt marshes, coral reefs, lagoons, and mangrove swamps (United States Environmental Protection Agency [USEPA] 2010, http://www.epa.gov/bioiweb1/aquatic/marine.html).

Water pollution by Ni is often indicated by the accumulation of Ni in bottom sediments. The objective of this book chapter is to review Ni in the sediments of freshwater and marine ecosystems. To date, reviews about Ni in aquatic sediments are few and usually limited to specific marine or freshwater ecosystems or portions of ecosystems. Also, many papers on Ni in sediments date back to the 1970s and 1980s and report work done largely in the United States and European countries. There is a need to review and upgrade the current state of knowledge on Ni in aquatic sediments including recent work done worldwide. This review discusses sources, accumulation, mobility, and bioavailability of Ni in the sediments of rivers, lakes, seas, bays, ports, basins, oceans, and deep ocean around the world including previously underreported areas in Africa, South America, Asia, and elsewhere.

7.2 General Properties of Nickel

Nickel was discovered in 1751 by the Swedish chemist, Axel Fredrik Cronstedt, in a mineral called niccolite. The name originated from the German word "kupfernickel" meaning Devil's copper or St Nicholas's (Old Nick") copper. Nickel (atomic number 28) is a hard, silvery-white metal whose strength, ductility, and resistance to heat and corrosion make it extremely useful for the development of a wide variety of materials. This extremely useful metal belongs in Group VIII of the periodic table, called the iron–cobalt group, and it is located between the elements cobalt and copper. Nickel is a transition metal, meaning it has valence electrons in two shells instead of one, allowing it to form several different oxidation states. Generally, Ni occurs in the 0 and II oxidation states, but the I and III states can exist under certain environmental conditions. Nickel is present as Ni^{2+} in common water-soluble compounds, for example, chlorides, sulfides, bromides, and others. Nickel has five stable isotopes in nature, with ^{58}Ni being the most abundant (68.27%).

Currently, nickel is extracted from sulfide and oxide ores, mainly pyrrhotite and pentlandite. In 2014 and 2015, the primary producers of Ni in the world in decreasing order were Philippines, Canada, Australia, and Russia. World production of Ni in 2015 was approximately 2,530,000 tons. The estimated 2015 world total Ni reserves were 79 million tons. Identified land-based resources averaging 1% Ni or greater contain at least 130 million tons of Ni with about 60% in laterites and 40% in sulfide deposits. Extensive nickel resources are also found in manganese crusts and nodules on the ocean floor. The decline in discovery of new sulfide deposits in traditional mining districts has led to exploration in more challenging locations such as east-central Africa and the Subarctic (U.S. Geological Survey 2016).

7.3 Natural and Anthropogenic Sources of Nickel in Sediments

Nickel constitutes about 55 mg kg^{-1} of the Earth's crust, making it the 23rd most common element (Table 7.1; Liu et al. 1984). Most Ni is in igneous rocks,

TABLE 7.1

Nickel Concentrations (in mg/kg Unless Otherwise Specified) in Various Environmental Materials

Material	Mean	Range	References
Igneous rocks	75	2–3600	Cannon (1978)
Shales and clays	68	20–250	
Black shales	50	10–500	
Coal	15	3–50	Trudinger et al. (1979)
Sedimentary rocks	20		Polemio et al. (1982)
Global crust	55		Liu et al. (1984)
Chinese crust	38		Institute of Geochemistry Chinese Academy of Sciences (1998)
Soil over serpentine rocks in Australia	770		Severne (1974)
Soil over serpentine rocks in New Zealand	1700–5000		Lyon et al. (1970)
Marlborough Creek sediments over serpentine rocks, Australia		160–1250	Duivenvoorden et al. (2017)
Shallow sediment (world)	35		Salomons and Forstner (1984)
Deep-sea nodules in Atlantic Ocean/Canary Basin	8280		Menendez et al. (2017)
Soils (world)	20		McGrath (1995)
Freshwater (μg/L)	0.5	0.02–27	Bowen (1979)
Seawater (μg/L)	0.56	0.13–43	
Air—Urban/industrial areas (ng/m^3)		4–120	Kabata-Pendias and Szteke (2015)
Air—Remote regions (ng/m^3)	0.9		
Cosmic grains from Greenland (%)	>1		Robin et al. (1988)
Tamentit iron meteorite (%)		5–25	https://en.wikipedia.org/wiki/Iron_meteorite
American nickel (five-CENT coin) (%)	25		https://en.wikipedia.org/wiki/Nickel_(United_States_coin)

of which Ni comprises approximately 75 mg kg^{-1} on average with a very broad range from 2 to 3600 mg kg^{-1} (Table 7.1; Cannon 1978). Serpentine rocks contain high levels of Ni; for example, soil over serpentine rocks in New Zealand has 1700 to 5000 mg kg^{-1} of Ni (Table 7.1; Lyon et al. 1970). Similarly, sediment over serpentine rocks will have elevated Ni, for example, Marlborough Creek sediment with a range from 160 to 1250 mg kg^{-1} (Table 7.1; Duivenvoorden et al. 2017). Deposits that are commercially exploitable contain at least 10,000 mg kg^{-1} Ni, and Ni in these deposits is associated mostly with mafic and ultramafic igneous rocks (e.g., the deposits in Sudbury, Ontario; Thompson, Manitoba; and Kambalda, Australia). The smallest amounts of Ni are in sedimentary rocks, on average about 20 mg kg^{-1} (Table 7.1). The sediment content of Ni is also variable (Table 7.1), with the world's average reported around 35 mg kg^{-1} for shallow sediments (Salomons and Forstner 1984). Therefore, relative to Ni, sediments can be grouped into two categories: those derived from sandstone or limestone, containing <50 mg kg^{-1} Ni, and those derived from basic igneous rocks or ultrabasic igneous rocks with higher levels of Ni (>50 mg kg^{-1}).

Nickel in sedimentary deposits can be categorized in accordance with its predominant source of origin, either lithogenic (natural) or anthropogenic. Nickel belongs to a group of metals that have become enriched in surface soils and sediments, for the most part, because of man's activities. Eroded soil particles account for 77% of all natural Ni emissions, followed by volcanogenic particles (15%) (Moore and Ramamoorthy 1984). However, according to Nriagu (1979), the rate of anthropogenic emission exceeds the natural rate by 180%. Nickel concentrations in ambient air vary considerably, and the highest values have been reported from highly industrialized areas. Nickel content in air in urban/industrial areas is 4 to 133 times higher than that in remote regions (Table 7.1). The major sources of Ni pollution in aquatic ecosystems, including the ocean, are domestic wastewater effluents, mining and smelting of Ni-bearing ores, the automotive combustion of nickel-containing diesel oil, industrial applications, waste incineration, wood combustion, iron and steel production, coal combustion, fertilizer production, and agricultural runoff. To determine the extent of Ni pollution in sediments, it is important to establish the natural level of Ni and then subtract it from existing Ni concentrations to derive the total anthropogenic enrichment. The rock standard is a world standard in general use and satisfies the basic requirement of being uncontaminated and based, for most elements, on a large number of sediment samples (Forstner and Wittmann 1979).

Source of Ni in the ocean are river transport, atmospheric input, industrial and municipal discharges, and release from ocean sediments. Generally, Ni enrichment of sediments, especially nearshore, reflects anthropogenic inputs. The nickel burden of the world's oceans is 10^4 times higher than that of freshwater (Nriagu 1980).

7.4 Factors Affecting the Mobility, Bioavailability, and Toxicity of Nickel

Aqueous speciation reactions, adsorption, and precipitation control the partitioning of Ni to sediments, and hence its mobility, bioavailability, and toxicity (Drever 1997; Langmuir 1997; Stumm and Morgan 1981; Truex et al. 2011). There are a number of factors that contribute to the chemical speciation, adsorption, and precipitation of Ni including pH, Eh, natural organic matter, clay minerals, and Fe and Al oxides. At pH <8.2, the dominant Ni species is mobile and bioavailable Ni^{2+}; at pH 8.2–13.5, Ni primarily precipitates as sparsely soluble NiO; and at pH >13.5, the dominant Ni species is mobile and bioavailable $Ni(OH)_4^{2-}$. The pH also affects the surface charge of sediment minerals that accumulate Ni. For oxides and silicates, the charge is mostly the result of negatively charged, partially bonded oxygen atoms that attract positively charged ions like Ni^{2+}. As pH decreases, increasingly abundant hydrogen ions neutralize this negative charge. At pH values below the zero point of charge (ZPC), the mineral surface becomes positively charged, while at pH values above the ZPC, the mineral surface becomes negatively charged. This variable surface charge behavior is an important control on the adsorption and mobility of Ni^{2+} and other metals. For aqueous Ni species carrying an electrical charge, partitioning is driven by electrostatic attraction and then potential bonding to mineral surfaces. Ni^{2+} tends to adsorb more strongly to minerals and to become less mobile as pH increases.

The Eh or redox potential is an important factor that can contribute to Ni speciation, mobility, and bioavailability. Under oxidizing conditions, Ni is largely mobile and bioavailable Ni^{2+}. However, under the reducing conditions prevailing in most sediments, Ni precipitates as immobile NiS_2 of very low solubility. Anoxic sediments contain iron monosulfides that react with metal cations to form insoluble metal sulfides that are not absorbed through the respiratory surfaces of aquatic organisms (DiToro et al. 1992). Thus, the measurement of acid volatile sulfide and simultaneously extracted metals (including nickel) in sediments can be used to evaluate the toxicity of metals to benthic organisms. Factors that control the partitioning of metals between these binding phases and the dissolved form have a large influence on metal uptake and toxicity.

Organic carbon and other solid- and dissolved-phase ligands present in sediments can also bind free metals and reduce their bioavailability and mobility (USEPA 2001). In Ni-spiked soils, Ni adsorbed on mineral phases (i.e., clays, Fe and Al oxides/hydroxides), but concurrent addition of Ni and dissolved organic matter (DOM) resulted in a strong (32%–144%) increase of adsorption capacity, indicating that DOM enhances the adsorption and immobilization of Ni^{2+} on sediment minerals (Refaey et al. 2017). In non-amended sediments, Ni associated with clay minerals and organic matter, but the addition of biosolid to these sediments provided a source of reactive organic matter and iron oxide that further increased Ni retention. Thus, the long-term mobility of Ni can be affected by organic matter degradation (Mamindy-Pajany et al. 2014). Conversely, organic matter can increase the soluble organic ligands in sediments and enhance the solubility and mobility of Ni as shown by the treatment of natural and metal-amended soils with disodium ethylenedi-amine tetraacetate (EDTA) solution. After 5 months of incubation, Ni was found predominantly in the exchangeable and organic matter fractions, with some Ni in the residual fraction. The EDTA increased Ni concentration in the exchangeable fraction of the metal-amended soils and enhanced its solubility in these soils because EDTA, a strong complexing agent, removed Ni from the organic matter and other Mn and Fe mineral fractions and redistributed it into the exchangeable fraction (Li and Shuman 1996).

Adsorption is an important process in immobilizing Ni^{2+} and other metals in aquatic environments (Knox et al. 2016; Uddin 2017). At low to neutral pH and low Ni concentrations, the mobility of Ni in sediments is mainly determined by adsorption on minerals (e.g., clays, Fe and Al oxyhydroxides) (Tipping et al. 2010; Weng et al. 2004). The main modes of Ni uptake by pyrophyllite, montmorillonite, and a 1:1 pyrophyllite–montmorillonite mixture were adsorption on montmorillonite and surface precipitation on pyrophyllite. For a clay mixture, adsorption on montmorillonite and surface precipitation on pyrophyllite competed for Ni uptake (Elzinga and Sparks 1999). In illite suspensions at lower ion strength (i.e., 0.003 M), Ni sorption increased over a pH range of 4.5–8.0. At higher ionic strength (i.e., 0.1 M), the formation of Ni–Al layered double hydroxide (LDH) phases occurred at pH values >6.25. Nickel–Al LDH precipitation became more important than other Ni adsorption mechanisms at higher pH and higher Al content (Elzinga and Sparks 2001).

Because Fe oxyhydroxides (e.g., ferrihydrite, goethite, and hematite) are widespread, the adsorption of Ni^{2+} on them is an important route for Ni immobilization in the environment. The adsorption of Ni on Fe oxyhydroxides in aqueous media gradually increases with increasing pH from 5 to 8 (Arai 2008; Bryce et al. 1994; Buerge-Weirich et al. 2002). There are multiple Ni surface species or different NiO_6 coordination environments on Fe octahedra depending on their crystallinity and structure. The formation of LDH might be suppressed by the presence of iron oxyhydroxides, which, along with other clay minerals and soil organic matter, may compete for the sorption of dissolved Ni (Arai 2008). Green-Pedersen and Pind (2000) reported that Ni sorption on montmorillonite surfaces was significantly enhanced at pH 6.5–8 when the clay mineral surfaces are coated or mixed with iron oxyhydroxide.

Water hardness can have a mitigating effect on the bioavailability (hence toxicity) of Ni and other metals to aquatic organisms. Cations responsible for water hardness compete with metals such as Ni for binding sites on the respiratory surfaces of aquatic organisms (DiToro et al. 2001). Specifically, water hardness reduces metal toxicity by saturating gill surface binding sites with Ca^{2+} and Mg^{2+} ions, thereby excluding toxic metal cations. For example, the exposure of fathead minnows to nickel in soft and very hard water resulted in acute lethal concentrations that differed by nearly a factor of 10 (Pickering 1974).

Generally, Ni is relatively nontoxic to benthic organisms in both freshwater and marine sediments because of its limited bioavailability, and total Ni in the muscle tissue of freshwater and marine fish is typically low, <0.5 mg kg^{-1} (Moore 1991). However, Ni levels can be substantially higher in organisms from ecosystems contaminated by Ni mining, Ni smelters, Ni-cadmium battery plants, electroplating plants, and sewage outfalls, for example, the Sudbury, Ontario, region where freshwater species had Ni body burdens of 9.5 to 13.8 mg kg^{-1} WW (Hutchinson et al. 1975). Ni bioconcentration factors (mg Ni per kg fresh sample weight divided by mg Ni per liter of water) for aquatic macrophytes ranged from 6 in pristine areas to 690 near a Ni smelter; comparable values for crustaceans, mollusks, and fishes were 10–39, 2–191, and 2–52, respectively (Sigel and Sigel 1988). Nickel toxicity affects the gills of fish and can destroy gill lamellae, eventually resulting in hypoxia and death (Ellgaard et al. 1995). Ni concentrations of 11–113 µg/L can kill fish embryos (Eisler 1998). Lethal concentrations for more resistant species were 150 µg/L for mysid shrimp, 237 µg/L for freshwater snails, and 410 µg/L for salamander embryos (Eisler 1998).

Sediment quality guidelines have been issued to prevent potentially adverse effects from Ni pollution in areas affected by Ni mining and related industrial activities (Table 7.2). The USEPA, Region V sets guidelines for metal pollution in freshwater sediment and divided sediment into three categories

TABLE 7.2

Sediment Quality Criteria for Nickel

Guideline	Acronym	Ni (mg kg^{-1}) Dry Weight	References
Probable effect level	PEL	36	Smith et al. (1996)
Severe effect level	SEL	75	Persaud et al. (1993)
Toxic effect level	TET	61	EC & MENVIQ (1992)
Effect range low	ERL	30	Long and Morgan (1991)
Effect range median	ERM	50	Long and Morgan (1991)
Probable effect level for *Hyalella azteca*	PEL-HA28	33	USEPA (1996)
Concentrations above which harmful effect are likely to be observed	PEC	48.6	MacDonald et al. (2000)
Not polluted		<20	USEPA (1995)
Moderately polluted		20–50	USEPA (1995)
Heavily polluted		>50	USEPA (1995)
Interim Sediment Quality Guideline	ISQG	22.7	CCME (1991)

based on Ni concentration: not polluted, ≤ 20 mg kg^{-1}; moderately polluted, 20–50 mg kg^{-1}; and heavily polluted, >50 mg kg^{-1}, respectively (www.epa .gov). These levels are in general agreement with sediment quality criteria developed by other researchers; for example, Long and Morgan (1991) indicated an effect range low (ERL) of 30 mg kg^{-1} that would protect most benthic organisms and an effect range median (ERM) of 50 mg kg^{-1}, above which toxic effects would be expected.

7.5 Nickel in Freshwater Ecosystem Sediments

7.5.1 Lakes

Lakes are large, lentic ecosystems that are surrounded by land. They are typically freshwater, but some are filled with saltwater (e.g., Great Salt Lake, Utah, USA). Although usually shallower than seas and oceans, lakes come in all shapes and sizes, and some are notable for surface area, depth, volume, or other characteristics (e.g., Lake Baikal). Rivers are connected to a sea or ocean; however, lakes are generally landlocked and fed by streams from melting glaciers or rivers. Some lakes are considered temporary water bodies and may dry up with changing geological or climatological conditions.

Despite its name, the Caspian Sea is sometimes regarded as the world's largest lake, though it contains an oceanic basin rather than being entirely over continental crust. For this reason, and because of its salinity (about one-third that of saltwater), Ni concentrations in it are discussed under marine environments/seas. The largest lakes in the world are Superior, Victoria, Huron, Michigan, Tanganyika, Baikal, Great Bear Lake, Malawi, and Great Slave Lake. Lake sediments are good indicators of pollution with heavy metal, including Ni, because freshwater lakes have been at the centers of important cultural developments for thousands of years. Generally, with increasing population and industrial densities, we observe increasing levels of Ni in lake sediments around the world (Table 7.3). In some cases, river discharge into a lake is a key factor responsible for the input of Ni into lake sediment, for example, Lake Symsar in Northern Poland (Table 7.3).

Apart from industrialized areas, Ni in the surface sediments of lakes (Table 7.3) is seldom above natural background levels for shallow sediments worldwide (35 mg kg^{-1}) (Salomons and Forstner 1984). Examples of lakes with background levels of Ni are Lake Michigan (USA), Symsar Lake (Poland), and East Lake (China) (Table 7.3). However, lakes in the near vicinity of mines, nickel deposits, and nickel smelters have higher Ni levels in surface sediments. For example, Ni concentrations in the surface sediment of Thomson Lake, Canada, are higher (average, 45 mg kg^{-1}; range, 30–85 mg kg^{-1}) than the average level for worldwide shallow sediments due to proximity to a major gold mine (Moore 1981). The Sudbury region of Ontario, Canada, has one of

TABLE 7.3

Nickel Concentrations in Lake Sediments

Lake	Ni (mg/kg)		n	Water Depth (m)	Year	Location	References
	Range	Mean					
North America							
Michigan		34		0–0.15		USA	Frye and Shimp (1973)
Richard		1632	3	1	1998	Ontario, Canada	Shuhaimi-Othman (2008)
		2368	3	2	1998	Ontario, Canada	
		4721.1	3	3	1998	Ontario, Canada	
		2438.2	3	4	1998	Ontario, Canada	
		1972.6	3	5	1998	Ontario, Canada	
		2017.6	3	6	1998	Ontario, Canada	
		1891.0	3	7	1998	Ontario, Canada	
		2435.6	3	8	1998	Ontario, Canada	
		1571.8	3	9	1998	Ontario, Canada	
Superior		82				Northern part	Mothersill and Fung (1972)
Europe							
Symsar		26.03	13	0–0.1	2012–2014	Lake center, northern Poland	Kuriata-Potasznik et al. (2016)

(Continued)

TABLE 7.3 (CONTINUED)

Nickel Concentrations in Lake Sediments

Lake	Ni (mg/kg)			Water Depth (m)	Year	Location	References
	Range	Mean	n				
Asia							
East	10.3–66.2	28.4	29	0–0.15	2012	Niuchao Hu/lakelet of East Lake	Liu et al. (2014)
	15.9–55.7	27.5	31			Guozheng Hu/lakelet of East Lake	
	13.2–49.5	26.9	13			Guandu Hu/lakelet of East Lake	
	12.4–44.5	26.9	16			Tangling Hu/lakelet of East Lake	
	11.3–32.2	24.7	17			Hou Hu/lakelet of East Lake	
		27.1				East Lake overall	
Longgan	26.4–51.4				2002–2008	Hubei Province, China	Bing et al. (2013)

the world's largest sulfide ore deposits of nickel and copper and has been the center of nickel–copper smelting in North America for the past 100 years. Sempkin (1975) and Sempkin and Kramer (1976) reported average Ni concentrations of 120 mg kg^{-1} (range, 3–630 mg kg^{-1}) for 65 lakes in the Sudbury area. These relatively high levels in the Sudbury lakes, including Richard Lake, are due to aerial fallout and point sources from nickel smelters that have raised background levels (Table 7.3). Metal analysis shows that Ni concentrations in the shallow sediment of Richard Lake peaked at 4721.1 mg kg^{-1} at 3 m followed by steadily decreasing concentrations at greater depths (Table 7.3; Shuhaimi-Othman 2008). Comparison of the Ni concentrations in Richard Lake with USEPA standards showed that these sediments were heavily polluted with Ni, suggesting that sediments from Richard Lake may pose a hazard to aquatic biota (Table 7.2).

7.5.2 Rivers

A river is a naturally flowing watercourse, usually freshwater, that discharges into an ocean, sea, lake, or another river. In some cases, rivers flow into the ground and become dry without reaching another body of water. A small river is referred to as a stream, creek, brook, rivulet, or rill; these names are geographically specific. Rivers are part of the hydrological cycle. Water generally collects in a river through surface runoff from precipitation on the drainage basin and from other sources such as groundwater recharge, springs, and the release of stored water in natural ice and glaciers.

Rivers have been used for drinking water, food, bathing, transport, as a means of defense, for hydropower to drive machinery, and as a means of waste disposal. Rivers have been essential for navigation for thousands of years. Riverine navigation provides cheap transport and is still used extensively on most major rivers of the world like the Amazon, Ganges, Nile, Mississippi, and Indus. Some rivers have contributed to the development of important civilizations and have been anthropogenically influenced for centuries or even millennia. Below, Ni in sediments will be discussed in rivers around the world by continent including Africa, Asia, Australia, Europe, North America, and South America (Figure 7.1 and Tables 7.4 through 7.9).

The chemistry of rivers is complex and depends on inputs from the atmosphere, the geology through which it travels, and inputs from man's activities. The chemical composition of the water has a large impact on riverine ecology, affecting both plants and animals as well as the uses that can be made of the river water. In rivers, nickel is transported mainly as a precipitated coating on particles and in association with organic matter. Ni content in the river sediment depends mainly on the river flow (river energy), erosion of the river bed, geology, and human activities. Generally, Ni concentrations in river sediments increase along the river, with the highest concentration observed in the mouth, for example, River Karoon, South of Iran (Diagomanolin et al. 2004).

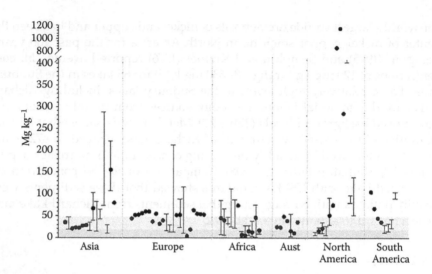

FIGURE 7.1
Nickel concentrations in river sediments (circles = mean, bars = range) (not polluted = gray shading, moderately polluted = light gray shading, and heavily polluted = no shading). (Sediment Quality Criteria from USEPA. 1995. QA/QC Guidance for Sampling and Analysis of Sediments, Water, and Tissues for Dredged Material Evaluations—Chemical Evaluations. EPA 823-B-95-001. Office of Water, Washington, DC, and Department of the Army, U.S. Army Corps of Engineers, Washington, DC.)

7.5.2.1 Rivers in Africa

Africa's longest rivers are the Nile, Congo, Niger, and Zambezi. Table 7.4 shows Ni concentrations in the sediments of these rivers as well as a few smaller rivers such as Kafue and Luangwa. Many parts of these rivers remain untamed and wild compared with rivers in Europe and the United States. However, the Nile, one of the largest rivers on Earth, has been densely inhabited for thousands of years, especially in its delta and lower valley where 95% of the 92 million people in Egypt reside (State Information Service 2014). Considerable industrial activity occurs in the Nile delta near Cairo. According to Wahaab and Badawy (2004), the annual consumption of Nile water by Egyptian industry is 6.38×10^8 m³, of which 5.49×10^8 m³ is discharged to the Nile River after use. Mean Ni concentration in the sediments of the delta and Aswan-Asyut Valley are 45 and 40 mg kg⁻¹, respectively (Table 7.4; Arafa et al. 2015). According to USEPA (1995) guidelines, these sediments are moderately polluted. They also exceed the probable effect level of 36 mg kg⁻¹ recommended by Smith et al. (1996) (Table 7.2). Arafa et al. (2015) suggested that elevated Ni concentrations in the Nile delta could also be related to weathering in the Ethiopian highlands and transport of the weathered material by the Blue Nile, the Nile main tributary.

Ni concentrations in sediments are elevated in the Congo River Basin in the vicinity of the large city of Kinshasa (Table 7.4). Ni concentration in the

TABLE 7.4

Nickel Concentrations in African River Sediments

River	Length (km)	Ni (mg/kg)			Year	Location	References
		Range	Mean	n			
Nile	6853	8–85	45	20		Riverbank/Delta/Egypt	Arafa et al. (2015)
		6–67	40	20		Riverbank/Aswan-Asyut Valley/Egypt	
Congo	4700	32.4–47.4		5	2013	Kinshasa/Democratic Rep. of the Congo	Mwanamoki et al. (2015)
		25.0–111.7[a]		5	2013	Kinshasa/Democratic Rep. of the Congo	
		22.1–86.8[b]		5	2013	Kinshasa/Democratic Rep. of the Congo	
			74[a]			Kinshasa/Democratic Rep. of the Congo	McLennan (1993)
Niger	4200	1.2–14.6	7.0	20	2009	Cross River, eastern Niger Delta Basin/Nigeria	Ekwere et al. (2013)
		1.2–28.0	5.4	16	2009	Qua Iboe River, eastern Niger Delta Basin/Nigeria	
		6.0–28.0	15.7	16	2009	Imo River, Eastern Niger Delta Basin/Nigeria	
Zambezi	2574	3–39	13	8	2008	Above Mambova/Zambia	Ikenka et al. (2010)
Kafue	1600	2–75	46	8	2008	Downstream from Copperbelt Area/Zambia	
Luangwa	770	9–17	17	5	2008	Eastern Zambia	

[a] 0–3 cm depth.
[b] 3–6 cm sediment depth.

TABLE 7.5

Nickel Concentrations in Asian River Sediments

River	Length (km)	Ni (mg/kg)			Year	Location	References
		Range	Mean	n			
Yangtze	6300		35.2	30	2010	Delta/China	Wang et al. (2015)
		19.4–47.3			2010	Delta/China	
Yellow	5464		20.9		2014	Upper reach/Gansu/China	Zuo et al. (2016)
			23.2		2014	Upper reach/Ningxia	
			22.5		2014	Upper reach/Inner Mongolia	
			27.4		2007	Delta/unvegetated sediment/China	Nie et al. (2010)
			28.7		2007	Delta/rhizosphere sediment/China	
Lena	4400	16–35		27	1991	Delta/Russia	Nolting et al. (1996)
Euphrates	2800	40.0–104.0	67.1	14	2012	From Albagoz to Ramadi/Iraq	Salah et al. (2012)
Ganges	2525	14.6–82.5		180	2012	15 stations, Samne Ghat to Ganga/Varuna confluence/India	Pandey et al. (2014)
Pearl	2400	42.1–65.1				Delta/Guangzhou/China	Cheung et al. (2003)
Tigris	1850	74–288				Syria, Turkey, Iraq	Varol (2011)
Periyar	244	10–30		7		Along river/India	Suraj et al. (1996)
Turag	160	108–221.6	155.4	15	2014	Upstream/Dhaka/Bangladesh	Mohiuddin et al. (2016)

TABLE 7.6
Nickel Concentrations in Australian River Sediments

River	Length (km)	Ni (mg/kg)			Year	Location	References
		Range	Mean	n			
Murray	2375		25	100		Between Locks 2 and 4/Lower Murray	Thoms (2007)
Hunter	186		24	3	1995–1996	Upper river	Lottermoser (1998)
			49	4	1995–1996	Lower river, Throsby Creek, Newcastle Harbor	
Richmond	170		39	18	1995–1996		
Hawkesbury–Nepean	120	3–54	16	90	1998–1999	Upper river, central New South Wales	Simonovski et al. (2003)
Williams	75		11	3	1995–1996		Lottermoser (1998)
Wilson	28		36	4	1995–1996	Near Lismore	

TABLE 7.7

Nickel Concentrations in European River Sediments

River	Length (km)	Ni (mg/kg)			Year	Location	References
		Range	Mean	n			
Volga	3692		44		1940	Delta/Russia	Winkels et al. (1998)
			52		1970	Delta/Russia	
			53		1990	Delta/Russia	
Danube	2860		56		1940	Delta/Romania and Ukraine	
			60		1970	Delta/Romania and Ukraine	
			60		1990	Delta/Romania and Ukraine	
Rhine	1236		37		1940	Delta/Netherlands	
			55		1970	Delta/Netherlands	
			32		1990	Delta/Netherlands	
Vistula	1047		40			Middle section/Poland	Helios-Rybicka (1986)
Elbe	1091	81–159		78	1996	Germany	Van der Veen et al. (2006)
		58–87		50	2005	Germany	Baborowski et al. (2007)
			114	22	1998–2002	Germany	Zerling et al. (2006)

(Continued)

TABLE 7.7 (CONTINUED)

Nickel Concentrations in European River Sediments

River	Length (km)	Ni (mg/kg)		n	Year	Location	References
		Range	Mean				
Odra	854		51.6			Poland	Adamiec and Helios-Rybicka (2002)
		23.5–89.3	52.7	115	1997–2000	Entire river system	Helios-Rybicka et al. (2005)
Olt	615	10.5–116.4			2010	Lower part/Romania	Iordache et al. (2016)
Segura	325	4.1				Ojos de Archivel/Spain	Garcia-Alonso et al. (2015)
		18.8				Orihuela/Spain	
Wupper	113	63		34	2013	Floodplain/Germany	Frohne et al. (2014)
Waal	80		55		1958	Floodplain/Netherlands	Japenga et al. (1990)
			54		1970	Floodplain/Netherlands	
			53		1981	Floodplain/Netherlands	

Nickel in Soils and Plants

TABLE 7.8
Nickel Concentrations in North American River Sediments

River	Length (km)	Ni (mg/kg)			Year	Location	References
		Range	Mean	n			
Missouri	3768	5.7–13.4		19	2002	Lower Missouri R. from Omaha to Jefferson City, Missouri/USA	Echols et al. (2008)
Mississippi	3734	10–23	16	8	1981	Above Lake Pepin, upper Mississippi River/USA	Bailey (1983)
		11–35	22	15	1979	Weaver Bottoms below Lake Pepin, upper Mississippi R./USA	
Columbia	1954	5.6–26.9		45		Mid-river between Vantage, Washington, and McNary Dam/USA	Youger and Mitsch (1989)
Ohio	1579	29.4–91.4	51.5	12	1987	Upper Ohio River/USA	Fitchko and Hutchinson (1975)
Grand River	406		74.8				
Sumas River	58	856.5–860.0			1993–1994	Canada–USA border	Smith et al. (2007)
		1159.6			2003–2004	Canada–USA border	
		285			1993–1994	Headwater/USA	
		413.2–403.3			2003–2004	Headwater/USA	
		80.1–97.2			1993–1994	Mouth, British Columbia/Canada	
		52.3–138.9			2003–2004	Mouth, British Columbia/Canada	
Lerma	750	~22–43		8		La Piedad, Mexico	Villalobos-Castaneda et al. (2016)

TABLE 7.9

Nickel Concentrations in South American River Sediments

River	Length (km)	Ni (mg/kg)			Year	Location	References
		Range	Mean	n			
Amazon	6400		105[a]		1977	Mainstem	Gibbs (1977)
Parana	4880		67[a]		1971–1973	Argentina	Depetris et al. (2003)
			45[a]		1997	Rosario/Argentina	
			36[a]		1998	Parana/Argentina	
Magdalena	1528	12.1–28.8		12	1995 (April–May, dry season)	Mouth/lagoons, Vienaga Grande de Santa Marta/Colombia	Perdomo et al. (1998)
		14.5–32.5		12	1995 (June–July, rainy season)	Mouth/lagoons, Vienaga Grande de Santa Marta/Colombia	
Piracicaba	115	90–105[a]			2005	Brazil	Mortatti and Probst (2010)

[a] Associated with total suspended solids.

vicinity of Kinshasa ranged from 22.1 to 111.7 mg kg^{-1} in surface sediment with a mean of 41.9 mg kg^{-1} (Table 7.4; Mwanamoki et al. 2015), indicating that the sediment is moderately polluted (Figure 7.1). Generally, concentrations of Ni in surface sediments (0–3 cm) in the Congo River Basin near Kinshasa were lower than in the deeper sediment profile (4 to 6 cm depth), suggesting slowdown of regional economic activities during the last decades (Mwanamoki et al. 2015). Elevated Ni concentrations (surface sediment mean of 46 mg kg^{-1}) also occur in the Kafue River, the largest tributary of the Zambezi due to mining activities in Kabwe and the Copperbelt (Table 7.4; Ikenka et al. 2010).

Rivers located in more remote parts of Africa, for example, the Luangwa River in eastern Zambia, are unpolluted, and Ni concentrations in surface sediment are below 20 mg kg^{-1} (Table 7.4; Ikenka et al. 2010). Unpolluted areas in Niger and Zambezi rivers also have low Ni levels in surface sediments, with means of 7.0–15.7 and 13.0, respectively (Table 7.4).

7.5.2.2 Rivers in Asia

Some of the world's longest rivers, such as the Yangtze, Yellow, and Euphrates, flow through Asia, the world's largest continent (Table 7.5). These waterways account for much of the region's fertility and provide important resources to sustain Asia's large population. Many of these rivers including the Tigris, Euphrates, and Indus have immense historical importance and have allowed some of the world's earliest civilizations to flourish around their banks.

The longest river of Asia and the third longest in the world, the Yangtze, flows through China. It originates at the Geladaindong Peak in Qinghai and covers a distance of 6300 km before joining the East China Sea at Shanghai. The average discharge of this gigantic river is 30,166 m^3 s^{-1}. The river not only has been important in the country's history but also is a source of economic prosperity for present-day China. The mean Ni concentration in the delta of the Yangtze River is 35.2 mg kg^{-1} (Table 7.5; Wang et al. 2015).

The Yellow River, the second largest river in China, has lower Ni concentrations in the upper reach that rise to slightly higher levels in the delta, 20.9 and 28.7 mg kg^{-1}, respectively (Table 7.5; Zuo et al. 2016 and Nie et al. 2010). These concentrations, especially in the upper reach, are not indicative of pollution. Ni concentrations in sediments of the Lena and Periyar Rivers (located in Siberia and India, respectively) are also low and do not exceed 35 mg kg^{-1}, the background value for the world's shallow sediments (Tables 7.1 and 7.5). The Lena River belongs to one of the world's largest river systems and is situated in an almost pristine environment; therefore, it is interesting to compare this river with other more anthropogenically influenced ones (Figure 7.1).

The Ni concentration gradient in rivers is rather simple when viewed on a large scale, but at finer scales, gradients are variable and influenced by local factors such as bank slumping, sediment deposition, human activities, and major events like floods and ice jams (Moore and Landrigran 1999). Therefore, there is a broad range of Ni concentrations in many river sediments, for example, the Euphrates, Ganges, Pearl, and Tigris: 40–104, 15–83, 42–65, and 74–288 mg of Ni kg⁻¹, respectively (Table 7.5; Figure 7.1). Ni pollution in river sediment is often more apparent in smaller rivers or tributaries, especially those in densely populated areas with diverse human and industrial activities such as the Turag River, an upper tributary of Buriganga, which is a major river in Bangladesh (Table 7.5; Figure 7.1) (Mohiuddin et al. 2016). The major pollutants of the Turag are soap and detergent, pharmaceuticals, dyes from the garment industry, aluminum, ink, textile, paint, and steel industry and waste discharges (Rahman et al. 2012). The average concentration of Ni in sediment of the Turag (155.4 mg kg⁻¹) is several times higher than the geochemical background (i.e., average worldwide shale and continental upper crust, 38–68 mg kg⁻¹; Tables 7.5 and 7.1) and exceeds the severe effect level (SEL) of 75 mg kg⁻¹ recommended by Persaud et al. (1993) (Table 7.2). Figure 7.1 shows that most of the Asian rivers presented in Table 7.5 are moderately or heavily polluted by Ni according to EPA quality standards for freshwater sediment (USEPA 1995).

7.5.2.3 Rivers in Australia

The continent of Australia is relatively dry, and sources of freshwater are few. In general, the rivers in Australia have a relatively low flow that varies along the course of the river. The Murray–Darling river system is the largest in Australia. It has a catchment of 1.07 million km², with the Murray catchment having an area of 310,000 km² above its confluence with the Darling. Most of the basin west of the Great Dividing Range is arid or semi-arid, with mean annual evaporation (1200 mm) considerably above mean annual rainfall (450 mm) (Thoms 2007). Both seasonal and annual stream flows in the basin are highly variable.

Because of a lack of freshwater resources in Australia, rivers are highly protected and regulated. The Murray–Darling system is regulated by 21 headwater dams and over 8000 low head weirs (Thoms 2007). Dams and weirs are efficient traps for sediments and associated pollutants. They interrupt the downstream movement of material leading to changes in sediment composition that could increase metal concentrations associated with agricultural and urban development. Peak loadings of heavy metals, including Ni, in the sediment were found in the depositional areas above each weir in the Murray–Darling system. Nickel and other heavy metals are amplified by changes in sediment texture, and the spatial concentrations of these pollutants reflect sediment-transport factors associated with the presence of weirs (Thoms 2007). Nickel concentrations in the fine fraction of the sediments in the lower Murray were 75 times higher than in other sediment fractions (Thoms 2007). However, the mean concentration of Ni in the Murray River is 25 mg kg⁻¹, comparable to natural Ni background levels reported by

Turekian and Wedepohl (1961) for similar rock types found in the catchment (Table 7.6).

Generally, Ni concentrations in river–stream sediments in Australia largely represent natural background values (Table 7.6 and Figure 7.1). However, Ni concentrations in Hunter River sediments within the heavily industrialized and urbanized Newcastle region exceed upstream background values by up to an order of magnitude: 49 and 24 mg kg^{-1}, respectively (Table 7.6; Lottermoser 1998). The Hawkesbury–Nepean River is similar: the upper part is not heavily polluted by Ni and other elements, but Ni concentrations in the downstream sediment near the Nepean Dam are relatively high (54 mg kg^{-1}) (Table 7.6, Figure 7.1; Simonovski et al. 2003).

7.5.2.4 Rivers in Europe

Europe, the second smallest continent, is home to almost 50 countries and is the third most populated continent in the world after Asia and Africa. The estimated number of rivers in Europe is 1352. Many of these are marked by conspicuous physical features, such as waterfalls and canyons, and are rich in dissolved minerals and organic compounds. The major rivers of Europe are the Volga and Danube (Table 7.7). The Volga is the longest, originating in the Valdai Hills and flowing for 3690 km before draining into the Caspian Sea. Forty percent of Russian people live near the Volga or its tributaries, and it used extensively for irrigation, fishing, commerce, and other purposes. As a likely result of anthropogenic influences, the mean Ni concentration in river sediment from the Volga delta (44–53 mg kg^{-1}) is higher than in sediment from the Lena River (16–35 mg kg^{-1}), a large Siberian river that traverses primarily undeveloped areas (Table 7.5). Nickel concentrations in sediment from the delta of the Danube (56–60 mg kg^{-1}), another river with a highly developed catchment, are even higher that in the Volga delta. These concentrations increased slightly from 1940 to 1970 and stayed at the same level for the next 20 years (Table 7.7; Winkels et al. 1998).

Most sediment in European rivers is moderately or highly polluted with Ni according to EPA sediment quality criteria (USEPA 1995) and other standards (Figure 7.1 and Tables 7.2 and 7.7). For example, Ni concentrations in fine sediments from the Elbe River ranged from 81 to 159 mg/kg compared with geogenic background levels of 53 mg kg^{-1} (van der Veen 2006). However, newly deposited sediments in the Elbe are less polluted than deeper, older ones (Schwartz 2006). Ni in the Danube delta correlates with the clay content of the sediment, but Ni in the Volga delta correlates with the clay fraction and organic carbon content (Winkels et al. 1998). Because the correlation between pollutants and the clay fraction is less pronounced in contaminated sediments, Ni in Volga delta sediments may represent natural background values (Winkels et al. 1998). In sediments of the Rhine delta, Ente (1981) found significant correlations between Ni and organic matter but not between Ni and the clay fraction similar to Winkels et al. concerning Danube delta sediment.

Ni concentration in the sediment of smaller European rivers are often high; for example, the mean Ni concentration for the entire Odra river system is 52.7 mg kg^{-1} (Table 7.7). The pattern of spatial and vertical Ni distribution in the river sediment indicated that a variety of sources might be responsible for this contamination: intensive, historical, and current mining and smelting activities are very likely the most important ones (Helios-Rybicka et al. 2005). Nickel concentrations in the Odra River varied widely from 23.5 to 89.3 mg kg^{-1}, with higher concentrations in areas of high mining and smelting activity (Table 7.7; Helios-Rybicka et al. 2005). The Sava River is another example of very elevated Ni in sediment, especially near Belgrade, the largest city of Serbia (Table 7.7; Milacic and Scancar 2010). In 2005, sediment quality typically corresponded to class 3 (Dutch regulation standard) with three sediment profiles corresponding to class 4 (Milacic and Scancar 2010) due to the high Ni concentrations. According to Dutch regulations (ref), class 3 sediments are heavily polluted, while class 4 sediments are extremely polluted and of unacceptable quality.

7.5.2.5 Rivers in North America

The Mississippi River is the largest river of North America and the United States at 3765 km in length. It flows from northwestern Minnesota south to the Gulf of Mexico and is a significant transportation artery. When combined with its major tributaries (the Missouri River and the Ohio River), it becomes the third largest river system in the world at 6236 km in length (Table 7.8). Ni concentrations in the upper Mississippi River above and below Lake Pepin are low and do not exceed background concentrations for shallow sediment (35 mg kg^{-1}) (Salomons and Forstner 1984). Ni concentrations in the upper Mississippi are correlated with the percentage of clay and organic matter in surface sediment (Bailey 1983). Low Ni concentrations are also found in the lower Missouri River from Omaha to Jefferson City (range, 5.7–13.4 mg kg^{-1}; Table 7.8; Echols et al. 2008), indicating no Ni pollution in the sediment (Figure 7.1). In contrast, Ni concentrations in the bottom sediment of the upper Ohio River varied over a broad range (29.4–91.4 mg kg^{-1}), indicating heavy Ni pollution, particularly downstream of East Liverpool, Ohio, and the Wheeling, West Virginia metropolitan area (Youger and Mitsch 1989). However, the data of Youger and Mitsch (1989) showed a reduction in Ni and other metal concentrations in the upper Ohio River valley over 1977–1987 due to a dramatic decrease in the iron and steel industry. In contrast, Ni concentrations in the Lerma River, one of the largest river of Mexico, have increased over the past 20 years due to population growth and industrial activities, especially in the area near La Piedad city. This river is moderately polluted according to USEPA sediment quality standards (Table 7.8 and Figure 7.1). Mid-river Ni concentrations in the Columbia River, which drains relatively undeveloped areas in the northwestern United States and parts of British Columbia, Canada, are in the range of natural background levels (Table 7.8).

Naturally occurring landslides of serpentinite materials contributed to high levels of Ni in sediments from the Sumas River in British Columbia, Canada (Table 7.8). Smith et al. (2007) reported that sediment samples were highly contaminated with Ni (above the SEL of 75 mg kg^{-1}, Table 7.2) in Swift Creek, a headwater tributary affected by a landslide, as well as the Sumas River (Table 7.8). Ni concentrations were higher in 2003/2004 than in 1993/1994, suggesting that the Ni-rich asbestos fiber material mobilized by the landslide had moved downstream over time. The authors found a strong relationship between bioavailable Ni and the concentration of Ni in Sumas River sediments.

7.5.2.6 Rivers in South America

Although the water chemistry of the Amazon River has been well documented, few studies have focused on fluvial sediments, and most scientist have studied suspended sediment composition rather than bottom sediments. Generally, there is no indication of Ni contamination in sediment of the mainstem section of the Amazon, the largest hydrographic basin in the world (Table 7.9; Duarte and Gioda 2014). The Ni concentration in the suspended sediment of the Amazon River is low, 31 mg kg^{-1}, and below the average Ni concentration in the suspended sediments from other rivers of South America (46 mg kg^{-1}), North America (50 mg kg^{-1}), Russia (123 mg kg^{-1}), China (68 mg kg^{-1}), Africa (78 mg kg^{-1}), and Europe (66 mg kg^{-1}) (Viers et al. 2009).

Nickel concentrations in other large South America rivers are generally near background levels. Nickel concentrations in bottom sediment from the Madeira River watershed in Brazil are extremely low: 3.8, 0.9, and 3.2 mg kg^{-1} in white water, clear water, and black water rivers, respectively (Lacerda et al. 1990). However, there are sometimes significant influxes from industrial effluents, rock erosion, and other sources. In mineral-rich zones of South America, mining is common, and surface water and sediments sometimes receive high levels of metals, including Ni. Metal contamination from mining operations generally affects smaller rivers (McClain 2002). Near large cities (e.g., Sao Paulo, Brazil), Ni concentrations can be increased by the release of large quantities of wastewater from domestic, industrial, and agricultural sources (Table 7.9). In the Piracicaba River, Brazil, the average Ni concentration showed a significant increase from upstream to downstream, reaching levels of moderate pollution near areas affected by the development of sugar cane plantations and urbanization (Mortatti and Probst 2010). In more industrialized tributaries near the mouth of the Parana River, Ni concentrations in suspended sediment were elevated relative to background levels based on samples collected during the summer and fall (48 and 36 mg kg^{-1}, respectively) (Table 7.9; near Parana, Argentina; Depetris et al. 2003). Ni concentrations in bottom sediments were also elevated in the mouth of the Magdalena River, reaching as high as 32.5 mg kg^{-1} in the rainy season (Table 7.9).

7.6 Nickel in Marine Ecosystem Sediments

Rivers transport eroded materials from the continents to the seas and oceans in dissolved and solid forms (Martin and Whitfield 1983). The distribution of Ni between the solute and particulate phases in transported material depends on its mobility and speciation, which is affected by dissolved mineral and organic ligands and the quantity and composition of mineral and organic particles. In turn, these factors are influenced by the recycling of continental crust (Goldstein and Jacobsen 1988; Taylor and McLennan 1985) and various anthropogenic activities that affect sedimentation rates and the chemistry of terrestrial runoff (Audry et al. 2006; Nriagu and Pacyna 1988). Material derived from soil erosion is carried in suspension by rivers, and the trace elements contained in this particulate matter are eventually transported to the oceans. Trace metals transported to seas and oceans subsequently accumulate in bottom sediments because minerals offer large specific surfaces for the sorption of these constituents.

Oceans and seas cover more than 70% of the Earth's surface and include a diversity of habitats from inshore estuarine and littoral zones that are strongly influenced by terrigenous and/or anthropogenic runoff to the abyssopelagic zone of the open ocean where such influences are weaker and the geochemical environment becomes the controlling factor affecting the disposition of Ni and other metals. These unique conditions facilitate the formation of polymetallic nodules, some more than 20 cm across, on the sediment surface over vast areas of the bottom of the ocean. The composition of these nodules varies but their Ni content may exceed 1% (International Seabed Authority 2010) and is sufficient to arouse commercial interests (Mero 1965). Nickel is also concentrated in much smaller (<5 mm) micronodules formed by a diagenetic process in some deep ocean sediments (Figure 7.2 and Table 7.10). Micronodules are common in red clay sediments in the Atlantic Ocean. These sediments may have a Ni content approaching 500 mg/kg (Menendez et al. 2017), which is substantially higher than Ni levels in other types of sediments from the deep Atlantic. Excluding micronodule-rich red clay, Ni levels in abyssopelagic sediments from the Atlantic Ocean are lower than those from the Pacific Ocean. Hutchinson et al. (1955) believed that the concentration of metals is higher in the Pacific than in the Atlantic because the ratio of hydrogenous sedimentation (i.e., chemical precipitation from seawater) to lithogenous sedimentation is greater in the former.

For the purposes of this paper, "sea" is defined as a body of saline water that is more-or-less landlocked and smaller than an ocean. The sediments of seas might be expected to differ from those of the open ocean because seas are strongly affected by surrounding lands and are geologically and chemically distinct from oceans. Nickel concentrations in sediment samples from the Caspian, Baltic, and Mediterranean seas were generally under 100 mg kg^{-1} (Table 7.10 and Figure 7.2). These concentrations were recorded in areas

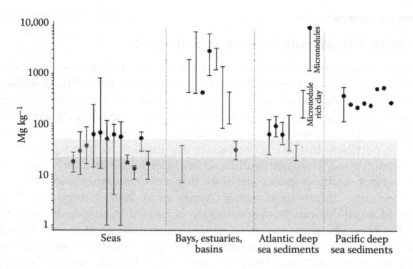

FIGURE 7.2
Nickel concentrations in marine sediments (circles = mean, bars = range) (not polluted = gray shading, moderately polluted = light gray shading, and heavily polluted = no shading). (Sediment Quality Criteria from USEPA. 1995. QA/QC Guidance for Sampling and Analysis of Sediments, Water, and Tissues for Dredged Material Evaluations—Chemical Evaluations. EPA 823-B-95-001. Office of Water, Washington, DC, and Department of the Army, U.S. Army Corps of Engineers, Washington, DC.)

not affected by point sources of pollution but possibly enriched by pollution from diffuse sources (Secrierus and Secrierus 2002). A Ni concentration of 815 mg kg^{-1}, along with elevated concentrations of other elements, was recorded in the Baltic Sea near a shipwreck (Emelyanov et al. 2010). However, Ni levels in the seas discussed herein were generally similar to or somewhat lower than found in Atlantic Ocean sediments and substantially lower than found in Pacific Ocean sediments.

The greatest Ni pollution would be expected in nearshore marine waters, especially bays and estuaries in populated areas subject to industrial development and mining or terrestrial runoff from Ni-rich soils. Ni concentrations were elevated in coastal Cuba near Moa, ranging from 900 to 6200 mg kg^{-1} (Cervantes-Guerra et al. 2017) (Table 7.10 and Figure 7.2). This area, with serpentine geology and a well-developed Ni and cobalt mining industry, was regarded by Cervantes-Guerra et al. (2017) as typical of aquatic environments in densely populated, industrialized regions. Comparable Ni levels were observed in Levisa Bay, Cuba, also subject to pollution from Ni mining (Gonzalez et al. 1997). The sediments of nearby Nipe and Cabonico bays were influenced by this pollution to a lesser degree, with Ni concentrations ranging between about 400 and 1900 mg/kg^{-1}. Louropoulou et al. (2015) reported Ni levels as high as 3207 mg kg^{-1} in Larymna Bay in the Evoikos Gulf near Greece, an area affected by discharge from an iron–nickel smelter, compared with 101–426 mg kg^{-1} in relatively unpolluted parts of the Evoikos

TABLE 7.10

Nickel Concentrations in Marine Sediments

Name	Mean	Ni (mg/kg) Range	n	Water Depth (m)	Sediment Depth (cm)	Fractions	Year	Location	References
Seas									
Baltic	64	10–815	375		0–5	All fractions	2006–2008	Bornholm Basin, bottom sediment	Emelyanov et al. (2010)
Baltic	18	11–27	18		0–5	Sand fraction	2006–2008	Bornholm Basin, bottom sediment	
Baltic	29	10–68	24		0–5	Coarse aleurite	2006–2008	Bornholm Basin, bottom sediment	
Baltic	37	16–88	15		0–5	Fine aleuritic mud	2006–2008	Bornholm Basin, bottom sediment	
Baltic	62	14–241	113		0–5	Aleuro-pelitic mud	2006–2008	Bornholm Basin, bottom sediment	
Baltic	67	13–815	205		0–5	Pelitic mud	2006–2008	Bornholm Basin, bottom sediment	
Black	49.8	1–117	24		0–20		1997	Northwestern Black Sea	Secrierus and Secrierus (2002)
	61.6	4–98	22		20–22			Northwestern Black Sea	
	55.4	1–110	19		30–32			Northwestern Black Sea	
Caspian	17.42	16.3–24.0					Winter 2012	Southeast coast/Iran	Bastami et al. (2014)
	13.11	7.5–15.26					Summer 2012	Southeast coast/Iran	

(Continued)

TABLE 7.10 (CONTINUED)

Nickel Concentrations in Marine Sediments

Name	Mean	Ni (mg/kg) Range	*n*	Water Depth (m)	Sediment Depth (cm)	Fractions	Year	Location	References
	51.6	29.4–67.8						Iran	De Mora et al. (2004)
Mediterranean	16.31	8–29							Moreno et al. (2009)
Bays/Ports/Basins									
Deception Bay		6.5–36.7						Southeast Queensland, Australia	Brady et al. (2014)
Nipe Bay		417–1893			Surface		1992–1993	Northern coast of Cuba	Gonzalez et al. (1997)
Levisa Bay		408–6787			Surface		1992–1993	Cuba	
Cabonico Bay	418				Surface		1992–1993	Cuba	
Cay Moa Bay	2800	900–6200	12		Surface		2009–2011	Cuba	Cervantes-Guerra et al. (2017)
Larymna Bay		1180–3207			Surface	<63 µm	2013	Northern Evoikos Gulf, Greece	Louropoulou et al. (2015)
		82.8–1362			Surface	>63 µm	2013	Northern Evoikos Gulf, Greece	
Evoikos Gulf		101–426			Surface	<63 µm	2013	Southern Evoikos Gulf, Greece	
Gdansk Basin	31	20–46			0–2.5	<63 µm	1996	Baltic Sea, Poland	Belzunce Segarra et al. (2007)

(*Continued*)

TABLE 7.10 (CONTINUED)

Nickel Concentrations in Marine Sediments

Name	Mean	Ni (mg/kg) Range	n	Water Depth (m)	Sediment Depth (cm)	Fractions	Year	Location	References
Open Oceans/Deep Sea									
Atlantic Ocean	63	25–122	18	5030		Red clay		Core A 160-8	Hutchinson et al. (1955)
	92	57–142	7	5080		Red clay		Core A 152-134	
	62	40–110					1951	Core Rotschi 273	
Atlantic Ocean		30.2–150				Red clay	1955–1988	Transect at 24°N across North Atlantic Ocean	Menendez et al. (2017)
Atlantic Ocean		18.7–38.9				Gray clay	1955–1988		
	8280	136–466				Micronodule-rich red clay	1955–1988	Transect at 24°N across North Atlantic	
Atlantic Ocean	364	1150–8440	49	6081		Micronodules	1955–1988	Core VM22-212/Canary Basin	
Pacific Ocean	247	114–535	9	4850				Core MP35.2	Hutchinson et al. (1955)
Pacific Ocean	214			0–2.5				Core MP35.2	
Pacific Ocean	259			46–53				Core MP35.2	
Pacific Ocean	235			100–110				Core MP35.2	
Pacific Ocean	503			155–161				Core MP35.2	
Pacific Ocean	526			259–268				Core MP35.2	
Pacific Ocean	267			374–376				Core MP35.2	
Pacific Ocean				30–786				Rotschi/all cores	

Gulf. They commented on the difficulty of discriminating between natural and anthropogenic sources of Ni in areas with naturally high Ni levels in the parent rock. In bays and basins not influenced by pollution or terrigenous runoff from Ni-rich catchments, including Deception Bay, Australia, and the Gdansk Basin, near Poland, Ni levels in sediment were under 50 mg kg^{-1}, lower than that in abyssopelagic ocean sediments and comparable to levels found in unpolluted seas.

7.7 Summary

Nickel is mobilized from geological sources by a variety of natural and anthropogenic processes and is ubiquitous in the Earth's aquatic ecosystems where it accumulates in sediment by a variety of mechanisms. Although potentially toxic to aquatic organisms via damage to respiratory surfaces and by other mechanisms, Ni poses less risk to aquatic ecosystems than most other heavy metals. Nickel toxicity is generally moderated by factors that reduce its bioavailability. It bioaccumulates to varying degrees in different types of organisms but does not biomagnify like mercury or some organic compounds. The sediments of most undisturbed aquatic ecosystems have relatively low concentrations of Ni, although Ni concentrations are higher in areas where the underlying geological parent material is rich in Ni. Additionally, Ni is concentrated in the polymetallic nodules and micronodules that naturally form in many of the world's deep ocean basins. Nickel is used by man for many purposes, and Ni pollution is not unusual. Sediment Ni levels are often elevated in highly developed regions, particularly near Ni mining and smelting operations.

References

Adamiec, E. and E. Helios-Rybicka. 2002. Distribution of pollutants in the Oda River system. Part IV. Heavy metal distribution in water of the upper and middle Odra River, 1998–2000. *Polish J. Environ. Stud.* 11(6):669–673.
Arafa, W.M., W.M. Badway, N.M. Fahmi et al. 2015. Geochemistry of sediments and surface soils from the Nile Delta and lower Nile valley studied by epithermal neutron activation analysis. *J. African Earth Sci.* 107:57–64.
Arai, Y. 2008. Spectroscopic evidence for Ni(II) surface speciation at the iron oxyhydroxides–water interface. *Environ. Sci. Technol.* 42:1151–1156.
Audry, S., G. Blan, and J. Schäfer. 2006. Solid state partitioning of trace metals in suspended particulate matter from a river system affected by smelting-waste drainage. *Sci. Total Environ.* 363:216–236.

Baborowski, M., O. Büttner, P. Morgenstern, F. Krüger, I. Lobe, H. Rupp, and W.V. Tümpling. 2007. Spatial and temporal variability of sediment deposition on artificial-lawn traps in a floodplain of the River Elbe. *Environ. Pollut.* 148:770–778.

Bailey, P.A. 1983. Distribution and enrichment of trace elements (Cd, Cr, Cu, Ni, Pb, Zn) in bottom sediments of navigation pools 4 (Lake Pepin), 5, and 9 of the upper Mississippi River. Master of Science in Biology Thesis, University of Wisconsin, La Crosse, Wisconsin, USA.

Bastami, K.D., H. Bagheri, V. Kheirabadi et al. 2014. Distribution and ecological risk of heavy metals in surface sediments along southeast coast of the Caspian Sea. *Mar. Pollut. Bull.* 81:262–267.

Belzunce Segarra, M.J., P. Szefer, M.J. Wilson et al. 2007. Chemical forms and distribution of heavy metals in core sediments from the Gdansk Basin, Baltic Sea. *Polish J. Environ. Stud.* 16(4):505–515.

Bing, H., Y. Hu, W-H. Nahm et al. 2013. Accumulation of heavy metals in the lacustrine sediment of Longgan Lake, middle reaches of Yangtze River, China. *Environ. Earth Sci.* 69:2679–2689.

Bowen, H.J.M. 1979. *Environmental Chemistry of Trace Elements*. New York: Academic Press, 333 pp.

Brady, J.P., A.A. Godwin, A. Ayoko, W.N. Martens, and A. Goonetilleke. 2014. Enrichment, distribution and sources of heavy metals in the sediments of Deception Bay, Queensland, Australia: A preliminary survey. *Mar. Pollut. Bull.* 81:248–255.

Bryce, A.L., W.A. Kornlcker, and A. W. Elzerman. 1994. Nickel adsorption to hydrous ferric oxide in the presence of EDTA: Effects of component addition sequence. *Environ. Sci. Technol.* 28:2353–2359.

Buerge-Weirich, D., R. Hari, H. Xue, P. Behra, and L. Sigg. 2002. Adsorption of Cu, Cd, and Ni on goethite in the presence of natural groundwater ligands. *Environ. Sci. Technol.* 36:328–336.

Cannon, H.L. 1978. Rocks: The geologic sources of most trace elements. *Geochem. Environ.* 3:17–31.

CCME (Canadian Council of Ministers of the Environment). 1991. Appendix IX—A protocol for the derivation of water quality guidelines for the protection of aquatic life (April 1991). In: Canadian Water Quality Guidelines, Canadian Council of Resource and Environment Ministers. 1987. Prepared by the Task Force on Water Quality Guidelines. [Updated and reprinted with minor revisions and editorial changes in Canadian Environmental Quality Guidelines, Chapter 4, Canadian Council of Ministers of the Environment, 1999, Winnipeg].

Cervantes-Guerra, Y., A. Pierra-Conde, Y. Almaguer-Carmenates et al. 2017. Metal accumulation in surface sediment of the urban and industrial coastal area of the municipality of Moa (Cuba); distribution and pollution assessment. *Mineria Geol.* 33:108–123.

Cheung, K.C., B.H.T. Poon, C.Y. Lan et al. 2003. Assessment of metal and nutrient concentrations in river water and sediment collected from the cities in the Pearl River Delta, South China. *Chemosphere* 52:1431–1440.

De Mora, S.J., M.R. Sheikholeslami, E. Wyse, S. Azemard, and R. Cassi. 2004. An assessment of metal contamination in coastal sediments of the Caspian Sea. *Mar. Pollut. Bull.* 48:61–77.

Depetris, P., J-L. Probst, A.I. Pasquini et al. 2003. The geochemical characteristics of the Parana River suspended sediments load: An initial assessment. *Hydrol. Process.* 17:1267–1277.

Diagomanolin, V., M. Farhang, M. Ghazi-Khansari, and N. Jafarzadeh. 2004. Heavy metals (Ni, Cr, Cu) in the Karoon waterway river Iran. *Toxicol. Lett.* 151(1):63–68.

DiToro, D.M., H.E. Allen, H.L. Bergman et al. 2001. Biotic ligand model of the acute toxicity of metals. I. Technical basis. *Environ. Toxicol. Chem.* 20:2383–2396.

DiToro, D.M., J.D. Mahony, D.J. Hansen et al. 1992. Acid volatile sulfide predicts the acute toxicity of cadmium and nickel in sediments. *Environ. Sci. Technol.* 26:96–101.

Drever, J.I. 1997. *The Geochemistry of Natural Waters: Surface and Groundwater Environments.* Upper Saddle River, NJ: Prentice Hall.

Duarte, A.F. and A. Gioda. 2014. Inorganic composition of suspended sediments in the Acre River, Amazon Basin, Brazil. *Latin Am. J. Sediment. Basin Anal.* 21(1):3–15.

Duivenvoorden, L.J., D.T. Roberts, and G.M. Tucker. 2017. Serpentine geology links to water quality and heavy metals in sediments of stream system in central Queensland, Australia. *Environ. Earth Sci.* 76:320–334.

Echols, K.R., W.G. Brumbaugh, C.E. Orazio et al. 2008. Distribution of pesticides, PAHs, PCBs, and bioavailable metals in depositional sediments of the lower Missouri River, USA. *Arch. Environ. Contam. Toxicol.* 55:1610172.

Eisler, R. 1998. Nickel hazards to fish, wildlife, and invertebrates: A synoptic review, Contaminant Hazards Review, Report No. 34, Biological Science Report USGS/BRD/BSR-1998-0001, Patuxent Wildlife Research Center, U.S. Geological Survey.

Ekwere, A., S. Ekwere, and V. Obim. 2013. Heavy metal geochemistry of stream sediments from parts of the Eastern Niger Delta Basin, South-Eastern Nigeria. *RMZ-M&G.* 60:205–210.

Ellgaard, E.G., S.E. Ashley, A.E. Langford, and D.C. Harlin. 1995. Kinetic analysis of the swimming behavior of the goldfish, *Carassius auratus*, exposed to nickel; hypoactivity induced by sublethal concentrations. *Bull. Environ. Contam. Toxicol.* 55:929–936.

Elzinga, E.J. and D.L. Sparks. 1999. Nickel sorption mechanisms in a pyrophyllite–montmorillonite mixture. *J. Colloid. Interface Sci.* 213:506–512.

Elzinga, E.J. and D.L. Sparks. 2001. Reaction condition effects on nickel sorption mechanisms in illite-water suspensions. *Soil Sci. Soc. Am. J.* 65:94–101.

Emelyanov, E., V. Kravtsov, Y. Savin et al. 2010. Influence of chemical weapons and warfare agents on the metal contents in sediments in the Bornholm Basin, the Baltic Sea. *Baltica* 23:77–90.

Ente, P.J. 1981. Remarks on Ijsselmeer deposits and historical fluctuations. *Flevobericht 177.* Lelystad: IJsselmeerpolders Development Authority, 69 pp.

Environment Canada and Ministere de l'Environnement du Quebec (EC & MENVIQ). 1992. *Interim Criteria for Quality Assessment of St. Lawrence River Sediment.* ISBN 0-662-19849-2. Ottawa, ON: Environment Canada.

Fitchko, J. and T.C. Hutchinson. 1975. A comparative study of heavy metal concentrations in river mouth sediments around the Great Lakes. *J. Great Lakes Res.* 1:46–78.

Forstner, U. and G.T.W. Wittmann. 1979. *Metal Pollution in the Aquatic Environment.* New York: Springer-Verlag, 486 pp.

Frohne, T., J. Rinklebe, and R.A. Diaz-Bone. 2014. Contamination of floodplain soils along the Wupper River, Germany, with As, Co, Cu, Ni, Sb, and Zn and the impact of pre-definite redox variations on the mobility of these elements. *Soil Sed. Contam.* 23:779–799.

Frye, J.C. and N.F. Shimp. 1973. Major, minor and trace elements in sediments of Late Pleistocene Lake Saline compared with those in Lake Michigan sediments. *Environ. Geol. Notes* No. 60. Illinois State Geol. Survey.

Garcia-Alonso, J., J. Gómez, F.R. Barboza et al. 2015. Pollution–toxicity relationships in sediments of the Segura River Basin. *Limnetica* 34(1):135–146.

Gibbs, R. 1977. Transport phases of transition metals in the Amazon and Yukon Rivers. *Geol. Soc. Am. Bull.* 88:829–843.

Goldstein, S.J. and S.B. Jacobsen 1988. Nd and Sr isotopic systematics of river water suspended material—Implications for crustal evolution. *Earth. Planet Sci. Lett.* 87:249–265.

Gonzalez, H., M. Ramirez, and I. Torres. 1997. Impact of nickel mining and metallurgical activities on the distribution of heavy metals in sediments of Levisa, Cabonico and Nipe Bays, Cuba. *Environ. Geochem. Health* 19:57–62.

Green-Pedersen, H. and N. Pind. 2000. Preparation, characterization, and sorption properties for Ni(II) of iron oxyhydroxide-montmorillonite. *Colloids Surface A* 168:133–145.

Helios-Rybicka, E. 1986. The role of clay minerals in the fixation of heavy metals in bottom sediments of the Upper Vistula River. *Quarterly Geologia*, AGH, Monograph, 32 [In Polish].

Helios-Rybicka, E., E. Adamiec, and U. Aleksander-Kwaterczak. 2005. Distribution of trace metals in the Odra River system: Water-suspended matter-sediments. *Limnologica* 35:185–198.

https://en.wikipedia.org/wiki/Iron_meteorite.

https://en.wikipedia.org/wiki/Nickel_(United_States_coin).

Hutchinson, G.E., R.J. Benoit, W.B. Cotter, and P.J. Wangersky. 1955. On the nickel, cobalt, and copper contents of deep-sea sediments. *Geology* 41:160–162.

Hutchinson, T.C., A. Fedorenko, J. Fitchko et al. 1975. Movement and compartmentation of nickel and copper in an aquatic ecosystem. In: D.D. Hemphill (Ed.), *Trace Substances in Environmental Health—IX. A Symposium*. Columbia: University of Missouri Press, pp. 89–105.

Ikenka, Y., S.M.M. Nakayama, K. Muzandu et al. 2010. Heavy metal contamination of soil and sediment in Zambia. *African J. Environ. Sci. Technol.* 4(11):729–739.

Institute of Geochemistry Chinese Academy of Sciences. 1998. *Advanced Geochemistry*. Beijing: Science Press, pp. 43–44.

International Seabed Authority (2010). A Geological Model of Polymetallic Nodule Deposits in the Clarion-Clipperton Fracture Zone and Prospector's Guide for Polymetallic Nodule Deposits in the Clarion Clipperton Fracture Zone. Technical Study: No. 6. ISBN 978-976-95268-2-2.

Iordache, M., I.V. Branzoi, L.R. Popescu et al. 2016. Evaluation of heavy metals pollution into a complex industrial area from Romania. *Environ. Eng. Manage. J.* 15(2):389–394.

Japenga, J., K.H. Zschuppe, A.J. De Groot et al. 1990. Heavy metals and organic micropollutants in floodplains of the river Waal, a distributary of the river Rhine, 1958–1981. *Netherlands J. Agric. Sci.* 38:3810397.

Kabata-Pendias, A. and B. Szteke. 2015. *Trace Elements in Abiotic and Biotic Environments*. Boca Raton, FL: CRC Press, 440 pp.

Knox, A.S., M.H. Paller, C.E. Milliken, T.M. Redder, J.R. Wolfe, and J. Seaman. 2016. Environmental impact of ongoing sources of metal contamination on remediated sediments. *Sci. Total Environ.* 563–564:108–117.

Kuriata-Potasznik, A., S. Szymczyk, A. Skwierawski et al. 2016. Heavy metal contamination in the surface layer of bottom sediments in a flow-through lake: A case study of Lake Symsar in Northern Poland. *Water* 8:358–373.
Lacerda, L.D., F.C.F. De Paula, A.R.C. Ovalle et al. 1990. Trace metals in fluvial sediments of the Madeira River watershed, Amazon, Brazil. *Sci. Total Environ.* 97/98:525–530.
Langmuir, D. 1997. *Aqueous Environmental Geochemistry*. Upper Saddle River, NJ: Prentice Hall.
Li, Z.B. and L.M. Shuman. 1996. Redistribution of forms of zinc, cadmium and nickel in soils treated with EDTA. *Sci. Total Environ.* 191:95–107.
Liu, J.Y., L.M. Cao, Z.L. Li et al. 1984. *Elementary Geochemistry*. Beijing: Science Press.
Liu, M., Y. Yang, X. Yun et al. 2014. Distribution and ecological assessment of heavy metals in surface sediments of the East Lake, China. *Ecotoxicology* 23:92–101.
Long, E.R. and L.G. Morgan. 1991. The potential for biological effects of sediment sorbed contaminants in the National Status and Trend Program. NOAA Technical Memorandum NOS OMA 52. Seattle, WA: National Oceanic and Atmospheric Administration, 175 pp. + Appendices.
Lottermoser, B.G. 1998. Heavy metal pollution of coastal river sediments, northeastern North South Wales, Australia: Lead isotope and chemical evidence. *Environ. Geol.* 36(1–2):118–126.
Louropoulou, E., F. Botsou, E. Koutsopoulou et al. 2015. Chromium and nickel distribution in sediments of a coastal area impacted from metallurgical activities: The case of the Larymna Bay. International Conference on the Environmental Perspectives of the Gulf of Elefsis, Elefsis, Greece, 11–12 September 2015.
Lyon, G.L., R.R. Brooks, P.J. Peterson et al. 1970. Some trace elements in plants from serpentine soils. *N.Z. J. Sci.* 13:133.
MacDonald, D.D., C.G. Ingersoll, and T. Berger. 2000. Development and evaluation of consensus-based sediment quality guidelines for freshwater ecosystem. *Arch. Environ. Cont. Toxicol.* 39:20–31.
Mamindy-Pajany, Y., S. Sayen, J.F.W. Mosselmans, and E. Guillon. 2014. Copper, nickel and zinc speciation in a biosolid-amended soil: pH adsorption edge, μ-XRF and μ-XANES investigations. *Environ. Sci. Technol.* 48:7237–7244.
Martin, J.M. and M. Whitfield. 1983. The significance of the river input of chemical elements to the ocean. In: C.S. Wong, E. Boyle, K.W. Bruland, J.D. Burton, and E.D. Goldberg (Eds.), *Trace Metals in Sea Water New York*. Plenum Publishing Corporation, pp. 265–296.
McClain, M.E. (Ed.). 2002. *The Ecohydrology of South American Rivers and Wetlands*. International Association of Hydrological Sciences (IAHS), IAHS Special Publication no.6., ISBN 1-901502-02-3. Wallingford, Oxfordshire, UK: IAHS Press.
McGrath, S.P. 1995. Chromium and nickel. In: B.J. Alloway (Ed.), *Heavy Metals in Soils*. London: Blackie.
McLennan, S.M., S. Hemming, D.K. McDaniel et al. 1993. Geochemical approaches of sedimentation, provenance, and tectonics. *Geol. Soc. Am. Spec. Pap.* 284:21–40.
Menendez, A., R.H. James, S. Roberts et al. 2017. Controls on the distribution of rare earth elements in deep-sea sediments in the North Atlantic Ocean. *Ore Geol. Rev.* 87:100–113.
Mero, J. 1965. The mineral resources of the sea. Elsevier Oceanography Series.

Milacic, R. and J. Scancar. 2010. A complex investigation of the extent of pollution in sediments of the Sava River. Part 1: Selected elements. *Environ. Monit. Assess.* 163:263–275.

Mohiuddin, K.M., M.S. Islam, S. Basak et al. 2016. Status of heavy metals in sediments of the Turag river in Bangladesh. *Progress. Agric.* 27(2):78–85.

Moore, J.N. and E.M. Landrigran. 1999. Mobilization of metal-contaminated sediment by ice-jam floods. *Environ. Geol.* 37:96–101.

Moore, J.W. 1981. Epipelic algal communities in a eutrophic northern lake contaminated with mine wastes. *Water Res.* 15:97–105.

Moore, J.W. 1991. *Inorganic Contaminants of Surface Water—Research and Monitoring Properties.* New York: Springer Verlag.

Moore, J.W. and S. Ramamoorthy. 1984. *Heavy Metals in Natural Waters.* New York: Springer-Verlag, 268 pp.

Moreno, M., G. Albertelli, and M. Fabiano. 2009. Nematode response to metal, PAHs and organic enrichment in tourist marinas of the Mediterranean Sea. *Mar. Pollut. Bull.* 58:1191–1201.

Mortatti, J. and J.-L. Probst. 2010. Characteristics of heavy metals and their evaluation in suspended sediments from Piracicaba River basin (Sao Paulo, Brazil). *Revista Brasileira de Geociencias.* 40(3):375–379.

Mothersill, J.S. and P.C. Fung. 1972. The stratigraphy, mineralogy and trace element concentrations of the Quaternary sediments of the northern Lake Superior basin. *Can. J. Earth Sci.* 9:1735–1755.

Mwanamoki, P.M., N. Devarajan, B. Niane et al. 2015. Trace metal distributions in the sediments from river-reservoir system: Case of the Congo River and Lake Ma Vallee, Kinshasa (Democratic Republic of Congo). *Environ. Sci. Pollut. Res.* 22:586–597.

Nie, M., N. Xian, X. Fu et al. 2010. The interactive effect of petroleum-hydrocarbon spillage and plant rhizosphere on concentrations and distribution of heavy metals in sediments in the Yellow River Delta, China. *J. Hazard. Mater.* 174:156–161.

Nolting, R.F., M. van Dalen, and W. Helder. 1996. Distribution of trace elements in sediment and pore waters of Lena delta and Laptev Sea. *Marine Chem.* 53:285–299.

Nriagu, J.O. 1979. Global inventory of natural and anthropogenic emissions of trace metals to the atmosphere. *Nature* 297:409–411.

Nriagu, J.O. 1980. Global cycle and properties of nickel. In: J.O. Nriagu (Ed.), *Nickel in the Environment.* New York: Wiley, pp. 1–26.

Nriagu, J.O. and J.M. Pacyna 1988. Quantitative assessment of worldwide contamination of air, water, and soils by trace metals. *Nature* 33:134–139.

Pandey, M., S. Tripathi, A.K. Pandey et al. 2014. Risk assessment of metal species in sediments of the river Ganga. *Catena.* 122:140–149.

Perdomo, L., I. Ensminger, L.F. Espinosa et al. 1998. The Mangrove ecosystem of the Cienaga Grande de Santa Marta (Colombia): Observations on regeneration and trace metals in Sediment. *Mar. Pollut. Bull.* 37(8–12):393–403.

Persaud, D., R. Jaagumagi, and A. Hayton. 1993. *Guidelines for the Protection and Management of Aquatic Quality in Ontario.* Toronto, ON: Water Resources Branch, Ontario Ministry of the Environment, 27 pp.

Pickering, Q.H. 1974. Chronic toxicity of nickel to the fathead minnow. *J. Water Pollut. Control Fed.* 46:760–765.

Polemio, M., S.A. Bufo, and N. Senesi. 1982. Minor elements in south-east Italy soils: A survey. *Plant Soil* 69:57–66.

Rahman, A.K.M.L., M. Islam, M.Z. Hossain, M.A. Ahsan. 2012. Study of the seasonal variations in Turag river water quality parameters. *African J. Pure and Appl. Chem.* 6(10):144–148.

Refaey, Y., B. Jansen, J.R. Parsons et al. 2017. Effects of clay minerals, hydroxides, and timing of dissolved organic matter addition on the competitive sorption of copper, nickel, and zinc: A column experiment. *J. Environ. Manage.* 187:273–285.

Robin, E., C. Jehanno, and M. Maurette. 1988. Characteristic and origin of Greenland Fe/Ni cosmic grains. Proceedings of the 18th Lunar and Planetary Science Conference, Lunar and Planetary Institute, Houston, TX, USA.

Salah, E.A.M., T.A. Zaidan, and A.S. Al-Rawi. 2012. Assessment of heavy metals pollution in the sediments of Euphrates River, Iraq. *J. Water Res. Protect.* 4:1009–1023.

Salomons, W. and U. Forstner. 1984. *Metals in Hydrocycle.* Berlin: Springer-Verlag, 349 pp.

Schwartz, R. 2006. Geochemical characterization and erosion stability of fine-grained field sediments of the Middle Elbe River. *Acta Hydrochim. Hydrobiol.* 34:223–233.

Secrierus, D. and A. Secrierus. 2002. Heavy metal enrichment of man-made origin of superficial sediment on the continental shelf of the north-western Black Sea. *Estuarine Coastal Shelf Sci.* 54:513–526.

Sempkin, R.G. 1975. A limnogeochemical study of Sudbury Area Lakes. M.Sc. thesis, McMaster University, Hamilton, Ontario, Canada, 248 pp.

Sempkin, R.G. and J.R. Kramer. 1976. Sediment geochemistry of Sudbury-Area Lakes. *Can. Mineral.* 14:73–90.

Severne, B.C. 1974. Nickel accumulation by *Hybanthus floribundus. Nature (London)* 248:807–808.

Shuhaimi-Othman, M. 2008. Metal concentrations in the sediments of Richard Lake, Sudbury, Canada and sediment toxicity in an amphipod *Hyalella azteca. J. Environ. Sci. Technol.* 1(1):34–41.

Sigel, H. and A. Sigel (Eds.). 1988. *Metal Ions in Biological Systems. Volume 23. Nickel and Its Role in Biology.* New York: Marcel Dekker, 488 pp.

Simonovski, J., C. Owens, and G. Birch. 2003. Heavy metals in sediments of the upper Hawkesbury-Nepean River. *Austral. Geogr. Stud.* 41(2):196–207.

Smith, I.M., K.J. Hall, L.M. Lavkulich et al. 2007. Trace metal concentrations in an intensive agricultural watershed in British Columbia, Canada. *J. Am. Water Res. Assoc.* 43(6):1455–1467.

Smith, S.L., D.D. MacDonald, K.A. Keenlysied et al. 1996. A preliminary evaluation of sediment quality assessment values for freshwater ecosystems. *J. Great Lakes Res.* 22:624–638.

State Information Service. 2014. <http://www.sis.gov.eg/en/>.

Stumm, W. and J.J. Morgan. 1981. *Aquatic Chemistry: An Introduction Emphasizing Chemical Equilibria in Natural Waters.* New York: John Wiley & Sons.

Suraj, G., M. Lalithambika, and C.S.P. Iyer. 1996. Clay minerals of Periyar river sediments and their role in the uptake of cadmium and nickel. *Ind. J. Mar. Sci.* 25:5–11.

Taylor, S.R. and S.M. McLennan. 1985. *The Continental Crust: Its Composition and Evolution.* Oxford: Blackwell, 312 pp.

Thoms, M.C. 2007. The distribution of heavy metals in a highly regulated river: The River Murray, Australia. In: *Water Quality and Sediment Behavior of the Future: Prediction for the 21st Century.* Proceeding of Symposium HS2005 at IUGG2007, Perugia, July 2007, IAHS Press, 314, pp. 145–154.

Tipping, E., J.J. Rothwell, L. Shotbolt, and A.J. Lawlor. 2010. Dynamic modelling of atmospherically-deposited Ni, Cu, Zn, Cd and Pb in Pennine catchments (Northern England). *Environ. Pollut.* 158:1521–1529.

Trudinger, P.A., D.J. Swaine, and G.W. Skyring. 1979. In: P.A. Trudinger and D.J. Swaine (Eds.), *Biogeochemical Cycling of Mineral-Forming Elements.* Amsterdam: Elsevier.

Truex, M., P. Brady, C.J. Newell et al. 2011. The scenarios approach to attenuation-based remedies for inorganic and radionuclide contaminants. Savannah River National Laboratory, U.S. Department of Energy.

Turekian, K.L. and K.H. Wedepohl. 1961. Distribution of elements in some major units of the earth crust. *Geol. Soc. Am. Bull.* 72:175–192.

Uddin, M.K. 2017. A review on the adsorption of heavy metals by clay minerals, with special focus on the past decade. *Chem. Eng. J.* 308:438–462.

USEPA (U.S. Environmental Protection Agency). 1995. QA/QC Guidance for Sampling and Analysis of Sediments, Water, and Tissues for Dredged Material Evaluations—Chemical Evaluations. EPA 823-B-95-001. Office of Water, Washington, DC, and Department of the Army, U.S. Army Corps of Engineers, Washington, DC.

USEPA. 1996. Calculation and evaluation of sediment effect concentrations for the amphipod *Hyalella azteca* and the midge *Chironomus riparius.* EPA 905/R-96/008, Chicago, IL.

USEPA. 2001. Method for assessing the chronic toxicity of marine and estuarine sediment-associated contaminants with the amphipod *Leptocheirus plumulosus.* 2nd edition. Washington, DC: U.S. Environmental Protection Agency; EPA 600/R-01/020.

USEPA. 2010. http://www.epa.gov/bioiweb1/aquatic/marine.html, EPA, February 20, 2010.

U.S. Geological Survey. 2016. Mineral commodity summaries, January 2016.

Van der Veen, A., C. Ahlers, D.W. Zachman, and K. Friese. 2006. Spatial distribution and bonding forms of heavy metals in sediments along the middle course of the River Elbe. *Acta Hydrochim. Hydrobiol.* 34:214–222.

Varol, M. 2011. Assessment of heavy metal contamination in sediments of the Tigris River (Turkey) using pollution indices and multivariate statistical techniques. *J. Hazard. Mater.* 195:355–364.

Viers, J., B. Dupre, and J. Gaillardet. 2009. Chemical composition of suspended sediments in world rivers: New insights from a new database. *Sci. Total Environ.* 407:853–868.

Villalobos-Castaneda, B., R. Cortes-Martinez, N. Segovia et al. 2016. Distribution and enrichment of trace elements and arsenic at the upper layer of sediments from Lerma River in la Piedad, Mexico: Case history. *Environ. Earth Sci.* 75:1490–1502.

Wahaab, R.A. and M.I. Badawy. 2004. Water quality assessment of the Nile river system: An overview. *Biomed. Environ. Sci.* 17:87–100.

Wang, J., R. Liu, H. Wang et al. 2015. Identification and apportionment of hazardous elements in the sediments in the Yangtze River estuary. *Environ. Sci. Pollut. Res.* 22:20215–20225.

Weng, L.P., A. Wolthoorn, T.M. Lexmond, E.J.M. Temminghoff, and W.H. Van Riemsdijk. 2004. Understanding the effects of soil characteristics on phytotoxicity and bioavailability of nickel using speciation models. *Environ. Sci. Technol.* 38:156–162.

Winkels, H.J., S.B. Kroonenberg, M.Y. Lychagin et al. 1998. Geochronology of priority pollutants in sedimentation zones of the Volga and Danube delta in comparison with the Rhine delta. *Applied Geochem.* 13(5):581–591.

Youger, J.D. and W.J. Mitsch. 1989. Heavy metal concentrations in Ohio River sediments—Longitudinal and temporal patterns. *Ohio J. Sci.* 89:172–175.

Zerling, L., C. Hanisch, and F.W. Junge. 2006. Heavy metal inflow into the floodplains at the mouth of the river Weiße Elster (central Germany). *Acta Hydrochim. Hydrobiol.* 34:234–244.

Zuo, W., X. Ma, K. Yang et al. 2016. Distribution and risk assessment of metals in surface water and sediments in the upper reaches of the Yellow River, China. *Soil Sed. Contam.* 25(8):917–940.

8

Nickel in Serpentine Soils

Zeng-Yei Hseu and Zueng-Sang Chen

CONTENTS

8.1 Characteristics of Serpentine Soil

8.1.1 Ultramafic Rocks and Serpentine Minerals

Serpentine soils are derived from ultramafic rocks alone and also from serpentinites (Coleman 1977). Ultramafic rocks are generally dominated with peridotites, dunites, and pyroxenites. Anhydrous ultramafic rocks become more hydrous and Ca content decreases relative to the original rocks in the metamorphism of ultramafic rocks with water near the Earth's surface or in the upper part of the Earth's mantle during subduction events, resulting in the serpentinization to form the serpentine minerals (Alexander et al. 2007):

$$\text{Dunite} + H_2O \rightarrow \text{Serpentine} + \text{Brucite} + \text{Magnetite} \qquad (8.1)$$

Serpentine is the name of a class of minerals of which antigorite, lizardite, and chrysotile are the most recognized members (Wicks and O'Hanley
1988). The ultramafic rocks with abundant serpentine are defined as serpentinite, which is commonly associated with brucite, talc, calcite, chlorite,
magnetite, and chromite (Equation 8.1; O'Hanley 1996). Serpentine-rich rock
has an olive greenish-gray color, usually blotched with stripes of various
shades that resemble the skin of a snake (Figure 8.1), from which the name
serpentine was derived (Brooks 1987). Therefore, the terms "ultramafic soil"
(Garnier et al. 2009; Lee et al. 2003), "serpentinitic soil" (Gasser and Dahlgren
1994; Hseu 2006; Hseu et al. 2007), and "serpentine soil" (Bini et al. 2017;
Cheng et al. 2009, 2011; Hseu et al. 2017; Oze et al. 2004a,b) have been used in
the literature. However, "serpentine soil" is generally called in this chapter.
Nickel is one of the major elements in the Earth's mantle, and thus global
serpentine soils contain high concentrations of geogenic nickel in serpentine
soils derived from the Earth's mantle materials.

Most of ultramafic rocks are associated with assemblages of mafic and
ultramafic rocks that are referred to as ophiolite (O'Hanley 1996). However,
a global majority of ophiolite complex site related to serpentine soils are
normally close to convergent boundaries of tectonic plates (Coleman 1977).
The worldwide distribution of major ophiolitic serpentinite bodies and
their related soils was close to the convergent margins. Serpentinite outcrops have been found in every continent except Antarctica. In this sense,
serpentine soils have been studied across the globe: in Europe (Bani et al.

FIGURE 8.1
Morphological characteristics of serpentine collected from Pogradec, Albania.

2014; Caillaud et al. 2009; Chardot et al. 2007; Kierczak et al. 2008; Massoura et al. 2006), North America (Alexander et al. 2007; Lee et al. 2004; McGahan et al. 2008, 2009; Morrison et al. 2009; Oze et al. 2004a,b), South America (Garnier et al. 2006, 2009), and the South Pacific Ocean particularly on New Caledonia's complex (Becquer et al. 2003, 2006, 2010). However, serpentine soils are mainly present in the areas within the Mediterranean and the Circum-Pacific margin (Oze et al. 2004a).

The release of Ni into ecosystems during serpentine weathering suggests that serpentine landscapes provide a source of non-anthropogenic metal contamination (Chardot et al. 2007; Ünver et al. 2013), although geogenic heavy metals in serpentine soils tend to be fixed by the mineral structures rather than the particle surfaces. Therefore, the potential environmental risk of serpentine soils seems to be involved on the increase of the bioavailability of Ni (Hseu and Iizuka 2013; Hseu and Lai 2017; Kierczak et al. 2008).

8.1.2 Constraints of Serpentine Soil

The chemical weathering of serpentine minerals has been widely studied for their large Cr- and Ni-ore deposits (Hajjar et al. 2017; Zhou et al. 2017), but serpentine soils have attracted worldwide interest for other reasons such as environment quality and human health (Kanellopoulos et al. 2015; McClain et al. 2017; Miranda et al. 2009). Serpentine soils have severe fertility limitations because of low Ca/Mg ratios (McGahan et al. 2008, 2009), low concentrations of P and K (Bonifacio and Barberis 1999), and Cr and Ni enrichments (Becquer et al. 2010; Cheng et al. 2009; Hseu 2006; Hseu et al. 2015a). They further show a unique flora (Oze et al. 2008; van der Ent et al. 2015) and unique physical properties (Alexander et al. 2007).

8.2 Geogenic Enrichment of Trace Elements in Serpentine Soils

8.2.1 Nickel and Other Trace Elements

Past studies of serpentine soils paid more attention to Cr and Ni than to Co, because of the clearer toxicity of Cr and Ni to organisms in comparison with Co. Compared with other rocks, serpentinites are richer in Cr and Ni contents up to 3400 mg/kg of Cr and 3600 mg/kg of Ni; however, the global average contents of Cr and Ni in soils are approximately 84 mg/kg and 34 mg/kg, respectively (McGrath 1995). Adriano (1986) reported a Cr content of 125,000 mg/kg in serpentine soils. Additionally, Co is also enriched in serpentine soils compared to soils from other parent materials (Hsiao et al. 2009; Kierczak et al. 2016). Total content of Co in soil varied widely from 0.05 to 300 mg/kg, with an average content in the range 10–15 mg/kg

TABLE 8.1

Total Content of Ni in Serpentine Soils from Different Countries

Value (mg/kg)	Region	References
72–2341	Lower Silesia, Poland	Kierczak et al. (2016)
4018–5320	Haut Limousin, France	Caillaud et al. (2009)
125–2507	Vosges Mountains, France	Chardot et al. (2007)
512–4060	Coast Range Ophiolite, California	Morrison et al. (2009)
400–5800	Coastal Range, Taiwan	Hseu et al. (2009)
326–520	Hyogo, Japan	Hseu et al. (2015a)
60–3800	Trás-os-Montes, Portugal	Alves et al. (2011)
7000–8500	New Caledonia	Becquer et al. (2006)
80–1887	Susa Valley, Italy	Bonifacio et al. (2010)
1582–2057	Tuscany, Italy	Bini et al. (2017)
3156–73786	Niquelândia, Brazil	Garnier et al. (2009)
44–435	Atalanti, Greece	Kanellopoulos et al. (2015)
5.91–940	Galicia, Spain	Miranda et al. (2009)
1590–3960	Franciscan Complex, California	Oze et al. (2004b)
25.7–2680	Turkey	Ünver et al. (2013)
1140–2000	Malakand, Pakistan	Kfayatullah et al. (2001)
515–2520	Czech	Quantin et al. (2008)
680–3700	Marivan and Dizaj, Iran	Ghaderian et al. (2007)
130–470	Assopos Basin, Greece	Economou-Eliopoulos et al. (2011)

(Kabata-Pendias and Pendias 2001). Total content of Ni in serpentine soils from different countries varies widely from values lower than the global average to more than 70,000 mg/kg (Table 8.1). For example, Cheng et al. (2009) demonstrated that total Ni contents of serpentine soils developed in place from weathering of the underlying serpentinites with values in the range 400–5800 mg/kg. These levels of Ni were much greater than those in the soils derived from other parent materials in eastern Taiwan, and were within the expected range, but toward the higher end for global serpentine soils. However, Ni content is lower in alluvial serpentine soils than in soils developed in place from serpentines, such as in the Assopos basin, Greece (Economou-Eliopoulos et al. 2011), and Galicia, Spain (Miranda et al. 2009).

8.2.2 Distribution of Ni in Serpentine Terrain

A study of well-developed serpentine soils in Tehama County, California, by Gough et al. (1989) found that B horizons had the highest Ni content. However, Ni content did not show clear depth functions in profiles of soils sampled from the foothills of the Santa Cruz Mountains, California (Oze et al. 2004b). The distribution profile of Ni widely varies in serpentine soils (Cheng et al. 2009). Hseu et al. (2015a) found an irregular depth trend in total

Ni of paddy soil profiles derived from serpentines, but Ni increases with increasing soil depth in certain serpentine soils in forests (Chang et al. 2013). The difference in the profile distribution of Ni between forest and paddy soils is a result of tillage, which is used to generate plow layers for cultivating paddy fields to plant rice.

To elucidate the distribution of geogenic Ni in serpentine soils, three pedons on the shoulder (Entisol), backslope (Inceptisol), and footslope (Alfisol) along a toposequence in eastern Taiwan were examined (Cheng et al. 2011). In this toposequence, the Ni content, ranging from 1314 to 4048 mg/kg in all soil horizons, varied considerably; however, the average Ni of pedon increased from the shoulder to the footslope. Cheng et al. (2011) indicated that erosion processes and lateral flow occurred in these upland soils and the Ni level reflected pedogenesis across the toposequence. The loss of base cations due to leaching and the increase of clay and secondary Fe oxides were relatively more prominent on the footslope than on the shoulder and backslope. These pedogenic progressions increased the total labile Ni in the soils. However, this release across the toposequence occurred in the following order: footslope > backslope > shoulder, which confirmed that the pedogenic processes substantially affected the leaching and accumulation of Ni from the Entisol at shoulder and Inceptisol at backslope to the Alfisol at footslope in the toposequence.

8.3 Mineral Origin of Nickel in Serpentine Soils

8.3.1 Nickel-Bearing Minerals

The understanding of serpentinite weathering and associated environmental consequences has significantly contributed to better knowledge in the release of Ni into ecosystems (Hseu and Lai 2017; Hseu et al. 2015a,b, 2017). Because Ni is geogenic in serpentine soils and not anthropogenic, apart from being adsorbed to colloidal surfaces, Ni is commonly incorporated into the crystal lattices of clay, silt, and sand-sized particles of the parent material and secondary minerals (Alexander et al. 2007). Chrysotile, antigorite, and lizartide are major polymorphs of serpentine, which have similar $[Mg_3Si_2O_5(OH)_4]$ composition (Caillaud et al. 2006). Minerals common in serpentine soils, serpentinized peridotites, and ultramafic lateritic deposits that associate with Ni include those minerals such as lizardite, antigorite, smectite, vermiculite, goethite, and hematite (Cheng et al. 2011; Massoura et al. 2006; Siebecker et al. 2017). Ni is commonly substituted for Mg^{2+} into the silicate structures of minerals in serpentine soils (Becquer et al. 2006), and as chemical weathering progresses, these minerals destabilize and release Ni. The major mineral origin of Ni in serpentine soils is derived from Ni-bearing silicate lattices

TABLE 8.2

Mean and Range of Ni Content (g/kg) in Phyllosilicates and Oxides of Serpentine Soil

Mineral	Mean	Range	References
Serpentine	–	1.1–2.1	Kierczak et al. (2016)
Serpentine	3.6	–	Caillaud et al. (2009)
Smectite	–	4.8–8.6	Morrison et al. (2015)
Smectite	0.1	–	Raous et al. (2013)
Vermiculite	–	4.1–6.5	Morrison et al. (2015)
Olivine	–	2.4–3.4	Kierczak et al. (2016)
Chlorite	–	1.5–2.2	Kierczak et al. (2016)
Chlorite	1.4	–	Caillaud et al. (2009)
Talc	0.015	–	Raous et al. (2013)
Ferrihydrite	–	5.2–28	Morrison et al. (2015)
Magnetite	–	2.5–5.7	Kierczak et al. (2016)
Magnetite	1.6	–	Caillaud et al. (2009)
Goethite	0.05	–	Raous et al. (2013)
Hematite	0.04	–	Raous et al. (2013)
Cr-spinel	–	<0.02	Hseu and Iizuka (2013)
Cr-spinel	0.002	–	Raous et al. (2013)

such as serpentine, chlorite, smectite, and vermiculite (Table 8.2). These silicate minerals are less resistant to weathering, compared to Cr-bearing spinels (Hseu and Iizuka 2013). This is the likely reason that bioavailability of Cr was much lower than that of Ni in serpentine soils abovementioned.

8.3.2 Mineral Transformation

Iron from olivine and pyroxene usually goes into the formation of magnetite by metamorphism disseminated through serpentine (Hseu et al. 2007). The magnetite was gradually replaced by pedogenic Fe oxides such as goethite, which paved the way to free Fe accumulation during the genetic processes of serpentine soils. Serpentine has initially weathered to smectite and interstratified chlorite-vermiculite by the formation of interlayer OH sheet in the incipient state of serpentine soils (Hseu et al. 2007; Lee et al. 2003). Under the highly weathering conditions of serpentine soils, thick argillic horizons with high pedogenic Fe oxides can be found (Figure 8.2; Chang et al. 2013). The susceptibility of serpentine minerals to weathering has been linked to enhanced availability of Ni in serpentine soils (Cheng et al. 2011; Hseu et al. 2015b; Morrison et al. 2009). Nickel released by weathering of primary minerals can substitute for magnesium in clay minerals such as smectites and vermiculite during the early stages of serpentine soil development (Hseu et al. 2007). Hseu et al. (2015b) investigated pedogenic clay mineral transformations in a paddy field formed on serpentinites in eastern

FIGURE 8.2
Morphological characteristics of initially (left) and highly weathered (right) serpentine soils. The left profile is Regosol and from Austria; the right profile is Acrisol and from Taiwan.

Taiwan and link the mineralogical characteristics to the bioavailability of Ni in the soil. They found that a marked enrichment of available Ni in the surface soils relative to the subsoils and this depth trend of available content reflected the release of Ni upon weathering of serpentinites followed by their sorption on the surfaces of transformed clay minerals.

Calcium and Mg are preferentially leached during serpentine soil weathering progresses, whereas Ni remains in the profiles and accumulates along with iron and manganese (Alves et al. 2011). However, Ni is slightly retained on clay and Fe/Mn oxide surfaces in comparison with other transition elements, even though considerable Ni can be adsorbed on Fe and Mn oxide surfaces at pH ≥ 5 (Lee et al. 2004). Silicon-rich goethite was the major Fe oxide observed in a study of serpentine Ferralsols in New Caledonia. The goethite scavenged large proportions of Ni. The adsorbed silica retarded the crystal growth of goethite and favored the incorporation of Ni. The presence of needle-like goethite in tropical serpentine Ferralsols as a Ni sink is reasonable (Becquer et al. 2006).

8.4 Bioavailability and Mobility of Nickel in Serpentine Soils

8.4.1 Single Extraction

Bioavailability of Ni in soils is critically dependent on its chemical speciation, and thus plants respond only to the fraction that is bioavailable to them.

The readily soluble fraction of heavy metals is generally considered to be bioavailable, but there is growing awareness that the current methods for assessment of soluble and bioavailable fractions need re-evaluation because these fractions depend on plant species and soil types (Menzies et al. 2007). Alves et al. (2011) studied Ni availability in upland soils in Portugal that were characterized by serpentinites, demonstrating no significant relationship between available Ni extracted using ethylenediamine tetraacetic acid (EDTA) and amorphous Fe oxides. However, Alves et al. (2011) indicated a strong association between Ni and Mn oxides in the studied Portuguese serpentine soils. The most widely used methods for evaluating the bioavailability of heavy metals in soils are single extraction and sequential extraction methods (Meers et al. 2007). Additionally, several studies on the mobility of Ni are available for serpentine soils using single or sequential extractions, for example, 0.1 N HCl in Hseu et al. (2015a) and the BCR protocol of sequential extraction in Hseu et al. (2017).

Ho et al. (2013) determined the validation of using the diethylenetriamine pentaacetate (DTPA)-extractable concentration as a bioavailability index of Ni that can be used to estimate Ni uptake by plants growing in serpentine soils. Hseu et al. (2015a) found that the concentrations of Ni extracted using 0.005 M DTPA and 0.1 N HCl increased toward the soil surface in serpentine soil profiles, illustrating that Ni availability is higher in more weathered surface soils than in less weathered subsoils. However, pedogenic Fe and Mn oxides are regarded as crucial Ni sinks in serpentine soils (Becquer et al. 2006; Cheng et al. 2011; Hseu 2006), and thus the roles of Fe and Mn oxides on the Ni retention have been assessed by conducting selective extractions by Hseu et al. (2015a). Hseu et al. (2015a) found that the levels of free Fe-bound Ni (Ni_d) and amorphous Fe-bound Ni (Ni_o) were higher than Mn-bound Ni (Ni_{hh}) levels in paddy soils (Table 8.3), reflecting the fact that Ni was predominantly retained by nearly amorphous Fe oxides (e.g., ferrihydrite) and goethite. This Fe-retaining Ni was significantly correlated with the DTPA-extractable Ni in the soils. Therefore, the authors elucidated the importance of Fe oxides in the fate of potentially labile Ni, and repeated redox and leaching cycles caused the redistribution of Ni in the paddy soils. A high concentration of Ni-laden and redox-sensitive Fe oxides can affect wetland soils and the environment when Ni is released into the soil solution and becomes bioavailable (Antić-Mladenović et al. 2011; Rinklebe et al. 2016). Additionally, Bani et al. (2014) found that Ni availability in serpentine soils is the highest in surface horizons as a consequence of biogeochemical recycling through litter fall in an ultramafic toposequence of Albania because the rich organic matter and high cation exchange capacity (CEC) in the surface soils may contribute to host a significant fraction of available Ni after the decay of Ni-rich litter.

The availability of Ni in soils is determined by soil properties, notably mineral composition, organic matter, CEC, pH, Fe and Mn oxides, and redox potentials (Alves et al. 2011; Bani et al. 2014; Chardot et al. 2007; Hseu 2006; Hseu et al. 2010, 2015a,b), and thus Ni contents in crops are variable.

TABLE 8.3

Selective Extractions of Fe, Mn, and Ni (mg/kg) in a Pedon of Serpentine Soil by DTPA (Ni_{DTPA}), Hydrogen Chloride (Ni_h), Dithionite–Citrate–Bicarbonate (Fe_d, Mn_d, and Ni_d), Ammonium Oxalate (Fe_o, Mn_o, and Ni_o), and Hydroxylamine (Mn_{hh} and Ni_{hh}) Dissolutions

Horizon	Ni_{DTPA}	Fe_d	Mn_d	Ni_d	Fe_o	Mn_o	Ni_o	Mn_{hh}	Ni_{hh}
Ap1	8.99	7721	287	18.8	7208	271	14.7	153	5.66
Ap2	1.30	3302	696	17.8	3261	573	14.5	167	6.78
BA	1.03	2018	574	10.6	1807	452	9.61	178	7.04
Bt1	0.52	1402	316	6.30	1356	319	5.17	136	4.44
Bt2	0.41	1508	268	4.20	1147	204	3.47	109	2.88
Bt3	0.28	1682	383	4.90	1288	301	3.89	102	2.80
Bw1	0.26	1856	337	3.60	1149	209	3.27	89.7	2.58
Bw2	0.22	1498	351	3.40	1183	307	3.04	88.9	2.50
C	0.21	1317	332	4.00	1068	288	3.75	105	3.28

Source: Hseu, Z. Y., T. Watanabe, A. Nakao, and S. Funakawa. 2015a. Partition of geogenic nickel in paddy soils derived from serpentinites. *Water and Paddy Environment* 14:417–426.

The clay content was positively and significantly correlated with 0.1 N HCl- and DTPA-extractable Ni levels (Table 8.4), indicating that clay minerals favor the accumulation of labile Ni in soil because they are strong adsorbents of Ni. However, 0.1 N HCl- and DTPA-extractable Ni levels significantly increased with decreasing soil pH. Additionally, the DTPA-extractable Ni exhibited a positive and significant correlation with OC content in the study soils. After assessment through a sequential extraction procedure (Cheng et al. 2011), the OC-complexed Ni was determined to be approximately <12% of the weathered Ni in the serpentine soils of eastern Taiwan, indicating the weak presence of labile Ni from this form in the soils. Free Fe content was positively and significantly correlated only with 0.1 N HCl-extractable Ni level, supporting the assertion that pedogenic Fe oxides are the main contribution sinks of total labile Ni resulting from weathered Ni-bearing minerals and also that the Fe oxides became soluble under acidic conditions with

TABLE 8.4

Linear Correlation Matrix between Soil Properties and Ni in Soil in Paddy Fields from Serpentines of Eastern Taiwan ($n = 40$)

	Clay	pH	Organic Carbon	Free Fe
Total Ni	NS	NS	NS	NS
0.1 N HCl-extractable Ni	0.78[b]	−0.38[a]	NS	0.58[b]
DTPA-extractable Ni	0.65[b]	−0.42[b]	0.32[a]	NS

Source: Hseu, Z. Y. and Y. J. Lai. 2017. Nickel accumulation in paddy rice on serpentine soils containing high geogenic nickel contents in Taiwan. *Environmental Geochemistry and Health* 186:151–157.

[a] Significant at $p < 0.05$.
[b] Significant at $p < 0.01$; NS, not significant.

strong acid extraction. The DTPA-extractable Ni is reported to account for 3% of total Ni in serpentine soils of paddy fields (Hseu et al. 2015a). Kabata-Pendias and Pendias (2001) recommended the safe limit of Ni uptake content for paddy rice to be within 10–30 mg/kg. However, Susaya et al. (2010) documented that EDTA-extractable Ni levels ranged from 5.60 to 13.4 mg/kg in serpentine soils in the Philippines and that Ni content ranged from 4.52 to 11.1 mg/kg in the edible parts of eggplants and tomatoes grown in these soils. Hseu and Lai (2017) investigated the extractability of Ni in serpentine soils collected from rice paddy fields in eastern Taiwan and found that available Ni content only accounted for <10% of total soil Ni content and 0.1 N HCl-extractable Ni was the more suitable index for Ni bioavailability in the soil to rice than DTPA-extractable Ni; however, their brown and polished rice samples contained Ni contents in the range of 1.50–4.53 and 2.45–5.54 mg/kg, respectively.

8.4.2 Sequential Extraction

Selective sequential extraction approaches are typically used to partition metals in serpentine soils into different fractions, that is, acid soluble (F1), Fe/Mn oxide bound (F2), organically bound (F3), and residual (F4) (Cheng et al. 2011; Rinklebe and Shaheen 2017; Rinklebe et al. 2016). Fractionation of Ni strongly depends on the mineralogy of the bedrock (Garnier et al. 2009; Raous et al. 2013), as well as on the climatic conditions (Kierczak et al. 2007), the landscape position in the toposequence (Bani et al. 2014; Cheng et al. 2011; Hseu 2006), and more precision on thermodynamic conditions of soils (Rinklebe et al. 2016). In the three pedons on the shoulder, backslope, and footslope of a toposequence in eastern Taiwan reported by Cheng et al. (2011), the lowest Ni contents were noted in acid-extractable form (F1), and the concentration in the residual fraction (F4) was much higher than those in other fractions (Table 8.5). Ni primarily recovered in the F4 fraction, which reflected the major source of Ni-bearing silicate lattices such as serpentines, smectites, and vermiculites abovementioned (Becquer et al. 2006; Hseu et al. 2007, 2015b). Additionally, the depth trends of all Ni fractions were variable in these pedons. The potential mobile Ni phases (F1, F2, and F3) increased substantially from pedon at the shoulder to pedon at the footslope. The pedogenic progressions along the toposequence increased the total labile fractions (F1, F2, and F3) of Ni in the serpentine soils (Table 8.5), which correlated with the pattern of erosion at the shoulder and accumulation at the footslope in the toposequence and resulted in the decrease of soil pH to increase the lability of Ni. This lability of the metals across the toposequence occurred in the following order: footslope (Alfisol) > backslope (Inceptisol) > shoulder (Entisol) (Cheng et al. 2011).

TABLE 8.5

Chemical Fractions (mg/kg) of Nickel in the Serpentine Soils at the Shoulder, Backslope, and Footslope along a Toposequence in Taiwan

Horizon	Depth (cm)	F1	F2	F3	F4
Shoulder					
A	0–10	4.55	10.3	46.1	1314
C1	10–40	4.75	33.8	53.4	1311
C2	40–70	4.54	42.2	34.4	1295
C3	70–100	6.41	20.4	37.3	1208
C4	100–120	6.34	22.0	33.3	1173
Backslope					
A	0–10	14.3	215	138	3289
Bw	10–30	13.0	209	166	3212
C1	30–55	7.82	226	140	2812
C2	55–90	9.72	206	149	1658
C3	90–120	9.70	238	159	1367
Footslope					
A	0–15	32.9	327	275	2786
BA	15–50	35.0	435	220	1930
Bt	50–80	31.8	395	283	2494
BC1	80–110	35.2	384	297	2991
BC2	110–130	40.7	323	328	3333
C	130–150	49.9	220	314	3190

Source: Cheng, C. H., S. H. Jien, Y. Iizuka, H. Tsai, Y. S. Chang, and Z. Y. Hseu. 2011. Pedogenic chromium and nickel partitioning in serpentine soils along a toposequence. *Soil Science Society of American Journal* 75:659–668.

Note: F1: acid extractable, F2: reducible, F3: oxidizable, and F4: residual.

8.4.3 Leaching Potential

The dissolution of Ni from serpentine soils through chemical weathering is regulated by factors including solution pH, ionic strength, and the type and concentration of inorganic and organic acids. In the dissolution kinetics of minerals, two of the most important chemical weathering reactions in soil environments are proton-promoted (pH effect) and ligand-promoted dissolution reactions (Hamer et al. 2003). However, various processes of pedogenesis lead to input/formation of protons and ligands from acid precipitation and low-molecular-weight organic acids from humus and root exudates (Ramos et al. 2014). Using inorganic and organic acids to study weathering reactions under laboratory conditions, Rajapaksha et al. (2012) demonstrated that Ni release from serpentine soil was accelerated by complex-forming ligands and that both protons and ligands corroborated to accelerate the release of Ni from the soil.

An improved understanding of the factors and processes that control Ni leaching from serpentine soils is needed, because of the environmental risk assessment and management strategies to reduce Ni mobilization, bioavailability, and transfer to the surrounding environment. Therefore, the dissolution kinetics with three inorganic acids (HCl, HNO_3, and H_2SO_4) and three organic acids (citric, acetic, and oxalic acids) in concentrations ranging from 0.05 to 10 mM have been performed for two contrasting serpentine sites in Taiwan (TW) and Austria (AT) to evaluate the leaching potential of Ni in serpentine soils with respect to pH and acid types by Hseu et al. (2017). They found that the release rate of Ni from the soils increased with decreasing pH in all acids. However, the organic acids caused stronger pH dependences than the inorganic acids, because of ligand-promoted dissolution (Table 8.6). The maximum total rate of Ni dissolution occurred with citric acid in both soils. However, the total rate of dissolution (R_T) of AT soil was consistently higher than that of TW soil in all acids. The ligand-promoted rates at pH 3.5, $R_{L(pH3.5)}$, were obtained by subtracting R_T of hydrochloric acid from the total dissolution rates of the other acids. The so-called enhancement factor R_T/R_H was used to compare the proton- and ligand-promoted Ni dissolution rates in different acids in Table 8.6. Compared to hydrochloric acid, nitric acid and sulfuric acid showed lower values of R_T/R_H and met their low effects on anionic ligands compared to HCl. For organic acids, R_T/R_H decreased in the order citric acid > oxalic acid > acetic acid due to the decrease of carboxylic functional group number. Thus, acetic acid with monocarboxylic anions only had a slight effect on the release of Ni while bi- and tri-carboxylic anions in oxalic and citric acids significantly increased Ni release, respectively

TABLE 8.6

Rate Constant (k_T), Total Rate of Dissolution (R_T), Ligand-Promoted Dissolution (R_L), and Hydrogen Dissolution (R_H) of Ni in the Presence of Inorganic and Organic Acids in the TW and AT Serpentine Soils

Acid	Log k_T		$R_{T(pH\,3.5)}$		$R_{L(pH\,3.5)}$		R_T/R_H	
	TW	AT	TW	AT	TW	AT	TW	AT
				mmol m^{-2} d^{-1} $\times$ 10^{-4}				
Hydrochloric acid	−2.02	−0.84	8.55	8.91	0.00	0.00	1.00	1.00
Nitric acid	−1.42	−0.23	7.95	8.19	−1.60	−0.72	0.83	0.92
Sulfuric acid	−2.08	−1.42	6.76	7.32	−2.79	−1.59	0.71	0.82
Acetic acid	−3.27	−1.42	14.8	15.5	6.25	6.59	1.73	1.78
Oxalic acid	−6.60	−1.80	18.2	24.0	9.65	15.1	2.13	2.70
Citric acid	−8.35	−3.42	34.7	89.1	26.2	80.2	4.08	10.0

Source: Hseu, Z. Y. and Y. J. Lai. 2017. Nickel accumulation in paddy rice on serpentine soils containing high geogenic nickel contents in Taiwan. *Environmental Geochemistry and Health* 186:151–157.

(Hseu et al. 2017). Therefore, Hseu et al. (2017) contribute evidence that Ni release from serpentine soils is dependent on the pH and ionic strength of these acids and may be strongly accelerated by complex-forming ligands; both ligands and protons promote the release of Ni into the water bodies of the forest ecosystem.

8.5 Conclusions

Because of Ni enrichment, serpentine soils have attracted worldwide studies on environment quality and human health. The global average concentration of Ni in soils is below 50 mg/kg. However, Ni in serpentine soils from different countries varies widely from values lower than the global average to more than 70,000 mg/kg, depending on the parent material source. Nickel is commonly substituted for Mg^{2+} in the silicate structures of minerals in serpentine soils, and as chemical weathering progresses, these minerals destabilize and release Ni. Thus, the pedogenic progressions in clay formation and Fe oxide accumulation increased the total labile Ni in the soils. This Fe-retaining Ni was significantly correlated with the DTPA-extractable Ni in paddy soils from serpentines. A high concentration of Ni-laden and redox-sensitive Fe oxides can affect wetland soils and the environment when Ni is released into the soil solution and becomes bioavailable at serpentine sites. Sequential fractionation of Ni in serpentine soil strongly depends on the mineralogy of the bedrock, as well as on the climatic conditions, the landscape position in the toposequence, and more precision on the thermodynamic conditions of soils. However, Ni release from serpentine soils might be accelerated by complex-forming ligands and the fact that both protons and ligands corroborated to accelerate the release of Ni from the soil. This chapter suggests a better understanding of source, bioavailability, and mobility of Ni in serpentine landscapes for the evaluation of risk to ecosystems.

Acknowledgment

The authors would like to thank the Ministry of Science and Technology of the Republic of China, Taiwan, for financially supporting the research of serpentine soils under Grant Nos.: MOST 105-2313-B-020-009-MY3, NSC 102-2313-B-020-009-MY3, NSC 99-2313-B-020-010-MY3, and NSC 96-2313-B-020-010-MY3.

References

Adriano, D. C. 1986. *Trace Elements in the Terrestrial Environment*. New York: Springer Verlag.

Alexander, E. B., R. G. Coleman, T. Keeler-Wolf, and S. Harrison. 2007. *Serpentine Geoecology of Western North America*. New York: Oxford University Press.

Alves, S., M. A. Trancoso, M. L. S. Goncalves, and M. M. C. Santos. 2011. A nickel availability study in serpentinised areas of Portugal. *Geoderma* 164:115–163.

Antić-Mladenović, S., J. Rinklebe, T. Frohne, H. J. Stärk, R. Wennrich. Z. Tomić, and V. Ličina. 2011. Impact of controlled redox conditions on nickel in a serpentine soil. *Journal of Soils and Sediments* 11:406–415.

Bani, A., G. Echevarria, E. Montarges-Pelletier, F. Gjoka, S. Sulce, and J. L. Morel. 2014. Pedogenesis and nickel biogeochemistry in a typical albanian ultramafic toposequence. *Environmental Monitoring and Assessment* 186:4431–4442.

Becquer, T., C. Quantin, and J. P. Boudot. 2010. Toxic levels of metals in Ferralsols under natural vegetation and crops in New Caledonia. *European Journal of Soil Science* 61:994–1002.

Becquer, T., C. Quantin, S. Rotté-Capet, J. Ghanbaja, C. Mustin, and A.J. Herbillon. 2006. Sources of trace metals in Ferralsols in New Caledonia. *European Journal of Soil Science* 57:200–213.

Becquer, T., C. Quantin, M. Sicot, and J. P. Boudot. 2003. Chromium availability in ultramafic soils from New Caledonia. *Science of the Total Environment* 301:251–261.

Bonifacio, E. and E. Barberis. 1999. Phosphorus dynamics during pedogenesis on serpentinite. *Soil Science* 164:960–967.

Bonifacio, E., G. Falsone, and S. Piazza. 2010. Linking Ni and Cr concentrations to soil mineralogy: Does it help to assess metal contamination when the natural background is high? *Journal of Soils and Sediments* 10:1475–1486.

Bini, C., L. Maleci, and M. Wahsha. 2017. Potentially toxic elements in serpentine soils and plants from Tuscany (Central Italy). A proxy for soil remediation. *Catena* 148:60–66.

Brooks, R. R. 1987. *Serpentine and Its Vegetation: A Multidisciplinary Approach*. London: Croom Helm.

Caillaud, J., D. Proust, and D. Righi. 2006. Weathering sequences of rock-forming minerals in a serpentinite: Influence of microsystems on clay mineralogy. *Clays and Clay Minerals* 54:87–100.

Caillaud, J., D. Proust, S. Philippe, C. Fontaine, and M. Fialin. 2009. Trace metals distribution from a serpentinite weathering at the scales of the weathering profile and its related weathering microsystems and clay minerals. *Geoderma* 149:199–208.

Chang, Y. T., Z. Y. Hseu, Y. Iizuka, and C. D. Yu. 2013. Morphology, geochemistry, and mineralogy of serpentine soils under the tropical forest in southeastern Taiwan. *Taiwan Journal of Forest Science* 28:185–201.

Chardot, V., G. Echevarria, M. Gury, S. Massoura, and J. L. Morel. 2007. Nickel bioavailability in an ultramafic toposequence in the Vosges Mountains (France). *Plant and Soil* 293:7–21.

Cheng, C. H., S. H. Jien, H. Tsai, Y. H. Chang, Y. C. Chen, and Z. Y. Hseu. 2009. Geochemical element differentiation in serpentine soils from the ophiolite complexes, eastern Taiwan. *Soil Science* 174:283–291.

Cheng, C. H., S. H. Jien, Y. Iizuka, H. Tsai, Y. S. Chang, and Z. Y. Hseu. 2011. Pedogenic chromium and nickel partitioning in serpentine soils along a toposequence. *Soil Science Society of American Journal* 75:659–668.

Coleman, G. G. 1977. *Ophiolites*. New York: Springer-Verlag.

Economou-Eliopoulos, M., I. Megremi, and C. Vasilatos. 2011. Factors controlling the heterogeneous distribution of Cr(VI) in soil, plants and groundwater: Evidence from the Assopos basin, Greece. *Chemie der Erde* 71:39–52.

Garnier, J., C. Quantin, E. Guimarães, V. K. Garg, E. S. Martins, and T. Becquer. 2009. Understanding the genesis of ultramafic soils and catena dynamics. *Geoderma* 151:204–214.

Garnier, J., C. Quaitin, E. S. Martins, and T. Becquer. 2006. Solid speciation and availability of chromium in ultramafic soils from Niquelandia, Brazil. *Journal of Geochemical Exploration* 88:206–209.

Gasser, U. G. and R. A. Dahlgren. 1994. Solid-phase speciation and surface association of metals in serpentinitic soils. *Soil Science* 158:409–420.

Ghaderian, S. M., A. Mohtadi, R. Rahiminejad, R. D. Reeves, and A. J. M. Baker. 2007. Hyperaccumulation of nickel by two *Alyssum* species from the serpentine soils of Iran. *Plant and Soil* 293:91–97.

Gough, L. P., G. R. Meadows, L. L. Jackson, and S. Dudka. 1989. Biogeochemistry of a highly serpentinized, chromite rich ultramafic area, Tehama County, California. USGS Bulletin No. 1901. Washington, DC: U.S. Dept. of the Interior.

Hajjar, Z., F. Gervilla, A. Essaifi, and A. Wafik. 2017. Mineralogical and geochemical features of the alteration processes of magmatic ores in the Beni Bousera Ultramafic massif (north Morocco). *Journal of African Earth Sciences* 132:47–63.

Hamer, M., R. C. Graham, C. Amrhein, and K. N. Bozhilov. 2003. Dissolution of ripidolite in organic and inorganic acid solutions. *Soil Science Society of American Journal* 67:654–661.

Ho, C. P., Z. Y. Hseu, N. C. Chen, and C. C. Tsai. 2013. Evaluating heavy metal concentration of plants on a serpentine site for phytoremediation applications. *Environmental Earth Science* 70:191–199.

Hseu, Z. Y. 2006. Concentration and distribution of chromium and nickel fractions along a serpentinitic toposequence. *Soil Science* 171:341–353.

Hseu, Z. Y. and Y. Iizuka. 2013. Pedogeochemical characteristics of chromite in a paddy soil derived from serpentinites. *Geoderma* 202–203:126–133.

Hseu, Z. Y. and Y. J. Lai. 2017. Nickel accumulation in paddy rice on serpentine soils containing high geogenic nickel contents in Taiwan. *Environmental Geochemistry and Health* 186:151–157.

Hseu, Z. Y., S. W. Su, H. Y. Lai, H. Y. Guo, T. C. Chen, and Z. S. Chen. 2010. Remediation techniques and heavy metals uptake by different rice varieties in metals-contaminated soils of Taiwan: New aspects for food safety regulation and sustainable agriculture. *Soil Science and Plant Nutrition* 56:31–52.

Hseu, Z. Y., Y. C. Su, F. Zehetner, and H. C. Hsi. 2017. Leaching potential of geogenic nickel in serpentine soils from Taiwan and Austria. *Journal of Environmental Management* 186:151–157.

Hseu, Z. Y., H. Tsai, H. C. Hsi, and Y. C. Chen. 2007. Weathering sequences of clay minerals in soils along a serpentinitic toposequence. *Clays and Clay Minerals* 55:389–401.

Hseu, Z. Y., T. Watanabe, A. Nakao, and S. Funakawa. 2015a. Partition of geogenic nickel in paddy soils derived from serpentinites. *Water and Paddy Environment* 14:417–426.

Hseu, Z. Y., F. Zehetner, F. Otnner, and Y. Iizuka, Y. 2015b. Clay mineral transformations and heavy metal release in paddy soils formed on serpentinites in Eastern Taiwan. *Clays and Clay Minerals* 63:119–131.

Hsiao, K. H., K. H. Bao, S. H. Wang, and Z. Y. Hseu. 2009. Extractable concentrations of cobalt from serpentine soils with several single extraction procedures. *Communications in Soil Science and Plant Analysis* 40:2200–2224.

Kabata-Pendias, A. and H. Pendias. 2001. *Trace Elements in Soils and Plants*. 3rd ed. Boca Raton: CRC Press.

Kanellopoulos, C., A. Argyraki, and P. Mitropoulos. 2015. Geochemistry of serpentine agricultural soils and associated groundwater chemistry and vegetation in the area of Atalanti, Greece. *Journal of Geochemical Exploration* 158:22–33.

Kfayatullah, Q., M. Shah Tahir, and M. Arfan. 2001. Biogeochemical and environmental study of the chromite-rich ultramafic terrain of Malakand area, Pakistan. *Environmental Geology* 40:1482–1486.

Kierczak, J, C. Neel, U. Aleksander-Kwaterczak, E. Helios-Rybicka, H. Bril, and J. Puziewicz. 2008. Solid speciation and mobility of potentially toxic elements from natural and contaminated soils: A combined approach. *Chemosphere* 73:776–784.

Kierczak, J., C. Neel, H. Bril, and J. Puziewicz. 2007. Effect of mineralogy and pedoclimatic variations on Ni and Cr distribution in serpentine soils under temperate climate. *Geoderma* 142:165–177.

Kierczak, J., A. Pedziwiatr, J. Waroslaw, and M. Modelska. 2016. Mobility of Ni, Cr and Co in serpentine soils derived on various ultrabasic bedrocks under temperate climate. *Geoderma* 268:78–91.

Lee, B. D., R. C. Graham, T. E. Laurent, and C. Amrhein. 2004. Pedogenesis in a wetland meadow and surrounding serpentinic landslide terrain, northern California, USA. *Geoderma* 118:303–320.

Lee, B. D., S. K. Sears, R. C. Graham, C. Amrhein, and H. Vali. 2003. Secondary mineral genesis from chlorite and serpentine in an ultramafic soil toposequence. *Soil Science Society of American Journal* 67:1309–1317.

Massoura, S. T., G. Echevarria, T. Becquer, J. Ghanbaja, E. Leclere-Cessac, and J. L. Morel. 2006. Control of nickel availability by nickel bearing minerals in natural and anthropogenic soils. *Geoderma* 136:28–37.

McClain, C. N., S. Fendorf, S. M. Webb, and K. Maher. 2017. Quantifying Cr(VI) production and export from serpentine soil of the California Coast Range. *Environmental Science & Technology* 51:141–149.

McGahan, D. G., R. J. Southard, and V. P. Claassen. 2008. Tectonic inclusions in serpentinite landscapes contribute plant nutrient calcium. *Soil Science Society of American Journal* 72:838–847.

McGahan, D. G., R. J. Southard, and V. P. Claassen. 2009. Plant-available calcium varies widely in soils on serpentinite landscapes. *Soil Science Society of American Journal* 73:2087–2095.

McGrath, S. P. 1995. Chromium and nickel. In *Heavy Metals in Soils*. 2nd ed., B. J. Alloway (Ed.), pp. 152–178. London: Blackie Academic and Professional.

Meers, E., R. Samson, F. M. G. Tack, A. Ruttens, M. Vandegehuchte, J. Vangronsveld, and M. G. Verloo. 2007. Phytoavailability assessment of heavy metals in soils by single extractions and accumulation by *Phaseolus vulgaris*. *Environmental and Experimental Botany* 60:385–396.

Menzies, N. W., M. J. Donn, and P. M. Kopittke. 2007. Evaluation of extractants for estimating of the phytoavailable trace metals in soils. *Environmental Pollution* 145:121–130.

Miranda, M., J. L. Benedito, I. Blanco-Penedo, and C. López-Lamas. 2009. Metal accumulation in cattle raised in a serpentine-soil area: Relationship between metal concentrations in soil, forage and animal tissues. *Journal of Trace Element in Medicine and Biology* 23:231–238.

Morrison, J. M., M. B. Goldhaber, L. Lee, J. M. Holloway, R. B. Wanty, R. E. Wolf, and J. F. Ranville. 2009. A regional-scale study of chromium and nickel in soils of northern California, USA. *Applied Geochemistry* 24:1500–1511.

Morrison, J. M., M. B. Goldhaber, C. T. Mills, G. N. Breit, R. L. Hooper, J. M. Holloway, S. F. Diehl, and J. F. Ranville. 2015. Weathering and transport of chromium and nickel from serpentinite in the Coast Range ophiolite to the Sacramento Valley, California, USA. *Applied Geochemistry* 61:72–86.

O'Hanley, D. S. 1996. *Serpentinites: Records of Tectonic and Petrological History.* New York: Oxford University Press.

Oze, C., S. Fendorf, D. K. Bird, and R. G. Coleman. 2004a. Chromium geochemistry of serpentine soils. *International Geology Review* 46:97–126.

Oze, C., S. Fendorf, D. K. Bird, and R. G. Coleman. 2004b. Chromium geochemistry in serpentinized ultramafic rocks and serpentine soils from the Franciscan complex of California. *American Journal of Science* 304:67–101.

Oze, C., C. Skinner, A. Schroth, and R. G. Coleman. 2008. Growing up green on serpentine soils: Biogeochemistry of serpentine vegetation in the Central Coast Range of California. *Applied Geochemistry* 23:3391–403.

Rajapaksha, A. U., M. Vithanage, C. Oze, W. M. A. T. Bandara, and R. Weerasooriya. 2012. Nickel and manganese release in serpentine soils from the Ussangoda Ultramafic Complex, Sri Lanka. *Geoderma* 189–190:1–9.

Ramos, M. E., S. Garcia-Palma, M. Rozalen, C. T. Johnston, and F. J. Huertas. 2014. Kinetics of montmorillonite dissolution: An experimental study of the effect of oxalate. *Chemical Geology* 363:283–292.

Raous, S., G. Echevarria, T. Sterckeman, K. Hanna, F., Thomas, E. S. Martins, and T. Becquer. 2013. Potentially toxic metals in ultramafic mining materials: Identification of the main bearing and reactive phases. *Geoderma* 192:111–119.

Rinklebe, J., S. Antić-Mladenović, T. Frohne, H. J. Stärk, Z. Tomić, and V. Ličina. 2016. Nickel in a serpentine-enriched Fluvisol: Redox affected dynamics and binding forms. *Geoderma* 263: 203–214.

Rinklebe, J. and S. M. Shaheen. 2017. Redox chemistry of nickel in soils and sediments: A review. *Chemosphere* 179:265–278.

Siebecker, M. G., R. L. Chaney, and D. L. Sparks. 2017. Nickel speciation in several serpentine (ultramafic) topsoils via bulk synchrotron-based techniques. *Geoderma* 298:35–45.

Susaya, J. P., K. H. Kim, V. B. Asio, Z. S. Chen, and I. Navarrete. 2010. Quantifying nickel in soils and plants in an ultramafic area in Philippines. *Environmental Monitoring and Assessment* 167:505–514.

Ünver, I., S. Madenoğlu, A. Dilsiz, and A. Naml. 2013. Influence of rainfall and temperature on DTPA extractable nickel content of serpentine soils in Turkey. *Geoderma* 202–203:203–211.

van der Ent, A., R. Repin, J. Sugau, and K. M. Wong. 2015. Plant diversity and ecology of ultramafic outcrops in Sabah (Malaysia). *Australian Journal of Botany* 63:204–215.

Wicks, F. J. and D. S. O'Hanley. 1988. Serpentine minerals; structures and petrology. *Reviews in Mineralogy and Geochemistry* 19:91–167.

Zhou, S., Y. Wei, B. Li, H. Wang, B. Ma, C. Wang, and X. Luo. 2017. Mineralogical characterization and design of a treatment process for Yunnan nickel laterite ore, China. *International Journal of Mineral Processing* 159:51–59.

9

Methods of Ni Determination in Soils and Plants

Thomai Nikoli and Theodora Matsi

CONTENTS

9.1 Introduction

Nickel has been extensively studied in soils and plants as a heavy metal. However, Ni was also identified as an essential element for all higher plants in 1987 (Brown et al. 1987). This beneficial role of Ni was officially recognized by the Association of American Plant Food Control Officials relatively recently (Terry 2004). Based on the above, the majority of the published research on Ni soil availability and absorption by plants focuses on its role as a hazardous metal. Hence, considerable research has been done on the evaluation of several methods for the determination of plant-available Ni in a wide variety of metal-polluted soils in order to predict toxicity in plants grown in those soils. In a lesser extent, similar research has been done regarding unpolluted soils and plants cultivated in such soils.

In this chapter, an effort has been made to compile and evaluate existing knowledge and research, concerning methods applied for the extraction of Ni from soils and plant tissues. Focus has been given on Ni-specific papers concerning arable soils (unpolluted or polluted) and field crops. In all the reported cases in the present chapter, the analytical determination of Ni in soil and plant extracts was conducted by atomic absorption spectrometry (with flame or graphite furnace) or by inductively coupled plasma emission spectrometry. However, in certain cases reported in the literature, Ni was determined in solid samples by X-ray fluorescence (Diaz Rizo 2011; Saikat et al. 2007).

9.2 Methods of Ni Determination in Soils

9.2.1 Single Extraction Methods

Many single extraction methods, which have been used not only for the other four essential metal micronutrients for plants but also for macronutrients and non-essential trace elements, were widely employed in order to assess Ni availability for plant species grown in unpolluted or polluted soils. The most common methods used for Ni can be categorized into three groups, that is, methods employing extracting solutions containing complexing substances (among others), (diluted) acids, and (buffered or unbuffered) salts, and are summarized in Table 9.1. The rationale of the former group of methods is to complex the water-soluble Ni, forcing in this way different pools of the soil solid phase to release Ni to the solution. The target of the latter group of methods is to mainly extract the most mobile forms of Ni (water soluble and readily exchangeable). Certain characteristics of the methods, that is, either the substance's concentration in the extracting solution or the soil-to-solution ratio or the shaking time, were modified in several cases by researchers. For this reason, only the substances of the extracting solutions are presented in Table 9.1.

According to Mitchell et al. (1978), the basic qualifications of a method, which is proposed for the extraction of available amounts of an element from soil (i.e., an element soil test), are to (a) be equally effective in a wide soil pH range, (b) provide a significant correlation with plant concentration and uptake, and (c) extract the element in sufficient quantities to be analytically feasible to measure. Nowadays, the third qualification is questionable, since methods of instrumental chemical analysis with quite low detection limits have been developed. Between the other two qualifications, we believe that the most important is the second. Additional qualifications of a soil test are the significant correlation of the extracted element with other soil properties that affect its bioavailability and the extraction of high amounts of the element. However, the high amounts might not be correlated at all with the plant parameters.

In the following two sections, selected cases of different soil tests used for Ni extraction from unpolluted and polluted soils are analyzed. Emphasis is given on research papers, in which either Ni was extracted from a relatively large number of soils or its concentration was related to plant parameters or various Ni soil tests were used and compared.

TABLE 9.1

Substances That Have Been Used for the Extraction of Ni from Soil (Examples)

	References	
Substance	**Unpolluted Soils**	**Polluted Soils**
Complexing Substances		
DTPA[a]	Caridad-Cancela and Paz-Gonzalez (2005), Elkhatib (1994), Hseu (2006), L'Huillier and Edighoffer (1996), Liang and Schoenau (1995), Licina et al. (2010), McGrath (1996), Meers et al. (2007), Mellum et al. (1998), Narwal and Singh (1998), Nikoli and Matsi (2014), Nikoli et al. (2016), Pande et al. (2012), Rahmatullah et al. (2001), Rodak et al. (2015), Sadiq (1985), Sarkunan et al. (1989), Sims et al. (1991)	Adams and Kissel (1989), Antoniadis et al. (2010), Bahmanyar (2008), Barbarick and Workman (1987), Bhattacharyya et al. (2008), Cajuste et al. (2001), Gupta and Sinha (2006), Haq et al. (1980), Korcak and Fanning (1978), Kukier and Chaney (2001), Miller et al. (1995), Mitchell et al. (1978), Moral et al. (2002), Mulchi et al. (1991), Papadopoulos et al. (2007), Roca and Pomares (1991), Sanders et al. (1986a,b, 1987), Sauerbeck and Hein (1991), Sukkariyah et al. (2005), Tsadilas et al. (1995), Xiou et al. (1991), Zahedifar et al. (2017)
AB-DTPA	Nikoli and Matsi (2014), Nikoli et al. (2016), Pande et al. (2012), Shahid et al. (2014)	Barbarick and Workman (1987), Roca and Pomares (1991)
EDTA	Elkhatib (1994), McGrath (1996), Misra and Pande (1974), Susaya et al. (2010), Takeda et al. (2006)	Alegria et al. (1991), Cajuste et al. (2001), Haq et al. (1980), Lavado et al. (2005), Roca and Pomares (1991), Sanders et al. (1986a,b, 1987), Stalikas et al. (1999), Wang et al. (2009)
Mehlich-3[a]	Caridad-Cancela and Paz-Gonzalez (2005), Fontes et al. (2008), Nikoli and Matsi (2014), Nikoli et al. (2016), Pande et al. (2012), Sims et al. (1991)	Mulchi et al. (1991), Xiou et al. (1991)
Acids		
Mehlich-1	Pande et al. (2012), Rodak et al. (2015), Sims et al. (1991)	Korcak and Fanning (1978), Mulchi et al. (1991), Sukkariyah et al. (2005), Xiou et al. (1991)
HNO_3	Elkhatib (1994), Meers et al. (2007), Misra and Pande (1974), Nikoli and Matsi (2014), Pierce et al. (1982), Takeda et al. (2006)	
HCl	Meers et al. (2007), Misra and Pande (1974), Nikoli and Matsi (2014), Pande et al. (2012)	Roca and Pomares (1991), Xiou et al. (1991)
CH_3COOH	Meers et al. (2007), Misra and Pande (1974)	Davis (1979), Haq et al. (1980), Wang et al. (2009)

(Continued)

TABLE 9.1 (CONTINUED)

Substances That Have Been Used for the Extraction of Ni from Soil (Examples)

Substance	References	
	Unpolluted Soils	**Polluted Soils**
Salts		
CH_3COONH_4	Allinson and Dzialo (1981), Elkhatib (1994), Halstead et al. (1969), Hseu (2006), Meers et al. (2007), Misra and Pande (1974), Takeda et al. (2006)	Haq et al. (1980), Bisessar (1989), Sauerbeck and Hein (1991), Temple and Bisessar (1981), Wang et al. (2009)
NH_4NO_3	Meers et al. (2007), Mellum et al. (1998), Takeda et al. (2006)	Gupta and Sinha (2006),Wang et al. (2009)
$CaCl_2$	Guo et al. (2010), Halstead et al. (1969), Meers et al. (2007), Takeda et al. (2006)	Cioccio et al. (2016), Gupta and Sinha (2006), Moral et al. (2002), Sanders and Adams (1987), Sanders et al. (1986a,b, 1987), Sauerbeck and Hein (1991), Sukkariyah et al. (2005)

[a] In almost all cases, the procedures of Lindsay and Norwell (1978) and Mehlich (1984) were followed for the DTPA and Mehlich-3 methods, respectively.

In the majority of the reported cases, Ni was extracted from soils employing methods with complexing substances and especially the DTPA method suggested by Lindsay and Norvell (1978). Even though this method was initially designed to extract metals from natural soils, it has been widely used in studies with polluted soils, in order to assess mobility and bioavailability of metals. Since Ni is mostly studied as a pollutant, and less as a micronutrient, the DTPA method has dominated over others, and in many cases, it has been proven to be a reliable indicator for Ni availability to plants. Nevertheless, the Mehlich-3 method, which contains EDTA in its solution, has also succeeded in several cases in predicting Ni bioavailability. Among the rest of the methods tested, salts like CH_3COONH_4 and $CaCl_2$ have sometimes presented promising results, while acids proved to be poor Ni soil tests, with the exception of the CH_3COOH and Mehlich-1 method, occasionally.

Apart from chemical extraction methods, ion exchange resin methods have also been employed for the assessment of soil-available Ni (Becquer et al. 2002; Liang and Schoenau 1995). These methods are not included in the current chapter because further research is needed for their establishment as reliable soil tests for various metals. In addition, it is questionable if resins take into account the available amounts of a metal that are connected to the soil solid phase. For the same reason, neither water as Ni extractant is commented in the present chapter. However, water has been used as Ni extractant from polluted soils and was proven to be a reliable Ni soil test (Haq et al. 1980), probably because Ni in the added material to soil was mainly in water-soluble form.

9.2.1.1 Application to Unpolluted Soils

Application of simple extraction methods of Ni on unpolluted soils is not as commonly reported in the literature as those on polluted soils, probably because the role of Ni as a pollutant and not as an essential plant micronutrient was of major interest until recently. However, various methods have been used in common soils for the establishment of the most reliable Ni soil test, as well as for the determination of Ni toxicity limits (Guo et al. 2010; Khalid and Tinsley 1980; Pande et al. 2012) or even its (sufficiency) critical levels (Nikoli and Matsi 2014; Nikoli et al. 2016).

Except for the extractants reported in Table 9.1, other substances used for Ni extraction from unpolluted soils were the AAAc-EDTA (Meers et al. 2007; Nikoli and Matsi 2014; Nikoli et al. 2016), CaDTPA-$Na_2B_4O_7$, KNO_3 (Elkhatib 1994), H_2SO_4, $C_6H_8O_7$ (Misra and Pande 1974), $NaNO_3$, $MgCl_2$ (Meers et al. 2007), $K_4P_2O_7$ (Misra and Pande 1974; Pierce et al. 1982), and Grigg's method (Misra and Pande 1974). In several of the following reported studies, extractants containing complexing substances proved to be the most reliable Ni soil tests. However, there are also studies that demonstrated the suitability of acids or salts for the extraction of soil-available Ni.

In several cases, soils enriched with Ni due to pedogenetic reasons were studied (Caridad-Cancela and Paz-Gonzalez 2005; L'Huillier and Edighoffer 1996; Licina et al. 2010; Mellum et al. 1998; Narwal and Singh 1998). Caridad-Cancela and Paz-Gonzalez (2005) used the DTPA and Mehlich-3 methods for the extraction of available Ni from 10 soils of Galicia (Spain). They reported that although Ni concentrations extracted with the two methods were highly correlated with each other, significant relationships with other soil properties, like clay content and cation exchange capacity, were obtained only for the Mehlich-3-extracted Ni.

On the other hand, L'Huillier and Edighoffer (1996) reported that the DTPA method was successful in predicting soil-available Ni for 10 crops grown in 60 soils of New Caledonia (UK) and especially for tomato (*Solanum lycopersicum* L.) and eggplant (*Solanum melongena* L.), as this was proved by the significant and strong relationships between soil and plant Ni. However, Mellum et al. (1998), who tested DTPA and NH_4NO_3 as Ni soil tests for certain crops using 50 soils of Southeastern Norway, mentioned that although DTPA-extracted Ni was significantly correlated with soil clay and organic C content, it did not relate to plant Ni, whereas the opposite was evidenced for the NH_4NO_3-extracted Ni.

As far as soils not naturally enriched with Ni are concerned, Nikoli and Matsi (2014) investigated the suitability of six methods in predicting Ni availability for ryegrass (*Lolium perenne* L.) grown in 30 soils of North Greece. Nickel extracted with the AAAc-EDTA or Mehlich-3 methods was the most suitable followed by Ni extracted by the DTPA and AB-DTPA methods, as it was judged by the highly significant relationships found between Ni extracted by the four methods and plant Ni. Similar findings are reported by

Fontes et al. (2008) for Mehlich-3-extracted Ni, who used three Brazilian soils and lettuce (*Lactuca sativa* L.) and bean (*Phaseolous vulgaris* L.) as test plants.

Pande et al. (2012) employed DTPA, AB-DTPA, Mehlich-3, Mehlich-1, and 0.1 M HCl for the assessment of Ni toxicity levels in bean and concluded that the DTPA method was the most promising as Ni soil test. Elkhatib (1994) extracted Ni from 22 calcareous soils employing CaDTPA-$Na_2B_4O_7$, DTPA, EDTA, 1 M CH_3COONH_4, 0.1 M HNO_3, and 1 M KNO_3 and reported that the most reliable Ni soil test for the prediction of Ni uptake by wheat (*Triticum aestivum* L.) was the CaDTPA-$Na_2B_4O_7$-extracted Ni, probably because of its DTPA content. Rodak et al. (2015) reported that Ni in the leaves, but especially in the grain of soybean (*Glycine max* L.), was significantly and strongly correlated with Ni extracted from 14 soils by the DTPA and Mehlich-1 methods.

Meers et al. (2007) assessed the availability of Ni, among various heavy metals, to bean in 21 soils, which differed in chemical properties, employing DTPA, EDTA, 0.11 M CH_3COOH, 0.5 M HNO_3, 0.1 M HCl, 1 M CH_3COONH_4, 0.01 M $CaCl_2$, 1 M $MgCl_2$, 1 M NH_4NO_3, and 0.1 M $NaNO_3$. Nickel extracted with salt solutions primarily, like $CaCl_2$, NH_4NO_3, CH_3COONH_4, and $MgCl_2$, and with the DTPA and EDTA methods secondarily was significantly correlated with Ni in bean. Takeda et al. (2006) reported that among the eight methods they studied, Ni extracted with mild extractants, that is, 0.01 M HNO_3, 0.01 M $CaCl_2$, and 1 M NH_4NO_3, from 16 soils of Japan was significantly correlated with Ni uptake by buckwheat (*Fagopyrum esculentum* M.). However, all methods failed to predict Ni uptake by komatsuna (*Brassica rapa* var. *perviridis* L.). In addition, Misra and Pande (1974) reported that from the nine Ni soil tests they used in 32 soils of India, only seven were significantly correlated with Ni uptake by sorghum (*Sorghum bicolor* L.) in the following decreasing order: Grigg's method > 0.1 M HCl > 0.5 M CH_3COOH > 1 M CH_3COONH_4 > 0.1 M HNO_3 > 0.1 M H_2SO_4.

9.2.1.2 Application to Polluted Soils

Several simple extraction methods of Ni were applied to soils that were considered polluted, mainly because of sewage sludge addition (Barbarick and Workman 1987; Korcak and Fanning 1978; Miller et al. 1995; Mulchi et al. 1991; Roca and Pomares 1991; Sanders and Adams 1987; Sanders et al. 1986a,b, 1987; Sukkariyah et al. 2005; Xiou et al. 1991), but also due to the addition of fly ash derived from coal-fired power plants (Gupta and Sinha 2006) or of wastes from metals' smelting industry (Bisessar 1989; Temple and Bisessar 1981) and to irrigation with wastewater (Cajuste et al. 2001; Stalikas et al. 1999).

Apart from the extractants reported in Table 9.1, other substances used for Ni extraction from polluted soils were $(NH_4)_2EDTA$ (Ogbazghi et al. 2015; Sauerbeck and Hein 1991), $(COOH)_2$, $C_4H_6O_6$, $C_6H_8O_7$ (Wang et al. 2009), $MgCl_2$ (Sauerbeck and Hein 1991; Wang et al. 2009), KCl, $NaNO_3$, $CuCl_2$ (Sauerbeck and Hein 1991), NH_4Cl, $SrCl_2$ (Moral et al. 2002), and $Sr(NO_3)_2$ (Kukier and

Chaney 2001, 2004). In most of the following reported cases, methods containing complexing substances in their extracting solutions, especially the DTPA, were the most reliable or almost equally reliable to methods of acids or salts for the assessment of soil-available Ni.

Papadopoulos et al. (2007) extracted Ni, among other metals, with the DTPA method from 570 soil samples collected from an area of North Greece, where lignite power stations are operating. They reported that the DTPA-extracted Ni was significantly correlated with the soil properties that affect metals' bioavailability, especially pH. Moreover, Gupta and Sinha (2006) examined the capacity of Indian mustard (*Brassica juncea* L.) to uptake Ni from fly ash–amended soils, using DTPA, 0.01 M $CaCl_2$, and 1 M NH_4NO_3. Their results showed that the DTPA method was the most suitable for determining soil-available Ni for mustard plants.

As far as polluted soils due to sewage sludge application are concerned, Mitchell et al. (1978) reported that DTPA-extracted Ni, from two soils amended with sewage sludge enriched with heavy metals, succeeded in predicting Ni plant uptake. Sukkariyah et al. (2005) examined the residual effect of soil addition of biosolids on the availability of Ni and other heavy metals, by means of a field experiment. Among the DTPA, Mehlich-1, and 0.01 M $CaCl_2$ methods used for Ni extraction from soils, the DTPA method better predicted Ni availability to radish (*Raphanus sativus* L.). Xiou et al. (1991) studied the availability of sludge-born Ni and other metals to corn (*Zea mays* L.) and Sudan grass (*Sorghum sudanense* L.) in the field. Although Ni extracted with the DTPA, Mehlich-3, Mehlich-1, and 0.1 M HCl methods was significantly correlated with plant Ni, the most significant correlations were obtained for the former method.

Mulchi et al. (1991) employed the Mehlich-1, Mehlich-3, and DTPA methods for Ni extraction from two soils amended with municipal sludge and cultivated with tobacco (*Nicotiana tabacum* L.). They found that all methods were almost equally efficient in predicting soil-available Ni for tobacco. Roca and Pomares (1991) compared the effectiveness of the DTPA, AB-DTPA, EDTA, and 0.1 M HCl methods for the extraction of available Ni from a calcareous soil that had received sewage sludge, under field conditions. All methods were found to be equally effective in predicting soil-available Ni for potato (*Solanum tuberosum* L.) and lettuce grown in the soil.

Barbarick and Workman (1987) compared the effectiveness of the AB-DTPA and DTPA methods for extracting available Ni, among other trace elements, from two sludge-amended soils, by means of a greenhouse experiment with Swiss chard (*Beta vulgaris* var. *cicla* L.) and a field experiment with wheat. Nickel extracted with both methods significantly correlated with Ni in Swiss chard, but only AB-DTPA-extracted Ni significantly correlated with Ni in wheat.

Haq et al. (1980) tested the suitability of the DTPA, EDTA-$(NH_4)_2CO_3$, NTA–sodium citrate, HCl-$AlCl_3$, 0.5 M CH_3COOH, 1 M CH_3COONH_4, and $(COOH)_2$–$(COONH_4)_2$ methods in predicting Ni availability for Swiss chard, grown in 46 soils polluted with metals. Among all the tested methods, the CH_3COOH method was the best Ni soil test for Swiss chard.

Methods employing salts (especially the $CaCl_2$) have been proposed by certain researchers as the most reliable soil Ni tests. Sanders et al. (1986a) reported that Ni extracted from sludge-amended soils with the EDTA, DTPA, and 0.1 M $CaCl_2$ methods was significantly and strongly correlated with Ni concentration in ryegrass, with the $CaCl_2$ being superior than the other two extractants. Similar results are reported for barley (*Hordeum vulgare* L.) (Sanders et al. 1986b), red beet (*Beta vulgaris* var. *crassa* L.) (Sanders et al. 1986b, 1987), and white clover (*Trifolium repens* L.) (Sanders et al. 1987). Sauerbeck and Hein (1991) investigated the uptake of Ni by 11 species, grown in two soils after the addition of sewage sludge enriched with Ni as $NiCl_2$. In order to predict Ni availability, they compared the efficiency of several extractants, that is, 0.005 M DTPA, 0.005 M DTPA-TEA, 0.005 M DTPA–0.05 M $CaCl_2$-TEA, 0.005 M $(NH_4)_2$EDTA, 0.01 M $(NH_4)_2$EDTA, 1 M CH_3COONH_4, 0.05 M $CaCl_2$, 1 M $MgCl_2$, 2 M KCl, 1 M $NaNO_3$, 0.025 M $CuCl_2$, and 0.125 M $CuCl_2$. Among the methods tested, the $NaNO_3$ and chloride salt solutions (especially $CaCl_2$) were the most suitable ones for predicting Ni availability, followed by the DTPA method.

Certain researchers used uncommon solutions for Ni extraction from soils (Kukier and Chaney 2001, 2004). Kukier and Chaney (2001) extracted available Ni from two soils polluted by particulate emissions from a Ni refinery, employing the DTPA and 0.01 M $Sr(NO_3)_2$ methods. They found that $Sr(NO_3)_2$ was more successful in predicting Ni plant uptake, whereas the DTPA method was only efficient when pH and bulk density were taken into account.

Although rare in the literature, there are studies that support that none of the methods tested was appropriate as a Ni soil test. Wang et al. (2009) reported that all the eight methods tested [EDTA, CH_3COOH, $(COOH)_2$, $C_4H_6O_6$, $C_6H_8O_7$, CH_3COONH_4, $MgCl_2$, and NH_4NO_3] failed to predict Ni availability to wheat grown in ten soils.

9.2.2 Methods of Total Ni Determination

Total Ni concentration in soils, although a poor bioavailability indicator, has been used for the assessment of the risk or the extent of soils' pollution with Ni, due to natural reasons or anthropogenic activities, such are metals' smelting processes and disposal of inorganic (e.g., fly ash and slag) or organic (e.g., sewage sludge mainly and tannery sludge) wastes to soil (Abollino et al. 2002; Bahmanyar 2008; Bunzl et al. 1983; Cajuste et al. 2001; Frank et al. 1982; Hazlett et al. 1984; Mulchi et al. 1991; Sanders et al. 1986a,b; Santos et al. 2014; Sastre et al. 2001). In addition, soil total Ni content has been used for the determination of Ni background levels in soils (unpolluted or polluted) and in survey, monitoring, and pedogenetic studies (Caridad-Cancela and Paz-Gonzalez 2005; Chukwuma 1995; Gil et al. 2004; Hamner et al. 2013;

Hseu 2006; Mellum et al. 1998; Pierce et al. 1982; Sims et al. 1991; Skorbilowicz and Samborska 2014; Susaya et al. 2010).

Similar to other metals, acid digestion methods were employed upon heating on a hot plate or in a microwave oven for the determination of soil total Ni. Concentrated solutions of acids were used in these methods, often including HF to attack silicates. Several researchers employed a digestion method with a HNO_3-HF-HCl solution (Abollino et al. 2002; Gupta and Sinha 2006; Pierce et al. 1982; Stalikas et al. 1999). Hseu (2006), in a study of Ni distribution along a serpentinitic toposequence, determined total Ni after digestion of soils with a HNO_3-HF-$HClO_4$-H_2SO_4 solution, which was also used by Rahmatullah et al. (2001). In addition, solutions of HNO_3-HF-$HClO_4$ (Bunzl et al. 1983; Mulchi et al. 1991; Quian et al. 1996; Takeda et al. 2006) and $HClO_4$-HF (De Melo et al. 2007; Gil et al. 2004; Nogueira et al. 2009) have been used for soils' digestion, before total Ni determination.

However, methods employing solutions of acids in the absence of HF, like HNO_3 or HNO_3-$HClO_4$ or HNO_3-HCl, were more preferable by the researchers for the determination of total Ni in soils. These methods are considered to be effective in extracting the total amounts of heavy metals, despite the fact that the silicates are not destructed, since heavy metals are not usually silicate bound in soils. However, because of the latter, total heavy metals determined by the particular methods are quite often called "pseudo-total."

In such methods, concentrated acids like HNO_3 (Cajuste et al. 2001; Caridad-Cancela and Paz-Gonzalez 2005; Chukwuma 1995; Cioccio et al. 2016; Davis 1979; Frank et al. 1982; Hamner et al. 2013; Moral et al. 2002; Sims and Kline 1991), HNO_3-HCl (aqua regia in most cases) (Atta-Aly 1999; Bahmanyar 2008; Meers et al. 2007; Mellum et al. 1998; Narwal and Singh 1998; Ogbazghi et al. 2015; Roca and Pomares 1991; Rooney et al. 2007; Sanders et al. 1986a,b, 1987; Skorbilowicz and Samborska 2014), HNO_3-$HClO_4$ (Alegria et al. 1991; Barbafieri 2000; Hazlett et al. 1984; Miller et al. 1995; Sadiq 1985; Sastre et al. 2001; Wang et al. 2009), HNO_3-$HClO_4$-HCl (Lavado et al. 2005), and HNO_3-H_2SO_4-HCl (Susaya et al. 2010) have been used. In addition, acid digestion methods involving the use of H_2O_2, like that of HNO_3-H_2O_2-HCl (De Melo et al. 2007; Nogueira et al. 2009; Santos et al. 2014; Sims et al. 1991), were also applied for the determination of total Ni.

9.3 Methods of Ni Determination in Plants

Nickel in different plant tissues and other biological parameters, like yield and Ni plant uptake, have been used in several studies as criteria for the evaluation of Ni soil tests regarding their ability to extract plant-available Ni from various soils. The soils were either not enriched with Ni, enriched with Ni due to pedogenetic reasons, or polluted with Ni mainly as a result of sewage sludge application or other organic or inorganic waste addition (Bahmanyar 2008; Barbafieri 2000; Barbarick and Workman 1987; Gupta and Sinha 2006; L'Huillier and Edighoffer 1996; McGrath 1996; Meers et al. 2007; Miller et al. 1995; Mitchell et al. 1978; Mulchi et al. 1991; Nikoli and Matsi 2014; Quian et al. 1996; Roca and Pomares 1991; Sanders et al. 1986a,b, 1987; Sauerbeck and Hein 1991; Sims and Kline 1991; Sukkariyah et al. 2005; Takeda et al. 2006; Xiou et al. 1991).

Moreover, the aforementioned parameters have been used to elucidate Ni deficiency or toxicity conditions in various plant species for agricultural or environmental purposes (Ahmad et al. 2011; De Melo et al. 2007; Guo et al. 2010; Khalid and Tinsley 1980; Kukier and Chaney 2004; Miller et al. 1995; Mitchell et al. 1978; Nikoli and Matsi 2014; Nikoli et al. 2016; Pande et al. 2012; Parida et al. 2003; Sanders et al. 1987; Sauerbeck and Hein 1991; Shahid et al. 2014) and are also reported in survey and pedogenetic studies (Mellum et al. 1998).

Similar to other elements, total Ni concentrations in plant biomass were determined and are reported in the literature. Before Ni determination in plant material, the destruction of its organic matter is necessary. This is usually achieved by employing a wet acid digestion or a dry ashing method followed by the ash dissolution with acid solutions. Although Ni is not considered a volatile element, the wet acid digestion procedures seem to be more preferable than the dry ashing methods by the researchers.

The destruction of organic matter by the wet acid digestion methods is commonly conducted by using concentrated solutions of an acid or a mixture of acids, with heating on a hot plate or in a microwave oven. Various concentrations and acids have been used in the case of Ni and other metals. The simplest method is that of HNO_3, which has been used for the pretreatment of plant material before Ni determination in wheat (Barbarick and Workman 1987; Hamner et al. 2013), barley (Sanders et al. 1986b; Sukkariyah et al. 2005), oats (*Avena sativa* L.) (Cioccio et al. 2016), corn (Guo et al. 2010), rice (*Oryza sativa* L.) (Chukwuma 1995), and soybean (Cioccio et al. 2016; Rodak et al. 2015).

In addition, HNO_3 has been used for the digestion of romaine lettuce (*Lactuca sativa* var. *longifolia* L.), radish (Sukkariyah et al. 2005), red beet (Frank et al. 1982; Sanders et al. 1986b, 1987), Swiss chard (Barbarick and Workman 1987), cabbage (*Brassica oleracea* var. *capitata* L.), lettuce, celery (*Apium graveolens* L.), radish (Frank et al. 1982), Indian mustard (Gupta and Sinha 2006), white clover (Sanders et al. 1987), and ryegrass (McGrath 1996; Nikoli and Matsi 2014; Nikoli et al. 2016; Sanders et al. 1986a).

The use of a HNO_3-$HClO_4$ solution for the digestion of plant tissues seems to be quite common among researchers, although caution is needed in handling $HClO_4$, because it is flammable upon drying. This solution has been applied to wheat (Lavado et al. 2005; Mitchell et al. 1978; Quian et al. 1996; Sauerbeck and Hein 1991; Sims and Kline 1991; Wang et al. 2009), corn (De Melo et al. 2007; Roca and Pomares 1991; Sadiq 1985; Sauerbeck and Hein 1991; Xiou et al. 1991), barley (Miller et al. 1995; Sauerbeck and Hein 1991), oats (Allinson and Dzialo 1981; Rahmatullah et al. 2001), Swiss chard (Miller et al. 1995), lettuce (Fontes et al. 2008; Mitchell et al. 1978; Roca and Pomares 1991; Sauerbeck and Hein 1991), bean (Fontes et al. 2008; Pande et al. 2012; Sauerbeck and Hein 1991; Singh and Keefer 1989), and soybean (Adams and Kissel 1989; Sims and Kline 1991).

Moreover, this solution has been applied to potato (Roca and Pomares 1991), rice (Sarkunan et al. 1989), Indian mustard (Sauerbeck and Hein 1991), tobacco (Mulchi et al. 1991), alfalfa (*Medicago sativa* L.) (Barbafieri 2000; Quian et al. 1996), ryegrass (Allinson and Dzialo 1981; Khalid and Tinsley 1980; Sauerbeck and Hein 1991), Sudan grass (Xiou et al. 1991), Egyptian clover (*Trifolium alexandrinum* L.) (Shahid et al. 2014), fenugreek (*Trigonella corniculata* L.) (Parida et al. 2003), carrot (*Daucus carota* L.), radish (Sauerbeck and Hein 1991; Singh and Keefer 1989), oil rape (*Brassica napus* L.), cabbage, tomato, kale (*Brassica oleracea* var. *acephala* L.) (Singh and Keefer 1989), spinach (*Spinacia oleracea* L.), and kohlrabi (*Brassica oleracea* var. *gongylodes* L.) (Sauerbeck and Hein 1991).

Other wet digestion methods are reported by Takeda et al. (2006), who digested komatsuna and buckwheat tissue samples in a microwave oven, employing a HNO_3-HF-$HClO_4$ solution. A H_2SO_4-$HClO_4$ solution was used by Singh et al. (1990) for wheat and a H_2SO_4-HNO_3-$HClO_4$ solution was used by Davis (1979) for barley, oil rape, lettuce, and ryegrass and by Misra and Pande (1974) for sorghum. In addition, solutions including H_2O_2 were used, like a H_2SO_4-H_2O_2 solution for sunflower (*Helianthus annuus* L.) (Ahmad et al. 2011), oats, lettuce, and radish (Liang and Schoenau 1995); a HNO_3-H_2O_2 solution for rice (Tang and Miller 1991), wheat, and corn (Tsadilas et al. 1995); a HNO_3-$HClO_4$-H_2O_2 solution for grapevine (*Vitis vinifera* L.) (Licina et al. 2010); and a HNO_3-H_2O_2-HCl solution for corn (Nogueira et al. 2009).

As far as the dry ashing methods are concerned, such methods were applied to cotton (*Gossypium hirsutum* L.) (Antoniadis et al. 2010), corn (Korcak and Fanning 1978; Kukier and Chaney 2004; L'Huillier and Edighoffer 1996), wheat (Kukier and Chaney 2001, 2004; Narwal and Singh 1998), rice (Bahmanyar 2008), oats, red beet (Kukier and Chaney 2001, 2004), Swiss chard (Haq et al. 1980; Kukier and Chaney 2004), bean (Giordani et al. 2005; Kukier and Chaney 2004; Meers et al. 2007), and soybean (Kukier and Chaney 2004).

Furthermore, dry ashing methods were applied to cereals, potato, and cauliflower (*Brassica oleracea* var. *botrytis* L.) (Mellum et al. 1998), cabbage (Giordani et al. 2005; Mellum et al. 1998), carrot (L'Huillier and Edighoffer 1996), radish (Kukier and Chaney 2004; L'Huillier and Edighoffer 1996), tomato (Giordani et al. 2005; Kukier and Chaney 2004; L'Huillier and Edighoffer 1996; Susaya et al. 2010), eggplant (L'Huillier and Edighoffer 1996; Susaya et al. 2010), Chinese cabbage (*Brassica rapa* var. *chinensis* L), courgette (*Cucurbita pepo* var. *cylindrical* L.) (L'Huillier and Edighoffer 1996), barley (Giordani et al. 2005; Kukier and Chaney 2004), spinach, sorghum, castorbean (*Ricinus communis* L.) (Giordani et al. 2005), watermelon (*Citrullus vulgaris* L.) (Susaya et al. 2010), oats, alfalfa (Halstead et al. 1969), and ryegrass (Kukier and Chaney 2004).

In the aforementioned methods, the temperatures applied for ashing, as well as the concentrations of acid solutions used for the dissolution of ash, varied widely. However, the most common acids used were HCl (Antoniadis et al. 2010; Halstead et al. 1969; L'Huillier and Edighoffer 1996; Susaya et al. 2010), HNO_3 (Giordani et al. 2005; Meers et al. 2007), and HNO_3-HCl (Haq et al. 1980; Kukier and Chaney 2001, 2004; Mellum et al. 1998; Narwal and Singh 1998).

References

Abollino, O., M. Aceto, M. Malandrino, E. Mentasti, C. Sarzanini and R. Barberis. 2002. Distribution and mobility of metals in contaminated sites. Chemometric investigation of pollutant profiles. *Environmental Pollution* 119:177–193.

Adams, J. F. and D. E. Kissel. 1989. Zinc, copper, and nickel availabilities as determined by soil solution and DTPA extraction of a sludge-amended soil. *Communications in Soil Science and Plant Analysis* 20:139–158.

Ahmad, M. S. A., M. Ashraf and M. Hussain. 2011. Phytotoxic effects of nickel on yield and concentration of macro- and micro-nutrients in sunflower (*Helianthus annuus* L.) achenes. *Journal of Hazardous Materials* 185:1295–1303.

Alegria, A., R. Barberfi, R. Boluda, F. Errecalde, R. Farr and M. J. Lagarda. 1991. Environmental cadmium, lead and nickel contamination: Possible relationship between soil and vegetable content. *Fresenius Journal of Analytical Chemistry* 339:654–657.

Allinson, D. W. and C. Dzialo. 1981. The influence of lead, cadmium, and nickel on the growth of ryegrass and oats. *Plant and Soil* 62:81–89.

Antoniadis, V., C. D. Tsadilas and V. Samaras. 2010. Trace element availability in a sewage sludge-amended cotton grown Mediterranean soil. *Chemosphere* 80:1308–1313.

Atta-Aly, M. A. 1999. Effect of nickel addition on the yield and quality of parsley leaves. *Scientia Horticulturae* 82:9–24.

Bahmanyar, M. A. 2008. Cadmium, nickel, chromium, and lead levels in soils and vegetables under long-term irrigation with industrial wastewater. *Communications in Soil Science and Plant Analysis* 39:2068–2079.

Barbafieri, M. 2000. The importance of nickel phytoavailable chemical species characterization in soil for phytoremediation applicability. *International Journal of Phytoremediation* 2:105–115.

Barbarick, K. A. and S. M. Workman. 1987. Ammonium bicarbonate-DTPA and DTPA extractions of sludge-amended soils. *Journal of Environmental Quality* 16:125–130.

Becquer, T., F. Rigault and T. Jaffre. 2002. Nickel bioavailability assessed by ion exchange resin in the field. *Communications in Soil Science and Plant Analysis* 33:439–450.

Bhattacharyya, P., K. Chakrabarti, A. Chakraborty, S. Tripathy, K. Kims and M. A. Powell. 2008. Cobalt and nickel uptake by rice and accumulation in soil amended with municipal solid waste compost. *Ecotoxicology and Environmental Safety* 69:506–512.

Bisessar, S. 1989. Effects of lime on nickel uptake and toxicity in celery grown on muck soil contaminated by a nickel refinery. *The Science of the Total Environment* 84:83–90.

Brown, P. H., R. M. Welch and E. E. Cary. 1987. Nickel: A micronutrient essential for higher plants. *Plant Physiology* 85:801–803.

Bunzl, K., G. Rosner and W. Schmidt. 1983. Distribution of lead, cobalt and nickel in the soil around a coal-fired power plant. *Journal of Plant Nutrition and Soil Science* 146:705–713.

Cajuste, L. J., A. Vazquez-A. and E. Miranda-C. 2001. Long-term changes in the extractability and availability of lead, cadmium, and nickel in soils under wastewater irrigation. *Communications in Soil Science and Plant Analysis* 33:3325–3333.

Caridad-Cancela, R. and A. Paz-Gonzalez. 2005. Total and extractable nickel and cadmium contents in natural soils. *Communications in Soil Science and Plant Analysis* 36:241–252.

Chukwuma, C., Sr. 1995. Evaluating baseline data for copper, manganese, nickel and zinc in rice, yam, cassava and guinea grass from cultivated soils in Nigeria. *Agriculture Ecosystems and Environment* 53:47–61.

Cioccio, S., Y. Gopalapillai, T. Dan and B. Hale. 2016. Effect of liming on nickel bioavailability and toxicity to oat and soybean grown in field soils containing aged emissions from a nickel refinery. *Environmental Toxicology and Chemistry* 9999:1–10.

Davis, R. D. 1979. Uptake of copper, nickel, and zinc by crops growing in contaminated soils. *Journal of Science of Food and Agriculture* 30:937–947.

De Melo, W. J., P. S. Aguiar, G. M. P. De Melo and V. P. De Melo. 2007. Nickel in a tropical soil treated with sewage sludge and cropped with maize in a long-term field study. *Soil Biology and Biochemistry* 39:1341–1347.

Diaz Rizo, O., I. Coto Hernandez, J. O. Arado Lopez, O. Diaz Arado, N. Lopez Pino and K. De Alessandro Rondriguez. 2011. Chromium, cobalt and nickel contents in urban soils of Moa, Northeastern Cuba. *Bulletin of Environmental Contamination and Toxicology* 86:189–193.

Elkhatib, E. A. 1994. Evaluation of six soil extractants for assessing nickel availability to wheat. *Arid Soil Research and Rehabilitation* 8:137–145.

Fontes, R. L. F., J. M. N. Pereira, J. C. L. Neves and M. P. F. Fontes. 2008. Cadmium, lead, copper, zinc, and nickel in lettuce and dry beans as related to Mehlich 3 extraction in three Brazilian latossols. *Journal of Plant Nutrition* 31:884–901.

Frank, R., K. I. Stonefield and P. Suda. 1982. Impact of nickel contamination on the production of vegetables on an organic soil, Ontario, Canada 1980–1981. *The Science of the Total Environment* 26:41–65.

Gil, C., R. Boluda and J. Ramos. 2004. Determination and evaluation of cadmium, lead and nickel in greenhouse soils of Almeria (Spain). *Chemosphere* 55:1027–1034.

Giordani, C., S. Cecchi and C. Zanchi. 2005. Phytoremediation of soil polluted by nickel using agricultural crops. *Environmental Management* 36:675–681.

Guo, X. Y., Y. B. Zuo, B. R. Wang, J. M. Li and Y. B. Ma. 2010. Toxicity and accumulation of copper and nickel in maize plants cropped on calcareous and acidic field soils. *Plant and Soil* 333:365–373.

Gupta, A. K. and S. Sinha. 2006. Role of *Brassica juncea* (L.) Czern. (var. Vaibhav) in the phytoextraction of Ni from soil amended with fly ash: Selection of extractant for metal bioavailability. *Journal of Hazardous Materials B* 136:371–378.

Halstead, R. L., B. J. Finn and A. J. MacLean. 1969. Extractability of nickel added to soils and its concentration in plants. *Canadian Journal of Soil Science* 49:335–342.

Hamner, K., J. Eriksson and H. Kirchmann. 2013. Nickel in Swedish soils and cereal grain in relation to soil properties, fertilization and seed quality. *Acta Agriculturae Scandinavica Section B: Soil and Plant Science* 63:712–722.

Haq, A. U., T. E. Bates and Y. K. Soon. 1980. Comparison of extractants for plant-available zinc, cadmium, nickel, and copper in contaminated soils. *Soil Science Society of America Journal* 44:772–777.

Hazlett, P. W., G. K. Rutherford and G. W. Van Loon. 1984. Characteristics of soil profiles affected by smelting of nickel and copper at Coniston, Ontario, Canada. *Geoderma* 32:273–285.

Hseu, Z. Y. 2006. Concentration and distribution of chromium and nickel fractions along a serpentinitic toposequence. *Soil Science* 171:341–353.

Khalid, B. Y. and J. Tinsley. 1980. Some effects of nickel toxicity on rye grass. *Plant and Soil* 55:139–144.

Korcak, R. F. and D. S. Fanning. 1978. Extractability of cadmium, copper, nickel, and zinc by double acid versus DTPA and plant content at excessive soil levels. *Journal of Environmental Quality* 7:506–512.

Kukier, U. and R. L. Chaney. 2001. Amelioration of Ni phytotoxicity in muck and mineral soils. *Journal of Environmental Quality* 30:1949–1960.

Kukier, U. and R. L. Chaney. 2004. In situ remediation of nickel phytotoxicity for different plant species. *Journal of Plant Nutrition* 27:465–495.

L'Huillier, L. and S. Edighoffer. 1996. Extractability of nickel and its concentration in cultivated plants in Ni rich ultramafic soils of New Caledonia. *Plant and Soil* 186:255–264.

Lavado, R. S., M. B. Rodriguez and M. A. Taboada. 2005. Treatment with biosolids affects soil availability and plant uptake of potentially toxic elements. *Agriculture Ecosystems and Environment* 109:360–364.

Liang, J. and J. J. Schoenau. 1995. Development of resin membranes as a sensitive indicator of heavy metal toxicity in the soil environment. *International Journal of Environmental Analytical Chemistry* 59:265–275.

Lindsay, W. L. and W. A. Norvell. 1978. Development of a DTPA soil test for zinc, iron, manganese and copper. *Soil Science Society of America Journal* 42:421–428.

Licina, V., S. Antic-Mladenovic, M. Kresovic and J. Rinklebe. 2010. Effect of high nickel and chromium background levels in serpentine soil on their accumulation in organs of a perennial plant. *Communications in Soil Science and Plant Analysis* 41:482–496.

McGrath, D. 1996. Application of single and sequential extraction procedures to polluted and unpolluted soils. *The Science of the Total Environment* 178:37–44.

Mehlich, A. 1984. Mehlich 3 soil test extractant: A modification of Mehlich 2 extractant. *Communications in Soil Science Plant Analysis* 15:1409–1416.

Meers, E., R. Samson, F. M. G. Tack, A. Ruttens, M. Vandegehuchte, J. Vangronsveld and M. G. Verloo. 2007. Phytoavailability assessment of heavy metals in soils by single extractions and accumulation by *Phaseolous vulgaris*. *Environmental and Experimental Botany* 60:385–396.

Mellum, H. K., A. K. M. Arnesen and B. R. Singh. 1998. Extractability and plant uptake of heavy metals in alum shale soils. *Communications in Soil Science and Plant Analysis* 29:1183–1198.

Miller, R. W., A. S. Azzari and D. T. Gardiner. 1995. Heavy metals in crops as affected by soil types and sewage sludge rates. *Communications in Soil Science and Plant Analysis* 25:703–711.

Misra, S. G. and P. Pande. 1974. Evaluation of a suitable extractant for available nickel in soils. *Plant and Soil* 41:697–700.

Mitchell, G. A., F. T. Bingham and A. L. Page. 1978. Yield and metal composition of lettuce and wheat grown on soils amended with sewage sludge enriched with cadmium, copper, nickel, and zinc. *Journal of Environmental Quality* 7:165–171.

Moral, R., R. J. Gilkes and J. Moreno-Caselles. 2002. A comparison of extractants for heavy metals in contaminated soils from Spain. *Communications in Soil Science and Plant Analysis* 33:2781–2791.

Mulchi, C. L., C. A. Adamu, P. F. Bell and R. L. Chaney. 1991. Residual heavy metal concentrations in sludge-amended coastal plain soils—I. Comparison of extractants. *Communications in Soil Science and Plant Analysis* 22:919–941.

Narwal, R. P. and B. R. Singh. 1998. Effect of organic materials on partitioning, extractability and plant uptake of metals in an alum shale soil. *Water, Air, and Soil Pollution* 103:405–421.

Nikoli, T. and T. Matsi. 2014. Evaluation of certain Ni soil tests for an initial estimation of Ni sufficiency critical levels. *Journal of Plant Nutrition and Soil Science* 177:596–603.

Nikoli, T., T. Matsi and N. Barbayiannis. 2016. Assessment of nickel's sufficiency critical levels in cultivated soils employing commonly used calibration techniques. *Journal of Plant Nutrition and Soil Science* 179:566–573.

Nogueira, T. A. R., W. J. De Melob, L. R. Oliveira, I. M. Fonseca, G. M. P. De Melo, S. A. Marcussi and M. O. Marques. 2009. Nickel in soil and maize plants grown on an Oxisol treated over a long time with sewage sludge. *Chemical Speciation and Bioavailability* 21:165–173.

Ogbazghi, Z. M., E. H. Tesfamariam, J. G. Annandale and P. C. De Jager. 2015. Mobility and uptake of zinc, cadmium, nickel, and lead in sludge-amended soils planted to dryland maize and irrigated maize-oat rotation. *Journal of Environmental Quality* 44:655–667.

Pande, J., P. C. Srivastava and S. K. Singh. 2012. Plant availability of nickel as influenced by farmyard manure and its critical toxic limits in French bean. *Journal of Plant Nutrition* 35:384–395.

Papadopoulos, A., C. Prochaska, F. Papadopoulos, N. Gantidis and E. Metaxa. 2007. Determination and evaluation of cadmium, copper, nickel, and zinc in agricultural soils of western Macedonia, Greece. *Environmental Management* 40:719–726.

Parida, B. K., I. M. Chhibba and V. K. Nayyar. 2003. Influence of nickel-contaminated soils on fenugreek (*Trigonella corniculata* L.) growth and mineral composition. *Scientia Horticulturae* 98:113–119.

Pierce, F. J., R. H. Dowdy and D. F. Grigal. 1982. Concentrations of six trace elements in some major Minnesota soils series. *Journal of Environmental Quality* 11:416–421.

Quian, J., Z. Wang, X. Shan, Q. Tu, B. Wen and B. Chen. 1996. Evaluation of plant availability of soil trace metals by chemical fractionation and multiple regression analysis. *Environmental Pollution* 91:309–315.

Rahmatullah, B. U. Z., M. Salim and H. Khawer. 2001. Nickel forms in calcareous soils and influence of Ni supply on growth and N uptake of oats grown in soil fertilized with urea. *International Journal of Agriculture and Biology* 3:230–232.

Roca, J. and F. Pomares. 1991. Prediction of available heavy metals by six chemical extractants in a sewage sludge-amended soil. *Communications in Soil Science and Plant Analysis* 22:2119–2136.

Rodak, B. W., M. F. De Moraes, J. A. Lopes Pascoalino, A. De Oliveira Jr., C. De Castro and V. Pauletti. 2015. Methods to quantify nickel in soils and plant tissues. *Revista Brasileira de Ciencia* do Solo 39:788–793.

Rooney, C. P., F. J. Zhao and S. P. McGrath. 2007. Phytotoxicity of nickel in a range of European soils: Influence of soil properties, Ni solubility and speciation. *Environmental Pollution* 145:596–605.

Sadiq, M. 1985. Uptake of cadmium, lead and nickel by corn grown in contaminated soils. *Water Air and Soil Pollution* 26:185–190.

Saikat, S., B. Barnes and D. Westwood. 2007. A review of laboratory results for bioaccessibility values of arsenic, lead and nickel in contaminated UK soils. *Journal of Environmental Science and Health Part A: Environmental Science and Engineering* 42:1213–1221.

Sanders, J. R., S. P. McGrath and T. McM. Adams. 1986a. Zinc, copper and nickel concentrations in ryegrass grown on sewage sludge-contaminated soils of different pH. *Journal of Science of Food and Agriculture* 37:961–968.

Sanders, J. R., T. McM. Adams and B. T. Christensen. 1986b. Extractability and bioavailability of zinc, nickel, cadmium, and copper in three Danish soils sampled 5 years after application of sewage sludge. *Journal of Science of Food and Agriculture* 37:1155–1164.

Sanders, J. R. and T. McM. Adams. 1987. The effects of pH and soil type on concentrations of zinc, copper and nickel extracted by calcium chloride from sewage sludge-treated soils. *Environmental Pollution* 43:219–228.

Sanders, J. R., S. P. McGrath and T. McM. Adams. 1987. Zinc, copper and nickel concentrations in soil extracts and crops grown on four soils treated with metal-loaded sewage sludges. *Environmental Pollution* 44:193–210.

Santos, J. L., A. S. F. Araújo, L. A. P. L. Nunes, M. L. J. Oliveira and W. J. Melo. 2014. Chromium, cadmium, nickel, and lead in a tropical soil after 3 years of consecutive applications of composted tannery sludge. *Communications in Soil Science and Plant Analysis* 45:1658–1666.

Sarkunan, V., A. K. Misra and P. K. Nayar. 1989. Interaction of zinc, copper and nickel in soil on yield and metal content in rice. *Journal of Environmental Science and Health. Part A: Environmental Science and Engineering* 24:459–466.

Sastre, I., M. A. Vicente and M. C. Lobo. 2001. Behaviour of cadmium and nickel in a soil amended with sewage sludge. *Land Degradation and Development* 12:27–33.

Sauerbeck, D. R. and A. Hein. 1991. The nickel uptake from different soils and its prediction by chemical extractions. *Water, Air, and Soil Pollution* 57–58:861–871.

Shahid, M., M. Sabir, M. Arif Ali and A. Ghafoor. 2014. Effect of organic amendments on phytoavailability of nickel and growth of berseem (*Trifolium alexandrinum* L.) under nickel contaminated soil conditions. *Chemical Speciation and Bioavailability* 26:37–42.

Sims, J. T., E. Igo and Y. Skeans. 1991. Comparison of routine soil tests end EPA method 3050 as extractants for heavy metals in Delaware soils. *Communications in Soil Science and Plant Analysis* 22:1031–1045.

Sims, J. T. and J. S. Kline. 1991. Chemical fractionation and plant uptake of heavy metals in soils amended with co-composted sewage sludge. *Journal of Environmental Quality* 20:387–395.

Singh, R. N. and R. F. Keefer. 1989. Uptake of nickel and cadmium by vegetables grown on soil amended with different sewage sludges. *Agriculture Ecosystems and Environment* 25:27–38.

Singh, B., Y. P. Dang and S. C. Mehta. 1990. Influence of nitrogen on the behaviour of nickel in wheat. *Plant and Soil* 127:213–218.

Skorbilowicz, E. and A. Samborska. 2014. Content of copper and nickel in soils of Vistula river catchment. *Journal of Ecological Engineering* 15:53–59.

Stalikas, C. D., G. A. Pilidis and S. M. Tzouwara-Karayanni. 1999. Use of a sequential extraction scheme with data normalization to assess the metal distribution in agricultural soils irrigated by lake water. *The Science of the Total Environment* 236:7–18.

Sukkariyah, B. F., G. Evanylo, L. Zelazny and R. L. Chaney. 2005. Cadmium, copper, nickel, and zinc availability in a biosolids-amended piedmont soil years after application. *Journal of Environmental Quality* 34:2255–2262.

Susaya, J. P., K. H. Kim, V. B. Asio, Z. S. Chen and I. Navarrete. 2010. Quantifying nickel in soils and plants in an ultramafic area in Philippines. *Environmental Monitoring and Assessment* 167:505–514.

Takeda, A., H. Tsukada, Y. Takaku, S. Hisamatsu, J. Inabu and M. Nanzyo. 2006. Extractability of major and trace elements from agricultural soils using chemical extraction methods: Application for phytoavailability assessment. *Soil Science and Plant Nutrition* 52:406–417.

Tang, T. and D. M. Miller. 1991. Growth and tissue composition of rice grown in soil treated with inorganic copper, nickel, and arsenic. *Communications in Soil Science and Plant Analysis* 22:2037–2045.

Temple, P. J. and S. Bisessar. 1981. Uptake and toxicity of nickel and other metals in crops grown on soil contaminated by a nickel refinery. *Journal of Plant Nutrition* 3:473–482.

Terry, D. 2004: AAPFCO Official Publication 57. Association of American Plant Food Control Officials, West Lafayette, Indiana.

Tsadilas, C. D., T. Matsi, N. Barbayiannis and D. Dimoyiannis. 1995. Influence of sewage sludge application on soil properties and on the distribution and availability of heavy metal fractions. *Communications in Soil Science and Plant Analysis* 26:2603–2619.

Wang, S., Z. Nan, X. Liu, Y. Li, S. Qin and X. Ding. 2009. Accumulation and bioavailability of copper and nickel in wheat plants grown in contaminated soils from the oasis, northwest China. *Geoderma* 152:290–295.

Xiou, H., R. W. Taylor, J. W. Shuford and W. Tadesse. 1991. Comparison of extractants for available sludge-borne metals: a residual study. *Water, Air, and Soil Pollution* 57–58:913–922.

Zahedifar, M., S. Dehghani, A. A. Moosavi and E. Gavili. 2017. Temporal variation of total and DTPA-extractable heavy metal contents as influenced by sewage sludge and perlite in a calcareous soil. *Archives of Agronomy and Soil Science* 63:136–149.

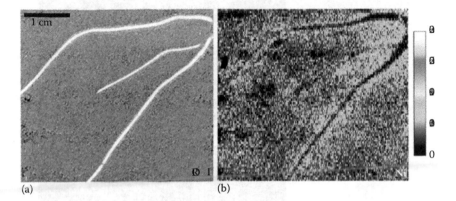

(a)   (b)

FIGURE 2.1
Photograph of a willow (*S. smithiana*) root (a) and related nickel depletion zone chemically imaged by combining diffusive gradients in thin films (DGT) deployment with subsequent laser ablation (LA)–ICP-MS. The scale is showing nickel fluxes in pg m^{-2} s^{-2}. (Modified with permission from Hoefer C., J. Santner, S. M. Borisov et al. 2017. Integrating chemical imaging of cationic trace metal solutes and pH into a single hydrogel layer. *Analytica Chimica Acta* 950:88–97.)

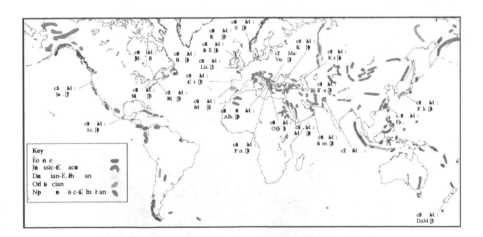

FIGURE 3.3
The global distribution of ophilites with approximate ages provides a proxy for the patchy, global distribution of ultramafic soils. (From Vaughan, A. P. M. and J. H. Scarrow. 2003. Ophiolite obduction pulses as a proxy indicator of superplume events? *Earth Planet. Sci. Lett.* 213(3–4):407–416. doi: http://dx.doi.org/10.1016/S0012-821X(03)00330-3.)

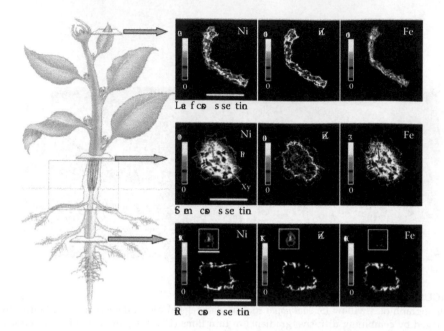

Ni | Zn | Fe
Leaf cross section

Ni | Zn | Fe
Stem cross section

Ni | Zn | Fe
Root cross section

FIGURE 3.6

Ni, Fe, and Zn fluorescence computed microtomography images of a leaf, stem, coarse and fine root cross sections from *A. murale* "Kotodesh." Inset in root tomogram is of a finer root. The colorimetric scale maps region-specific relative metal concentrations ($\mu g\ g^{-1}$) for each element, with brighter colors indicating areas of higher enrichment. The yellow scale bar represents ~500 μm; the white scale bar (root inset) represents ~100 μm. [Plant figure adapted from *Plant Physiology*, 3rd ed., Tiaz and Zeiger (eds.) with permission from Sinauer Associates, Inc., Publishers.] (From McNear, D. H., E. Peltier, J. Everhart, R. L. Chaney, S. Sutton, M. Newville, M. Rivers, and D. L. Sparks. 2005a. Application of quantitative fluorescence and absorption-edge computed microtomography to image metal compartmentalization in *Alyssum murale*. *Environ. Sci. Technol.* 39(7):2210–2218. doi: 10.1021/es0492034.)

FIGURE 4.4

Ultramafic rock in the process of altering to serpentine. (From Hayes, G. [http://geotripper .blogspot.com/2015/12/a-netherworld-incompatible-with.html].)

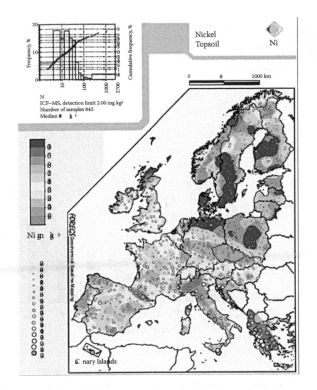

FIGURE 5.1
Geochemical map of Europe showing the distribution of total Ni in topsoils.

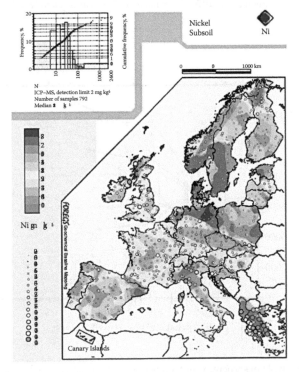

FIGURE 5.2
Geochemical map of Europe showing the distribution of total Ni in subsoils.

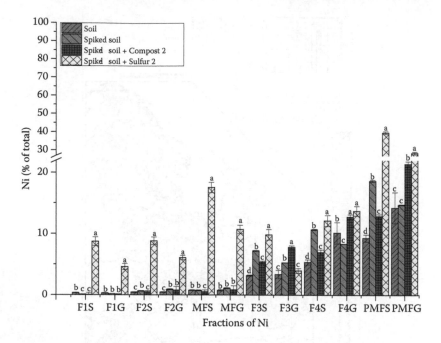

FIGURE 12.7

Impact of application of sulfur and compost on the fractions and potential mobilization of Ni under sorghum and barnyard grass cultivation. S, sorghum; G, barnyard grass; F1 = soluble + exchangeable fraction; F2 = sorbed and carbonate fraction; F3 = Fe/Mn oxide fraction; F4 = organic fraction; MF = mobile fraction (=$\sum$F1 – F2); PMF = potential mobile fraction (=$\sum$F1 – F4). (Reproduced from Shaheen et al., 2017c. *Environmental Geochemistry and Health* 39(6), 1305–1324.)

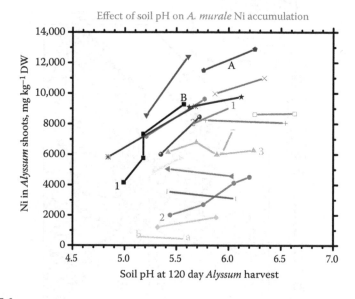

FIGURE 15.6

Effect of variation in pH of serpentine or Ni refinery contaminated soils (1, 2) on Ni concentration in shoots of *A. murale*.

10

Macroscale and X-ray Absorption Spectroscopic Studies of Soil Nickel Speciation

Yohey Hashimoto and Shan-Li Wang

CONTENTS

10.1 The Concept of Speciation

Speciation refers to the specific chemical forms of an element in soil solids, pore water, and gaseous phases. The following speciation will be discussed mainly by considering solid phase species. Speciation of an element in soil solids can be conceptually grouped into five principal mechanisms: (i) outer-sphere surface complexation, (ii) inner-sphere surface complexation, (iii) multinuclear surface complexation, (iv) homogeneous precipitation, and (v) lattice diffusion (Figure 10.1) (Manceau et al. 2002). Outer-sphere surface complexation involves electrostatic attraction between the sorbed metal ion and sorbent surface while the primary water of hydration around the metal ion remains. Inner-sphere surface complexation is formed when surface functional groups of the sorbent displace hydration water from the coordination

FIGURE 10.1
Basic processes of adsorbate molecules at the mineral–water interface: (a) physisorption; (b) chemisorption; (c) detachment; (d) absorption or inclusion; (e) occlusion; (f) attachment; (g) hetero-nucleation; (h) organo-mineral complexation; (i) complexation to bacterial exopolymer. (From Manceau, A., Marcus, M.A., Tamura, N., 2002. Quantitative speciation of heavy metals in soils and sediments by synchrotron X-ray techniques. *Reviews in Mineralogy and Geochemistry* 49, 341–428. With permission.)

sphere of the sorbed metal ion. Cations with multinuclear surface complexation undergo polymerization on the sorbent surface and form a hetero-epitaxial overgrowth. Homogeneous precipitation involves polymerization of cations on solid surface and in solution. Lattice diffusion occurs when the sorbed ion diffuses into the structure of sorbent by filling vacancies or substituting for sorbed ions. These reactions determine the distributions of metal species in soil, which possess different solubility. Thus, understanding metal speciation can bring insight into the key roles of different reactions in determining the mobility and bioavailability of metals in soils. The information is of great importance in understanding the fates of the metals in soils, which further determine their risks to ecosystems and public health.

Macroscale and molecular-scale research techniques have been employed to qualitatively and quantitatively characterize chemical species of metals in soils, which are essential but challenging tasks. A representative method of macroscopic techniques is sequential extraction, which classified the soil metals based on their solubility in reagents with different chemical natures (e.g., pH, ionic strength, and chelating power). Because the quantitative results of sequential extraction provides almost no information about the chemical identity of metal species, this method can only operationally define the speciation (or more precisely fractionation) of metals in soils and is not the method to determine "actual" metal species in soils. One of the drawbacks of using sequential extraction techniques for soil metal fractionation is lack of standardization in their methods and procedures, and the results reported in literature are not reasonably compared. However, information obtained from sequential extraction or other macroscale research has been used to infer the existence of different metal species in soils. In accordance to an IUPAC (International Union of Pure and Applied Chemistry) report on sequential extraction of trace elements in soil and sediment (Hlavay et al. 2004), it is stated that "Despite all limitations, sequential extraction schemes can provide a valuable tool to distinguish among trace metal fractions of different solubility." Sequential extraction is a useful tool for deducing possible metal species in soils, if the method and outcome are properly used and interpreted, respectively.

Spectroscopic techniques target intact species of soil elements without drastic physical and chemical alternations to the original soil conditions. Accordingly, the composition and structure of elements are maintained in their integrity in the soil, although sample pretreatments may be needed in some cases. Spectroscopic techniques that have been applied to determine elemental speciation in soils include X-ray diffraction (XRD), X-ray absorption spectroscopy (XAS), nuclear magnetic resonance (NMR), infrared spectroscopy, and electron microscopy. For determining Ni species in soils, however, the low Ni concentration and matrix interference of multiple elements in the samples significantly restrict their analytical accuracy. XRD is for example useful to identify the target species (i.e., minerals or crystalline precipitates) of elements when they are abundant and present as well-ordered crystals in the soil. However, it is rare to have the soil Ni concentration over several percent even in severely contaminated soils and ultramafic soils, and adsorption of Ni on mineral surfaces and Ni associated with organic substance cannot be identified by XRD. For the same reason, the low natural abundance of [61]Ni isotope also restricts the use of NMR spectroscopy on soil Ni studies (Ure and Davidson 2002).

Synchrotron XAS is the only nondestructive method that overcomes these drawbacks for determination of Ni speciation in soils. XAS is an element-specific technique that can be applied to determine the local structure of an absorbing element, even when the element is at a low concentration in materials with complex chemical compositions, such as soils. Thus, much of our current understanding of metal speciation and related reactions or processes in soils or other natural materials (e.g., rocks, sediments and living organisms) has been obtained by X-ray absorption fine structures (XAFS) of metals. To interpret the spectroscopic results, model compounds of Ni species are measured and the data are processed by deconvolution algorithm to gain information contributing to individual chemical species of soil Ni. The information obtained by XAFS represents an average of all Ni species present in the soil (bulk XAFS) and targets Ni species in a single micrometer particle and their spatial distribution (micro XAFS). One challenge is to apply the information and principles developed from microscale reaction mechanisms to macroscale analyses such as sequential extraction to assess Ni reaction in soils. In this chapter, we review studies on macroscale and XAFS research techniques for Ni speciation in soils. The chapter starts with a review of macroscopic research to demonstrate the importance of chemical speciation as an approach for understanding the behavior of Ni, focusing mainly on soil retention, reaction kinetics, and potential mobility. In Section 10.3, the state of synchrotron XAFS techniques that have been used to determine Ni speciation in soils is described, with emphasis on the differences between naturally and anthropogenically derived soil Ni.

10.2 Chemical Speciation of Soil Ni—Macroscale Studies

10.2.1 Minerals

Ni is ubiquitous in the environment. Generally, the identification of Ni-associated minerals in environmental solid phases is accomplished using a combination of mineralogical (such as XRD and petrographic and electron microscopic examinations) and chemical methods (total elemental composition and chemical extraction) with thermodynamic calculation sometimes. The complementary information from these different methods allows the identification of Ni-associated minerals and understanding their chemical transformations under certain conditions in the environment. Ni content in igneous rocks can be as high as 2000 mg kg^{-1} (Kabata-Pendias and Pendias 2011). As a siderophile element with chalcophile and lithophile characteristics, Ni is present in igneous rocks as several sulfide minerals, such as pentlandite (Fe,Ni)$_9$S$_8$, nickeline (NiAs) and ullmannite (NiSbS), or substitute Fe^{2+} ions in Fe-Mg silicate minerals, such as olivine (3000–4500 mg kg^{-1}), spinel (3000–3500 mg kg^{-1}), and orthopyroxene (650–1000 mg kg^{-1}) minerals (Trescases 2012). Ni-containing silicates are serpentinized as a result of hydrothermal alternation, and part of Fe and Ni from the minerals form oxides during the serpentinization process (Trescases 2012). In surface environment, the Ni-containing sulfide and silicate minerals in igneous rocks are unstable and undergo rapid weathering. During this weathering process, Ni is released and depletes due to its relative high mobility, while a part of released Ni can re-precipitate with other elements, mainly Fe and S, to form secondary minerals. As a consequence, Ni in sedimentary rocks is down to the range of 5–90 mg kg^{-1} and is associated with Fe-containing minerals or forms sulfide under reducing conditions (Kabata-Pendias and Pendias 2011). Rocks and geological sediments are the parent materials of soils. Thus, soils inherit minerals (i.e., primary minerals) directly from their parent materials during the soil formation process. The content of Ni in soils primarily depends on its content in soil parent materials, although its content in surface soils is also determined by the soil formation processes and anthropogenic activities. The mean concentration of Ni in uncontaminated soils varies from 0.2 to 450 mg kg^{-1} (Gonnelli and Renella 2013), but can reach a value as high as 10,000 mg kg^{-1} in serpentine-derived soils (Hseu 2006). The high Ni content of serpentine-derived soils can be traced back to the geological and pedogenic processes of ultramafic rocks, in which the Ni content is particularly high.

Soil development processes lead to gradual transformation of primary minerals to secondary minerals in soils. Continuous percolation of water through soil profiles leaches basic cations and silicic acid, and the resultant soil chemical conditions favor the formation of clay minerals like kaolinite and Fe/Al hydroxides and oxides in the soils. For example, the ultramafic pedogenesis leads to the transfer of Fe from primary silicate minerals (e.g., olivine) to secondary Fe oxides and oxyhydroxides and the alternation of serpentine and chlorite to vermiculite and smectite (Butt and Cluzel 2013; Trescases 2012; Uren 1992). During these processes, Ni associated with the primary minerals are released into soil solutions and may be either sorbed on these secondary minerals (Dahn et al. 2003; Ford et al. 1999; Scheidegger et al. 1996) or incorporated into the structures of these minerals (Hseu 2006; Manceau and Calas 1986; Scheidegger et al. 1997). Consequently, most Ni is associated with secondary minerals in the clay fraction such as Fe and Mn oxides, garnierites, chlorite, vermiculite, and saponite; comparatively, Ni content is insignificant in the primary minerals persisting in the sand fraction of soils (such as quartz and feldspars) (Butt and Cluzel 2013; Trescases 2012; Uren 1992). In serpentine soils, Ni is commonly found to be associated with minerals such as lizardite, antigorite, smectite, vermiculite, goethite, and hematite (Al-Khirbash 2016; Caillaud et al. 2006; Cheng et al. 2011; Echevarria et al. 2006; Gaudin et al. 2005; Massoura et al. 2006; Trescases 2012).

The anthropogenic sources of Ni in soil include agricultural activities, such as the application of fertilizer or sewage sludge, and industrial activities, such as steel works, metal plating and coinage, and fuel combustion (Alloway 2013). The Ni species depend on the nature of their origins and determine the releasing rate of Ni into soil solution. After releasing into soil solution, Ni undergoes different reactions with the solution and solid components of soils, leading to the formations of Ni complexes in soil solution and on soil surface, and inorganic precipitates, which are determined using sequential extraction or spectroscopic methods.

10.2.2 Ni Adsorption on Soil Solid Phases

Adsorption is one of the key processes in determining the bioavailability and mobility of contaminants in soils (Violante et al. 2010). The ability of soils to adsorb metal ions are dependent on the compositions of soil colloids (such as clays, metal oxides and hydroxides, metal carbonates, and phosphates, and humic substances), and the compositions and properties of soil solutions, such as pH, ionic strength, and the types and concentrations of cations and anions. It has been well understood that soil colloids play a key role in the overall adsorption process of Ni in soil due to their high surface area and adsorptive affinity toward Ni. Such an understanding is based on the results of extensive kinetics and/or equilibrium studies on the Ni adsorption for humic substances (Green-Pedersen et al. 1997; Strathmann and Myneni 2004) and for major inorganic colloids, including clay minerals, such as kaolinite (Bansal 1982; Eick and Fendorf 1998; Eick et al. 2001; Mattigod et al. 1979; Scheidegger et al. 1997; Sen Gupta and Bhattacharyya 2006), montmorillonite (Bradbury and Baeyens 2011; Dahn et al. 2003; Demirkiran et al. 2016; Green-Pedersen et al. 1997; Marcussen et al. 2009; Scheidegger et al. 1997; Sen Gupta and Bhattacharyya 2006; Sheikhhosseini et al. 2013, 2014), vermiculite (Bourliva et al. 2012; Malamis and Katsou 2013; Sen Gupta and Bhattacharyya 2006; Vijayaraghavan and Raja 2015), and illite (Bradbury and Baeyens 2011; Echeverria et al. 2003; Elzinga and Sparks 2001; Gu and Evans 2007; Zhao et al. 2017); the oxides and hydroxides of Al (Rajapaksha et al. 2012b; Scheidegger et al. 1997; Trivedi and Axe 2001), Fe (Arai 2008; Ford et al. 1997; Marcussen et al. 2009; Rajapaksha et al. 2012b; Tamura and Furuichi 1997; Trivedi and Axe 2001), and Mn (Green-Pedersen et al. 1997; Tamura and Furuichi 1997; Trivedi and Axe 2001; Trivedi et al. 2001), and the minerals of phosphate (Elouear et al. 2009; Ivanets et al. 2017; Perrone et al. 2001) and carbonate (Belova et al. 2007, 2014; Green-Pedersen et al. 1997; Lakshtanov and Stipp 2007; Pokrovsky and Schott 2002; Zachara et al. 1991). The results of kinetics or equilibrium approaches have also been applied to reveal the dependence of the Ni adsorptions of different soil colloids on solution pH, ionic strength, temperature, reaction time, solid/solution ratios, and co-existing cations and anions (Bradl 2004; Shaheen et al. 2013).

Cation or anion sorption on inorganic adsorbents proceeds generally fast initially and gradually decreases over time. The former has been attributed to adsorption reaction, whereas the latter has been attributed to surface precipitation or diffusion of adsorbate into the pores of adsorbent (Sparks 1998; Sposito 2004). Accordingly, kinetics studies have been conducted to distinguish adsorption and surface precipitation due to the significant difference between the time scales of adsorption (minutes or hours) and surface precipitation (days or weeks) (Sparks 1998). For example, Ni sorption on kaolinite at pH 7.5 exhibited two distinctive linear sections in the first-order kinetics plots, corresponding to a fast and a slow reaction with the rate constants of $1.25–1.82 \times 10^{-4}$ and $1.73–1.93 \times 10^{-5}$ min^{-1}, respectively (Eick et al. 2001). The fast and slow reactions were attributed to chemisorption and surface nucleation/precipitation, respectively (Eick et al. 2001). Similar phenomena were also observed for Ni adsorption on kaolinite (Scheidegger and Sparks 1996) and palygorskite and sepiolite (Sheikhhosseini et al. 2014). With the evidence of Ni K-edge EXAFS, the Ni adsorption on pyrophyllite occurred through the formation of bidentate inner-sphere Ni complexation via edge sharing of Ni and Al octahedra, and the slow reaction was attributed to the formation of polynuclear Ni complexes and Ni-Al hydroxide on the surface of pyrophyllite (Scheidegger et al. 1996, 1998). The formation of Ni-Al hydroxide was also observed when Ni was reacted with kaolinite, montmorillonite, and gibbsite (Eick and Fendorf 1998; Eick et al. 2001; Scheidegger et al. 1997). The formation of Ni-Al hydroxides increased with pH, Ni loading, and reaction time (Elzinga and Sparks 2001). As indicated in Peltier et al. (2010), the formation of Ni-Al hydroxide surface precipitate resulted in the reduction of desorbed Ni and therefore lowered Ni bioavailability in soils.

Adsorption isotherm is an equilibrium approach to determine Ni absorptivity of soils or soil components. The adsorptions of Ni on various soil minerals generally exhibit an *L*- or *H*-type isotherm according to Giles's classification (Giles and Smith 1974). These types of isotherms reveal that the adsorption affinity decreases with increasing surface coverage while the *H*-type isotherm reflects a higher affinity between adsorbate and adsorbent as compared to the *L*-type isotherm. As seen in Figure 10.2, for example, the Ni adsorption isotherms of illite are *L*-type at lower pH values and *H*-type at pH 8 and 9, indicating an increasing Ni affinity of illite with increasing pH (Echeverria et al. 2003). The formation of hydrolytic Ni(II) species [such as $Ni(OH)^+$ and $Ni(OH)_2^0$] also resulted in

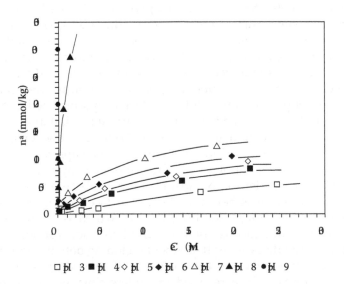

$\square$ pI 3 $\blacksquare$ pI 4 $\lozenge$ pI 5 $\blacklozenge$ pI 6 $\triangle$ pI 7 $\blacktriangle$ pI 8 $\bullet$ pI 9

FIGURE 10.2
Adsorption isotherms of Ni(II) on illite at different pH values (temperature = 25°C and background electrolyte = 0.01 M NaNO₃). (From Echeverria, J., Indurain, J., Churio, E., Garrido, J., 2003. Simultaneous effect of pH, temperature, ionic strength, and initial concentration on the retention of Ni on illite. *Colloids and Surfaces A—Physicochemical and Engineering Aspects* 218, 175–187. With permission.)

the increasing Ni(II) adsorption at higher pH ranges. However, the macroscopic observation did not take into account the different sites on the surface of an adsorbent for Ni uptake as well as the formation of surface precipitates, such as $Ni(OH)_2$ or Ni-Al layered double hydroxides (LDHs). Adsorption isotherms can be fitted with adsorption models, such as the Langmuir or Freundlich equation, leading to the determination of various model coefficients. These coefficients are convenient for comparing different adsorption data as functions of different experimental parameters, such as pH, ionic strength, and types of coexisting ions, to understand their effects on the adsorption reaction in a quantitative manner. However, the experimental conditions may not be well defined and thus yield poor model coefficients that are often not comparable among different studies. Meanwhile, this approach is considered empirical due to the lack of mechanistic significance in the adsorption isotherm models, which has long been recognized (Veith and Sposito 1977). In particular, these models do not take into account the electrostatic interactions between ions in solution and a charged solid surface as it is the case in soil colloids (Bradl 2004). Thus, mechanistic interpretation of these adsorption isotherms based on the goodness of fit cannot represent the real adsorption mechanism, unless it is supported by spectroscopic evidence.

To take into account the electrostatic interactions between ions in solution and a charged solid surface, various mechanistic models, termed "surface complexation models," have been developed and interested readers can be referred to the review articles in the literature (Bradl 2004; Merdy et al. 2006). In these models, the extent of the interactions between dissolved ions and surface functional groups controls surface charge buildup and adsorption properties of minerals in aqueous solutions. To account for the dependence of surface charge on pH, the modeling of Ni sorption is often carried out with the 2-pKa approach in which the surface is hydroxylated in two steps (Reactions 1 and 2). When the formation of surface precipitation is excluded from modeling or avoided in an experimental setting, the adsorption can be classified as outer-sphere or inner-sphere surface complexes. For example, Rajapaksha et al. (2012b) applied the diffuse double-layer model with the 2-pKa approach to model the Ni adsorption of gibbsite and goethite. The adsorption of Ni^{2+} forms either monodentate surface complexes (Reaction 3) or bidentate surface complexes (Reaction 4), which is analogous to the hydrolysis of Ni^{2+} in aqueous solution. The Ni(II) adsorptions of gibbsite and goethite were fitted well with a model including both monodentate and bidentate complexes, whereas the inclusion of monodentate or bidentate Ni surface complexes alone resulted in underestimation and overestimation of the Ni sorption, respectively (Rajapaksha et al. 2012b). Thus, Ni(II) is adsorbed on gibbsite and goethite through the formation of both monodentate and bidentate inner-sphere complexes (Rajapaksha et al. 2012b). In Gu and Evans (2008), the surface sites with different binding affinities toward Ni(II) were considered in modeling the Ni adsorption of kaolinite

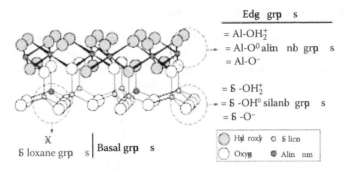

Edg grp s

= Al-OH$_2^+$
= Al-O^0 alin nb grp s
= Al-O$^-$

= S -OH$_2^+$
= S -OH^0 silanb grp s
= S -O$^-$

Hyd roxy   S lico
Oxyg   Alin nm

X
S loxane grp s | Basal grp s

FIGURE 10.3
The surface binding sites of metal ions on kaolinite proposed. (From Gu, X.Y., Evans, L.J., 2008. Surface complexation modelling of Cd(II), Cu(II), Ni(II), Pb(II) and Zn(II) adsorption onto kaolinite. *Geochimica et Cosmochimica Acta* 72, 267–276. With permission.)

(Figure 10.3). The proton binding constants were first determined for the edge and basal surface sites of kaolinite and subsequently used in the modeling of Ni adsorption of kaolinite. The results suggested that Ni ions were bound to permanent negatively charged sites on basal surfaces in the lower pH range through nonspecific ion-exchange reactions. With increasing pH, the variable charged edge sites (= SOH) became the major adsorption sites and Ni ions were adsorbed through the formation of monodentate complexes with the sites (Gu and Evans 2008). Echeverria et al. (2003) reported that the main sorption mechanism of Ni(II) on illite was ion exchange at lower pH values and the contributions of forming inner-sphere complex and precipitate to Ni sorption became relatively significant at higher pH values. Similar mechanisms of Ni(II) sorption on illite were proposed in Gu and Evans (2007), indicating that nonspecific ion-exchange reaction at lower pH and specific adsorption at higher pH occurred on the basal surfaces and mineral edges of illite, respectively. Although surface complexation modeling has provided reasonable prediction of metal ion sorption, the correlation of model parameters can result in uncertainty in the prediction. This uncertainty can be resolved by constraining the model calibration process, which can be accomplished using the information provided by spectroscopic methods, as discussed in Section 10.3.

$$= S - OH_2^+ \Rightarrow = S - OH + H^+ \quad K_{a1} \qquad \text{(Reaction 1)}$$

$$= S - OH \Rightarrow = S - O^- + H^+ \quad K_{a2} \qquad \text{(Reaction 2)}$$

$$= S - OH + Ni^{2+} \Rightarrow = S - ONi^+ + H^+ \qquad \text{(Reaction 3)}$$

$$2 = S - OH + Ni^{2+} \Rightarrow (= S - O)_2 Ni + 2H^+ \qquad \text{(Reaction 4)}$$

10.2.3 Chemical Extraction of Soil Ni

Nickel present in soils can be associated with soil solid components through a variety of reactions, mainly precipitation, coprecipitation, adsorption, and ion exchange. These reactions result in the formation of diverse Ni species with different extents of mobility and bioavailability, ranging from water-soluble to recalcitrant forms. Accordingly, extensive efforts have been devoted to develop methods to assess or determine Ni speciation in soils. One approach to the determination of Ni speciation is stepwise chemical extractions involving multiple extraction reagents (i.e., sequential extraction). Generally, a set of reagents is chosen on the basis of their selectivity toward particular soil solid phases and arranged with progressively increasing extraction capacity. The speciation of Ni (or other elements) is then classified into fractions that are associated with the target soil solid components of the extraction reagents. For example, the reagents of the Tessier method (Tessier et al. 1979), which has been widely applied in the literature, include (i) $MgCl_2$ (pH 7), (ii) NaOAc/HOAc (pH 5), (iii) NH_2OH HCl in 25% HOAc (pH 2), (iv) H_2O_2 in HNO_3 (pH 2) and subsequent NH_4OAc, and (v) HF and $HClO_4$ for (1) exchangeable, (2) bound to carbonate, (3) bound to iron and manganese oxides, (4) bond to organic matter, and (5) residual phases, respectively. Most methods follow a similar extraction scheme with variations in extraction reagents and conditions and interested readers can refer to previous reviews in the literature for the comparison of the sequential extraction methods (Gleyzes et al. 2002; Hass and Fine 2010; Hlavay et al. 2004; Rao et al. 2008; Tack and Verloo 1995; Vodyanitskii 2006; Zimmerman and Weindorf 2010).

With a careful design of the sequential extraction procedure in terms of reagents and extraction conditions, ideally, the reagent in each extraction step can selectively extract metals bound to a specific soil solid phase with minimal disturbance on the other phases. However, due to the lack of knowledge about the specificity of reagents and the existence of actual metal phases in soils, the results of sequential extraction methods can only operationally define metal fractions rather than metal species in samples. It is therefore inappropriate to refer the fractions to species because the term "chemical species" is defined as a specific form of an element with a distinct isotopic composition, electronic or oxidation state, and/or complex or molecular structure (Templeton et al. 2000). In most cases, a further characterization of the nickel species within a fraction is not accomplished, unless spectroscopic methods, such as XAS, are applied to determine the predominant Ni species. Meanwhile, the metal fractionation results of different sequential extraction methods may not be comparable due to the differences in the number of reagents, temperature, extraction time, and recovery of the residual extractant. Even so, if a sequential extraction method is chosen and applied systematically to a set of soil samples, the information about metal fractionation can still provide useful information to study the fates of metals in soils or assess the mobility and bioavailability of metals under the effects of certain soil conditions or managements.

Sequential extraction methods have been applied to provide information about biogeochemical cycling of Ni and pedogenic transformation of Ni-associated minerals in soils. Soils developed over ultramafic rocks often contain high levels of Ni and the release of geogenic Ni can have serious threat to public health. Therefore, Ni fractionation in serpentine soils has received extensive attention. Previous studies on serpentine soils in different regions in the world generally revealed that a significant portion of Ni (>50%) in serpentine soils is associated with residual phase, attributed to the geogenic origin of Ni (Antic-Mladenovic et al. 2011; Cheng et al. 2011; Hseu 2006; Kelepertzis and Stathopoulou 2013; Rajapaksha et al. 2012a; Rinklebe et al. 2016). Ni in residual phase is also predominant in alluvial soils (Barman et al. 2015). With the Ni fractionation results of sols collected from toposequences, Ni becomes more mobile with the level of soil development and weathering, and the mobility of Ni depended on pH, ionic strength, and organic acids (Cheng et al. 2011; Hseu 2006; Hseu et al. 2017). Nickel availability was correlated with total organic carbon in serpentine soils (Bani et al. 2014; Vithanage et al. 2014). The mobilization and availability of Ni have been primarily correlated with soil organic matter due to the formation of Ni–DOC complexes, whereas the immobilization of Ni was attributed to the formation of Fe/Mn (hydro)oxides and Ni coprecipitates (Antic-Mladenovic et al. 2011; Rinklebe et al. 2016). Accordingly, Ni fractionation by sequential extraction methods can provide an improved understanding of the processes and factors that control Ni mobility and availability in soils.

The anthropogenic input of Ni into soils generally results in the accumulation of Ni in the top soils, and a higher ratio of Ni tends to reside in more labile or mobile fractions as opposed to geogenic Ni predominantly associated with the residual fraction. For example, Rajaie et al. (2008) reported that Ni added into calcareous soils was predominantly associated with soluble, exchangeable, carbonate-bound, and organically bound fractions, which are considered more labile in comparison with those associated with Mn and Fe oxides and residual fraction. The effect of soil pH was revealed in Tewari et al. (2010), showing that Ni occurred predominantly in the water-soluble and exchangeable fractions in an acidic soil and in less labile fractions in an alkaline soil. Thomasi et al. (2015) reported that Ni added into Brazilian Oxysols was predominantly associated with soluble, exchangeable, and organic fractions. Although the distribution of added Ni varied with soil types, Ni generally transformed from more labile fractions into less labile fractions with time (Han and Bannin 2000; Rajaie et al. 2008; Tewari et al. 2010; Thomasi et al. 2015). This is supported by the results of a recent field study showing that the extractable Ni in acidic, neutral, and alkaline soils all decreased after 6 years (Jiang et al. 2018).

Sequential extraction methods have also been applied to study the effects of remediation methods on the mobility and bioavailability of Ni in contaminated soils by comparing the Ni fractionations before and after the remediation treatments. In Giannis et al. (2010), for example the sequential extraction method was applied to determine the distribution of Ni and other metals during the electrokinetic experiments with varying pH, redox potential, and electrolyte compositions. The results revealed that the metals became easier to be extracted after the treatment (Giannis et al. 2010). Sequential extraction methods have also been applied to reveal positive effects on the immobilization of Ni in soils for the amendments of several low-cost materials, such as biochars (Bandara et al. 2017) and coal fly ash (Saffari et al. 2015). For example, a field study of 3 years revealed that biochar amendment increased the residual fractions of Ni in the soils, which can result in a lower mobility of Ni in the soils (Shen et al. 2016). Thus, the results of sequential extraction methods can improve the understanding of Ni fractionation upon remediation treatment that is essential for developing an adequate strategy for remediating Ni-contaminated soils and assessing the effects of the strategy on reducing the risk of Ni contamination.

10.3 XAS for Ni Speciation

10.3.1 Principles of XAS

In soil science and geochemistry, XAS has been recognized as an essential tool for the determination of the oxidation state and chemical structure of target elements in geomaterials. XAS measures the absorption coefficient of X-ray of the element in the sample as a function of X-ray energy. The absorption edge occurs near the binding energy of core electrons that is specific to each element. This elemental specificity provides an important advantage in analyzing metal speciation of materials with complicated compositions. XAS measurement of an element yields XAFS, including both X-ray absorption near-edge structure (XANES) and extended X-ray absorption fine structure (EXAFS). XAFS is related to the average local molecular coordinating environment of the X-ray absorber element (Figure 10.4). In general, the position of the white-line peak (i.e., the energy at the highest X-ray absorption at ~8350 eV in Figure 10.4) on the XANES region can be used to determine the oxidation state of the element, and the EXAFS region encompasses the information about the type of neighbor elements and their coordination number and interatomic distance. The advantages of XAS for studying chemical species of elements in soils include high elemental specificity, high sensitivity to the oxidation state and coordination environment of an element, and minimum requirement of sample preparation. Kelly et al. (2008) summarized the application of synchrotron XAS in soil and geoscience, including sample preparation, measurement, and data interpretation. An excellent book providing a comprehensive description of XAS studies for soil and sediment samples has been published (Singh and Grafe 2010).

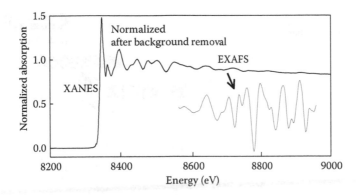

FIGURE 10.4
Nickel *K*-edge XAFS of nickel oxide (NiO) consisting of XANES and EXAFS regions.

10.3.2 Modes of Adsorption and Precipitation of Ni for Soil and Clay Minerals

The adsorption processes of Ni (and other metals) on soil colloids are generally studied through kinetics and/or equilibrium approaches, which determine the amount of adsorbate on the adsorbent surface as a function of the adsorption time and adsorbate concentration, respectively. With macroscopic observations alone, the distinction between adsorption and surface precipitation is not always clear due to their indirect nature with respect to molecular mechanism. The application of molecular spectroscopy has helped constrain the calibration process of mechanistic models of metal adsorption by limiting the range of potential surface complexes. A recent study of Zhao et al. (2017) has combined spectroscopic results with macroscopic experiments to demonstrate the change of surface Ni species and sorption mechanisms on illite at different surface coverage (Ni loading), pH, temperature, and reaction time (Figure 10.5). At low Ni concentration, Ni is adsorbed through ion-exchange reaction and inner-sphere surface complexation. With increasing pH or Ni concentration, surface precipitation of Ni occurs. With increasing reaction time and temperature, the formation of Ni-Al LDH becomes relatively more significant to $Ni(OH)_2$ phases. With spectroscopic evidence, it has been well recognized that Ni sorption on soil inorganic colloids facilitates the formation of surface precipitate phases, such as Ni hydroxides and Ni-metal LDH, even when the solution composition is under-saturated. A surface precipitation mechanism is generally favored because of the high sorptive concentration and long reaction time in the sorption process.

The molecular retention mechanisms of Ni with soil and clay minerals have been investigated using the corresponding XAFS (Table 10.1). Iron oxyhydroxides including hematite, ferrihydrite, and goethite are known to exhibit strong Ni retention on their surface via formation of inner-sphere surface complexes (Arai 2008; Xu et al. 2007). Arai (2008) revealed the difference in

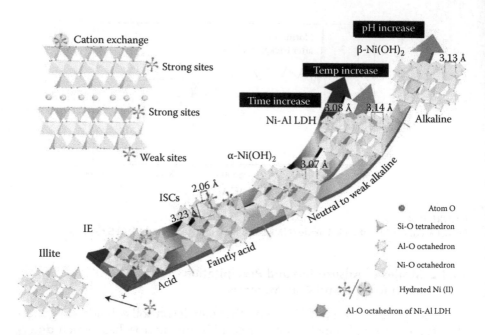

FIGURE 10.5
Ni sorption mechanisms of illite under different temperature and pH conditions proposed by Zhao et al. (2017) (Open access, CC by 4.0).

the distribution of monodentate and bidentate modes for inner-sphere surface complexes of Ni for hematite, ferrihydrite, and goethite at weak acid to neural solution pH. Nickel can form outer-sphere surface complexes with hydrous Mn oxide (Trivedi et al. 2001), illite (Elzinga and Sparks 2001), fulvic acid, and some organic acids (Strathmann and Myneni 2004), although it is rare for Fe oxyhydroxides. Precipitation of Ni occurs at a neutral to alkaline pH range. In solution free of Al or inert adsorbents, the retention of Ni is attributed to precipitation of a-type metal hydroxides (Scheckel et al. 2000; Scheinost et al. 1999). The precipitates are predominantly Al-containing LDH when the adsorbent phase releases Al during the reaction with Ni in the solution (Scheinost and Sparks 2000; Scheinost et al. 1999). Formations of these precipitates in solution rapidly decrease dissolved Ni and competes with surface adsorption of Ni on soil solid phases. Scheckel et al. (2000) compared the stability of several Ni sorption phases and decreased in the order of Ni-Al LDH on pyrophyllite, α-Ni(OH)$_2$ on talc, and Ni-Al LDH on gibbsite. Yamaguchi et al. (2001) found that the formation of α-Ni(OH)$_2$ and Ni-Al LDH on the mineral surface is affected by the presence of low-molecular-weight organic acids. Later, several studies revealed that the formation of Ni-Al LDH can be suppressed by the coating of humic acid on the surface of kaolinite via direct Ni sorption on functional groups of humic acid and/or by the formation of Ni–humus–metal ternary complexes (Nachtegaal and Sparks 2003; Strathmann and Myneni 2005).

TABLE 10.1

A Summary of Adsorption and Precipitation Mechanisms of Nickel in Soil
Clay Minerals

Adsorbent	pH	I (mol L^{-1})	Suggested Surface Species	References
Hematite	6.85	0.01	Inner-sphere bidentate (63%) and monodentate (37%)	Arai (2008)
Ferrihydrite	5.09–6.89	0.01	Inner-sphere bidentate (95%) and monodentate (5%)	Arai (2008)
Ferrihydrite	6–7	0.0028–0.1	Inner-sphere bidentate	Xu et al. (2007)
Goethite	5.09–6.89	0.01	Inner-sphere bidentate (80%) and monodentate (20%)	Arai (2008)
Pyrophyllite	7.5	0.1	Inner-sphere bidentate	Scheidegger et al.(1996)
Pyrophyllite	7.5	0.1	Surface precipitation	Elzinga and Sparks (1999)
Pyrophyllite	7.5	0.1	Ni-Al LDH	Elzinga and Sparks (2001)
Hydrous Mn oxide	5.0, 7.0		Outer-sphere	Trivedi et al. (2001)
Monocarboxylates[a]	6.0	0.44	Inner-sphere monodentate and possibly outer-sphere	Strathmann and Myneni (2004)
Di- and tricarboxylates[b]	3.5–7.0	0.1–0.56	Inner-sphere	Strathmann and Myneni (2004)
Kaolinite	7.5	0.02	Ni-Al LDH	Nachtegaal and Sparks (2003)
Montmorillonite	7.5	0.1	Surface adsorption (inner-sphere)	Elzinga and Sparks (1999)
Illite	6.25–8.00	0.003–0.1	Ni-Al LDH	Elzinga and Sparks (2001)
Illite	6.0	0.003–0.1	Outer-sphere	Elzinga and Sparks (2001)
Talc	7.5	0.1	a-Ni(OH)$_2$	Elzinga and Sparks (2001)
Gibbsite	7.5	0.1	Ni-Al LDH	Elzinga and Sparks (2001)
Fulvic acid	5.5–8.5	–	Inner-sphere and possibly outer-sphere	Strathmann and Myneni (2004)
Humic acid	7.5	0.02	Inner-sphere	Nachtegaal and Sparks (2003)

Note: I, Ionic strength of solution
[a] Acetate
[b] Oxalate, malonate, lactate, malate and citrate

10.3.3 Ni Species in Anthropogenically Contaminated Soils

There are so far few studies investigating Ni species in contaminated soils using XAS. McNear et al. (2007) studied Ni species of soils surrounding a historic refinery in Ontario, Canada. The concentration of Ni in the soils reached 4.9 g kg^{-1}, and they investigated the effect of soil type on the speciation of Ni in the organic soil (70% organic matter) and mineral soil. According

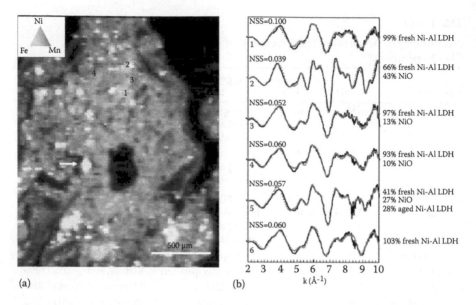

FIGURE 10.6
(a) μ-SXRF tricolor map of the Welland loam unlimed soil and (b) μ-XAFS spectra from selected points within the map (numbers 1–6). Solid lines represent the k^3 weighted χ-spectra and the dotted lines denote the best fits obtained using a linear least squares fitting approach. (From McNear, J.D.H., Chaney, R.L., Sparks, D.L., 2007. The effects of soil type and chemical treatment on nickel speciation in refinery enriched soils: A multi-technique investigation. *Geochimica et Cosmochimica Acta* 71, 2190–2208. With permission.)

to the XAFS investigation, NiO was the common Ni species in the organic and mineral soils, accounting for 30%–57% of total Ni. Microscale observations using electron microscopy and micro-X-ray fluorescence revealed that the particles of NiO had a spherical structure with 10–50 μm in diameter and were not bound to other metals. The other Ni species dominant in the mineral soil were found to be Ni-Al LDH phases. In the organic soil, Ni-Al LDH was not present, but Ni was associated with fulvic acid. Based on the XAFS results, the study of McNear et al. (2007) concluded that soluble Ni or persistent NiO particles deposited from the refinery into the soils have been partially transformed into secondary phases as precipitates and organic complexes (Figure 10.6).

10.3.4 Ni Species in Ultramafic Rocks and Soils

The nature of weathered ultramafic rocks and ultramafic soils is inherently different in locations where climatic conditions and other soil-forming factors affect weathering of parent materials. Such variability in soil mineralogy and chemistry results in the forms of Ni in ultramafic minerals and soils. Ni species in ultramafic geology have frequently been studied in lateritic soils. The primary minerals associated with Ni lateritic soils are phyllosilicates

(Manceau and Calas 1985; Wells et al. 2009), Fe oxyhydroxides (Dublet et al. 2012; Manceau et al. 2000), and Mn oxides (Dublet et al. 2012; Fan and Gerson 2011; Manceau et al. 2000), depending on the weathering stage of ultramafic rocks.

Phyllosilicates have been determined as a primary Ni-bearing phase in the early weathering stage of ultramafic rock (i.e., saprolite units). Perhaps the first application of XAFS to investigate the Ni-Mg substitution in weathering ultramafic rocks was the research of Manceau and Calas (1985) who studied intracrystalline distribution of Ni in phyllosilicates. They showed a hetero-geneous distribution of Ni in the octahedral sheet of phyllosilicates in an ore deposit in New Caledonia. Wells et al. (2009) also reported detailed informa-tion on mineralogy and crystal chemistry of hydrous Mg-Ni silicates (i.e., garnierites) in the Goro lateritic Ni deposit in New Caledonia. In a saprolite ore in the Philippines, the EXAFS investigation determined that up to 90% of Ni is associated with a serpentine mineral and a lesser extent of Ni is associ-ated with birnessite (Fan and Gerson 2011).

In contrast, Fe and Mn (oxyhydr)oxides serve as a primary host for Ni in the late or ultimate weathering stage of ultramafic rock. In a natural goethite col-lected from West Africa, about 75% of Ni is substituted for Fe in goethite and about 25% of Ni is associated with phyllomanganate impurities (Manceau et al. 2000). Fan and Gerson (2011) studied a lateritic Ni ore in the Philippines and found that 60% of total Ni was associated with phyllomanganate and 40% was incorporated in the goethite structure through substitution for Fe. Dublet et al. (2012) investigated Ni speciation in a 64-m vertical profile of saprolitic–lateritic regolith in New Caledonia using EXAFS. In the top 27 m of the lat-erite unit, Ni in primary minerals (olivine and serpentine) in the bedrock is incorporated mainly in goethite and to a lesser extent in Mn oxides (up to 30% of total Ni). In another vertical profile of New Caledonian saprolitic–lateritic regolith, incorporation of Ni in siderite ($FeCO_3$) was observed between 37 and 40 m depth where total C concentration was elevated to 6% (Dublet et al. 2014). Formation of siderite is uncommon in the oxidized environment, but is attrib-uted to the development of a swampy and organic-rich condition to create hydromorphic and reducing conditions in this specific profile.

Nickel species in ultramafic soils and sediments have recently been inves-tigated using the Ni K-edge XAFS. The concentration of Ni in soils derived from ultramafic parent materials is often lower than that of ultramafic rocks and ore deposits (Rinklebe et al., 2016), which is one of the constraint factors affecting the quality of the Ni K-edge XAFS. Noël et al. (2015) investigated Ni in mangrove sediments in New Caledonia and found differences in Ni species in a tidal zone with redox cycles and an oxidized laterite zone. In the oxidized laterite zone, Ni in the surface (2.5 cm) to subsurface (47.5 cm) soils is associ-ated with serpentine, talc, and goethite. In the tidal zone, contrarily, a large fraction of Ni is redistributed in a smectite pool throughout the soil profile, and a notable change in Ni species was found in a deeper soil profile where goethite-bearing Ni is absent and Ni-sorbed pyrite and Ni incorporation in pyrite are predominant. Siebecker et al. (2017) studied Ni species in the surface soils (<15 cm) from different ultramafic geologies in the United States. Nickel in these soils was found to be associated with primary serpentine minerals with layered octahedral and tetrahedral sheets, inosilicate minerals, and iron oxides. Compared to anthropogenically contaminated soils and ultramafic rocks, however, it seems difficult to identify specific minerals or adsorbent bearing Ni in the soils. This may be attributed to (i) structural similarity of Ni *K*-edge EXAFS spectra for several important standard compounds that are dominantly present in ultramafic soils (Siebecker et al. 2017) and (ii) complexity of weathering and soil-forming processes from ultramafic parent materials that eventually give multiple paths for the Ni-bearing phases.

To the best of our knowledge, however, no spectroscopic results have been reported to evidence the presence of Ni associated with humus in naturally occurring soils (Levard et al. 2009; Noël et al. 2015; Siebecker et al. 2017). Levard et al. (2009) studied Ni species in an Andisol derived from volcanic parent materials that naturally contain relatively high levels of Ni (575 mg kg^{-1}). They found that Ni in the soil is mainly associated with short-range-order aluminosilicates or their analogs, although the soils classified in Andisols are characterized by accumulation of humus (soil organic C concentration was not mentioned therein). This contradicts the result of soil Ni fractionation using sequential extraction methods that often found Ni in organically associated pools (Rajaie et al. 2008; Thomasi et al. 2015). Although the results between macroscopic and XAS studies are not directly comparable and often incompatible, Ni associated with humus is a minor species in uncontaminated soils or might be sometimes overlooked by XAS studies. Laboratory batch experiments found that soil humus has a high affinity for Ni and has a marked effect on Ni sorption via formations of ternary surface complexes with boehmite (i.e., humus–Ni–boehmite) (Nachtegaal and Sparks 2003; Strathmann and Myneni 2005). As mentioned previously, Ni associated with humus was identified in an anthropogenically contaminated soil (71% organic matter) where Ni was originally added as a form of NiO. It seems that the high concentration of organic C is a determinant factor for the formation of Ni–humus complexes in soils, but further studies on the identification of soil Ni species using XAFS are needed.

10.4 Conclusions

Characterizing chemical species of Ni in soils is a challenging task in soil science and geochemistry. The XAS analysis of soil Ni provides evidence that can be used to develop chemical speciation model studies to predict the bioavailability, transformations, and potential mobility of Ni in the terrestrial environment. Although XAS analysis is a useful and nondestructive technique, it should be noted that knowledge of the chemistry and mineralogy of soil is critical to determine Ni species. Macroscale and XAS investigations of soil Ni are complementary. The principles developed from macroscale investigations provide support to channel the XAS studies of chemical speciation, and the XAS studies on soils and model systems give a mechanistic understanding of macroscale phenomena.

Acknowledgment

The XAS experiments were conducted using Beamline BL01B1 at SPring-8, Japan Synchrotron Radiation Research Institute, Hyogo, Japan (Proposal Number: 2009B1280).

References

Al-Khirbash, S.A., 2016. Geology, mineralogy, and geochemistry of low grade Ni-lateritic soil (Oman Mountains, Oman). *Chemie Der Erde-Geochemistry* 76, 363–381.

Alloway, B.J., 2013. Sources of heavy metals and metalloids in soils. In: Alloway, B.J. (Ed.). *Heavy Metals in Soils*. Springer, Dordrecht, Heidelberg, New York, pp. 11–50.

Antic-Mladenovic, S., Rinklebe, J., Frohne, T., Stark, H.J., Wennrich, R., Tomic, Z., Licina, V., 2011. Impact of controlled redox conditions on nickel in a serpentine soil. *Journal of Soils and Sediments* 11, 406–415.

Arai, Y., 2008. Spectroscopic evidence for Ni(II) surface speciation at the iron oxyhydroxides—Water interface. *Environmental Science & Technology* 42, 1151–1156.

Bandara, T., Herath, I., Kumarathilaka, P., Seneviratne, M., Seneviratne, G., Rajakaruna, N., Vithanage, M., Ok, Y.S., 2017. Role of woody biochar and fungal–bacterial co-inoculation on enzyme activity and metal immobilization in serpentine soil. *Journal of Soils and Sediments* 17, 665–673.

Bani, A., Echevarria, G., Montarges-Pelletier, E., Gjoka, F., Sulce, S., Morel, J.L., 2014. Pedogenesis and nickel biogeochemistry in a typical Albanian ultramafic toposequence. *Environmental Monitoring and Assessment* 186, 4431–4442.

Bansal, O.P., 1982. Thermodynamics of K-Ni and Ca-Ni exchange reactions on kaolinite clay. *European Journal of Soil Science* 33, 63–71.

Barman, M., Datta, S.P., Rattan, R.K., Meena, M.C., 2015. Chemical fractions and bioavailability of nickel in alluvial soils. *Plant, Soil and Environment* 61, 17–22.

Belova, D.A., Lakshtanov, L.Z., Carneiro, J.F., Stipp, S.L.S., 2014. Nickel adsorption on chalk and calcite. *Journal of Contaminant Hydrology* 170, 1–9.

Belova, D.A., Lakshtanov, L.Z., Stipp, S.L.S., 2007. Nickel sorption on chalk and calcite. *Geochimica et Cosmochimica Acta* 71, A77.

Bourliva, A., Michailidis, K., Sikalidis, C., Filippidis, A., Betsiou, M., 2012. Nickel removal from aqueous solutions utilizing Greek natural bentonite and vermiculite. *Fresenius Environmental Bulletin* 21, 2466–2471.

Bradbury, M.H., Baeyens, B., 2011. Predictive sorption modelling of Ni(II), Co(II), Eu(IIII), Th(IV) and U(VI) on MX-80 bentonite and opalinus clay: A "bottom-up" approach. *Applied Clay Science* 52, 27–33.

Bradl, H.B., 2004. Adsorption of heavy metal ions on soils and soils constituents. *Journal of Colloid and Interface Science* 277, 1–18.

Butt, C.R.M., Cluzel, D., 2013. Nickel laterite ore deposits: Weathered serpentinites. *Elements* 9, 123–128.

Caillaud, J., Proust, D., Righi, D., 2006. Weathering sequences of rock-forming minerals in a serpentinite: Influence of microsystems on clay mineralogy. *Clays and Clay Minerals* 54, 87–100.

Cheng, C.H., Jien, S.H., Iizuka, Y., Tsai, H., Chang, Y.H., Hseu, Z.Y., 2011. Pedogenic chromium and nickel partitioning in serpentine soils along a toposequence. *Soil Science Society of America Journal* 75, 659–668.

Dahn, R., Scheidegger, A.M., Manceau, A., Schlegel, M.L., Baeyens, B., Bradbury, M.H., Chateigner, D., 2003. Structural evidence for the sorption of Ni(II) atoms on the edges of montmorillonite clay minerals: A polarized X-ray absorption fine structure study. *Geochimica et Cosmochimica Acta* 67, 1–15.

Demirkiran, A.R., Acemioglu, B., Gonen, T., 2016. Sorption of copper and nickel ions from solution by clay minerals. *Oxidation Communications* 39, 817–829.

Dublet, G., Juillot, F., Morin, G., Fritsch, E., Fandeur, D., Ona-Nguema, G., Brown Jr, G.E., 2012. Ni speciation in a New Caledonian lateritic regolith: A quantitative x-ray absorption spectroscopy investigation. *Geochimica et Cosmochimica Acta* 95, 119–133.

Dublet, G., Juillot, F., Morin, G., Fritsch, E., Noel, V., Brest, J., Brown, G.E., 2014. XAS evidence for Ni sequestration by siderite in a lateritic Ni-deposit from New Caledonia. *American Mineralogist* 99, 225–234.

Echevarria, G., Massoura, S.T., Sterckeman, T., Becquer, T., Schwartz, C., Morel, J.L., 2006. Assessment and control of the bioavailability of nickel in soils. *Environmental Toxicology and Chemistry* 25, 643–651.

Echeverria, J., Indurain, J., Churio, E., Garrido, J., 2003. Simultaneous effect of pH, temperature, ionic strength, and initial concentration on the retention of Ni on illite. *Colloids and Surfaces A—Physicochemical and Engineering Aspects* 218, 175–187.

Eick, M.J., Fendorf, S.E., 1998. Reaction sequence of nickel(II) with kaolinite: Mineral dissolution and surface complexation and precipitation. *Soil Science Society of America Journal* 62, 1257–1267.

Eick, M.J., Naprstek, B.R., Brady, P.V., 2001. Kinetics of Ni(II) sorption and desorption on kaolinite: Residence time effects. *Soil Science* 166, 11–17.

Elouear, Z., Amor, R.B., Bouzid, J., Boujelben, N., 2009. Use of phosphate rock for the removal of Ni^{2+} from aqueous solutions: Kinetic and thermodynamics studies. *Journal of Environmental Engineering-ASCE* 135, 259–265.

Elzinga, E.J., Sparks, D.L., 2001. Reaction condition effects on nickel sorption mechanisms in illite-water suspensions. *Soil Science Society of America Journal* 65, 94–101.

Fan, R., Gerson, A.R., 2011. Nickel geochemistry of a Philippine laterite examined by bulk and microprobe synchrotron analyses. *Geochimica et Cosmochimica Acta* 75, 6400–6415.

Ford, R.G., Bertsch, P.M., Farley, K.J., 1997. Changes in transition and heavy metal partitioning during hydrous iron oxide aging. *Environmental Science & Technology* 31, 2028–2033.

Ford, R.G., Scheinost, A.C., Scheckel, K.G., Sparks, D.L., 1999. The link between clay mineral weathering and the stabilization of Ni surface precipitates. *Environmental Science & Technology* 33, 3140–3144.

Gaudin, A., Decarreau, A., Noack, Y., Grauby, O., 2005. Clay mineralogy of the nickel laterite ore developed from serpentinised peridotites at Murrin Murrin, Western Australia. *Australian Journal of Earth Sciences* 52, 231–241.

Giannis, A., Pentari, D., Wang, J.Y., Gidarakos, E., 2010. Application of sequential extraction analysis to electrokinetic remediation of cadmium, nickel and zinc from contaminated soils. *Journal of Hazardous Materials* 184, 547–554.

Giles, C.H., Smith, D., 1974. A general treatment and classification of the solute adsorption isotherm I. Theoretical. *Journal of Colloid and Interface Science* 47, 755–765.

Gleyzes, C., Tellier, S., Astruc, M., 2002. Fractionation studies of trace elements in contaminated soils and sediments: A review of sequential extraction procedures. *TrAC—Trends in Analytical Chemistry* 21, 451–467.

Gonnelli, C., Renella, G., 2013. Chromium and nickel. In: Alloway, B.J. (Ed.). *Heavy Metals in Soils: Trace Metals and Metalloids in Soils and Their Bioavailability.* Springer, Dordrecht, pp. 313–333.

Green-Pedersen, H., Jensen, B.T., Pind, N., 1997. Nickel adsorption on MnO2, Fe(OH)3, montmorillonite, humic acid and calcite: A comparative study. *Environmental Technology* 18, 807–815.

Gu, X.Y., Evans, L.J., 2007. Modelling the adsorption of Cd(II), Cu(II), Ni(II), Pb(II), and Zn(II) onto Fithian illite. *Journal of Colloid and Interface Science* 307, 317–325.

Gu, X.Y., Evans, L.J., 2008. Surface complexation modelling of Cd(II), Cu(II), Ni(II), Pb(II) and Zn(II) adsorption onto kaolinite. *Geochimica et Cosmochimica Acta* 72, 267–276.

Han, F.X., Bannin, A., 2000. Long term transformations of Cd, Co, Cu, Ni, Zn, V, Mn and Fe in the native arid zone soils under saturated condition. *Soil Science Society of America Journal* 31, 943–957.

Hass, A., Fine, P., 2010. Sequential selective extraction procedures for the study of heavy metals in soils, sediments, and waste materials: A critical review. *Critical Reviews in Environmental Science and Technology* 40, 365–399.

Hlavay, J., Prohaska, T., Weisz, M., Wenzel, W.W., Stingeder, G.J., 2004. Determination of trace elements bound to soils and sediment fractions. *Pure and Applied Chemistry* 76, 415–442.

Hseu, Z.Y., 2006. Concentration and distribution of chromium and nickel fractions along a serpentinitic toposequence. *Soil Science* 171, 341–353.

Hseu, Z.Y., Su, Y.C., Zehetner, F., Hsi, H.C., 2017. Leaching potential of geogenic nickel in serpentine soils from Taiwan and Austria. *Journal of Environmental Management* 186, 151–157.

Ivanets, A.I., Srivastava, V., Kitikova, N.V., Shashkova, I.L., Sillanpaa, M., 2017. Kinetic and thermodynamic studies of the Co(II) and Ni(II) ions removal from aqueous solutions by Ca-Mg phosphates. *Chemosphere* 171, 348–354.

Jiang, B., Su, D., Wang, X.K., Liu, J., Ma, Y., 2018. Field evidence of decreased extractability of copper and nickel added to soils in 6-year field experiments. *Frontiers in Environmental Science and Engineering* 12.

Kabata-Pendias, A., Pendias, H., 2011. *Trace Elements in Soils and Plants*. CRC Press, Boca Raton, FL.

Kelepertzis, E., Stathopoulou, E., 2013. Availability of geogenic heavy metals in soils of Thiva town (central Greece). *Environmental Monitoring and Assessment* 185, 9603–9618.

Kelly, S.D., Hesterberg, D., Ravel, B., 2008. Analysis of soils and minerals using X-ray absorption spectroscopy. *Method of Soil Analysis, Part 5—Mineralogical Method*. Soil Science Society of America, Madison, WI.

Lakshtanov, L.Z., Stipp, S.L.S., 2007. Experimental study of nickel(II) interaction with calcite: Adsorption and coprecipitation. *Geochimica et Cosmochimica Acta* 71, 3686–3697.

Levard, C., Doelsch, E., Rose, J.E., Masion, A., Basile-Doelsch, I., Proux, O., Hazemann, J.-L., Borschneck, D., Bottero, J.-Y., 2009. Role of natural nanoparticles on the speciation of Ni in andosols of la Reunion. *Geochimica et Cosmochimica Acta* 73, 4750–4760.

Malamis, S., Katsou, E., 2013. A review on zinc and nickel adsorption on natural and modified zeolite, bentonite and vermiculite: Examination of process parameters, kinetics and isotherms. *Journal of Hazardous Materials* 252, 428–461.

Manceau, A., Calas, G., 1985. Heterogeneous distribution of nickel in hydrous silicates from New Caledonia ore deposits. *American Mineralogist* 70, 549–558.

Manceau, A., Calas, G., 1986. Nickel-bearing clay-minerals. 2. Intracrystalline distribution of nickel—An x-ray absorption study. *Clay Minerals* 21, 341–360.

Manceau, A., Marcus, M.A., Tamura, N., 2002. Quantitative speciation of heavy metals in soils and sediments by synchrotron X-ray techniques. *Reviews in Mineralogy and Geochemistry* 49, 341–428.

Manceau, A., Schlegel, M.L., Musso, M., Sole, V.A., Gauthier, C., Petit, P.E., Trolard, F., 2000. Crystal chemistry of trace elements in natural and synthetic goethite. *Geochimica et Cosmochimica Acta* 64, 3643–3661.

Marcussen, H., Holm, P.E., Strobel, B.W., Hansen, H.C.B., 2009. Nickel sorption to goethite and montmorillonite in presence of citrate. *Environmental Science & Technology* 43, 1122–1127.

Massoura, S.T., Echevarria, G., Becquer, T., Ghanbaja, J., Leclere-Cessac, E., Morel, J.L., 2006. Control of nickel availability by nickel bearing minerals in natural and anthropogenic soils. *Geoderma* 136, 28–37.

Mattigod, S.V., Gibali, A.S., Page, A.L., 1979. Effect of ionic strength and ion pair formation on the adsorption of nickel by kaolinite. *Clays and Clay Minerals* 27, 411–416.

McNear, J.D.H., Chaney, R.L., Sparks, D.L., 2007. The effects of soil type and chemical treatment on nickel speciation in refinery enriched soils: A multi-technique investigation. *Geochimica et Cosmochimica Acta* 71, 2190–2208.

Merdy, P., Huclier, S., Koopal, L.K., 2006. Modeling metal–particle interactions with an emphasis on natural organic matter. *Environmental Science & Technology*, 7459–7466.

Nachtegaal, M., Sparks, D.L., 2003. Nickel sequestration in a kaolinite–humic acid complex. *Environmental Science & Technology* 37, 529–534.

Noël, V., Morin, G., Juillot, F., Marchand, C., Brest, J., Bargar, J.R., Muñoz, M., Marakovic, G., Ardo, S., Brown, G.E., 2015. Ni cycling in mangrove sediments from New Caledonia. *Geochimica et Cosmochimica Acta* 169, 82–98.

Peltier, E., van der Lelie, D., Sparks, D.L. 2010. Formation and stability of Ni-Al hydroxide phases in soils. *Environmental Science & Technology*, 44, 302–308.

Perrone, J., Fourest, B., Giffaut, E., 2001. Sorption of nickel on carbonate fluoroapatites. *Journal of Colloid and Interface Science* 239, 303–313.

Pokrovsky, O.S., Schott, J., 2002. Surface chemistry and dissolution kinetics of divalent metal carbonates. *Environmental Science & Technology* 36, 426–432.

Rajaie, M., Karimian, N., Yasrebi, J., 2008. Nickel transformation in two calcareous soil textural classes as affected by applied nickel sulfate. *Geoderma* 144, 344–351.

Rajapaksha, A.U., Vithanage, M., Oze, C., Bandara, W., Weerasooriya, R., 2012a. Nickel and manganese release in serpentine soil from the Ussangoda Ultramafic Complex, Sri Lanka. *Geoderma* 189, 1–9.

Rajapaksha, A.U., Vithanage, M., Weerasooriya, R., Dissanayake, C.B., 2012b. Surface complexation of nickel on iron and aluminum oxides: A comparative study with single and dual site clays. *Colloids and Surfaces A—Physicochemical and Engineering Aspects* 405, 79–87.

Rao, C.R.M., Sahuquillo, A., Sanchez, J.F.L., 2008. A review of the different methods applied in environmental geochemistry for single and sequential extraction of trace elements in soils and related materials. *Water, Air and Soil Pollution* 189, 291–333.

Rinklebe, J., Antic-Mladenovic, S., Frohne, T., Stark, H.J., Tomic, Z., Licina, V., 2016. Nickel in a serpentine-enriched Fluvisol: Redox affected dynamics and binding forms. *Geoderma* 263, 203–214.

Saffari, M., Karimian, N., Ronaghi, A., Ghasemi-Fasaei, R., 2015. Stabilization of nickel in a contaminated calcareous soil amended with low-cost amendments. *Journal of Soil Science and Plant Nutrition* 15, 896–913.

Scheckel, K.G., Scheinost, A.C., Ford, R.G., Sparks, D.L., 2000. Stability of layered Ni hydroxide surface precipitates—A dissolution kinetics study. *Geochimica et Cosmochimica Acta* 64, 2727–2735.

Scheidegger, A.M., Lamble, G.M., Sparks, D.L., 1996. Investigation of Ni sorption on pyrophyllite: An XAFS study. *Environmental Science & Technology* 30, 548–554.

Scheidegger, A.M., Lamble, G.M., Sparks, D.L., 1997. Spectroscopic evidence for the formation of mixed-cation hydroxide phases upon metal sorption on clays and aluminum oxides. *Journal of Colloid and Interface Science* 186, 118–128.

Scheidegger, A.M., Sparks, D.L., 1996. Kinetics of the formation and the dissolution of nickel surface precipitates on phyrophyllite. *Chemical Geology* 132, 157–164.

Scheidegger, A.M., Strawn, D.G., Lamble, G.M., Sparks, D.L., 1998. The kinetics of mixed Ni-Al hydroxide formation on clay and aluminum oxide minerals: A time-resolved XAFS study. *Geochimica et Cosmochimica Acta* 62, 2233–2245.

Scheinost, A.C., Ford, R.G., Sparks, D.L., 1999. The role of Al in the formation of secondary Ni precipitates on pyrophyllite, gibbsite, talc, and amorphous silica: A DRS study. *Geochimica et Cosmochimica Acta* 63, 3193–3203.

Scheinost, A.C., Sparks, D.L., 2000. Formation of layered single- and double-metal hydroxide precipitates at the mineral/water interface: A multiple-scattering XAFS analysis. *Journal of Colloid and Interface Science* 223, 167–178.

Sen Gupta, S., Bhattacharyya, K.G., 2006. Adsorption of Ni(II) on clays. *Journal of Colloid and Interface Science* 295, 21–32.

Shaheen, S.M., Tsadilas, C.D., Rinklebe, J., 2013. A review of the distribution coefficients of trace elements in soils: Influence of sorption system, element characteristics, and soil colloidal properties. *Advances in Colloid and Interface Science* 201–202, 43–56.

Sheikhhosseini, A., Shirvani, M., Shariatmadari, H., 2013. Competitive sorption of nickel, cadmium, zinc and copper on palygorskite and sepiolite silicate clay minerals. *Geoderma* 192, 249–253.

Sheikhhosseini, A., Shirvani, M., Shariatmadari, H., Zvomuya, F., Najafic, B., 2014. Kinetics and thermodynamics of nickel sorption to calcium-palygorskite and calcium-sepiolite: A batch study. *Geoderma* 217, 111–117.

Shen, Z., Som, A.M., Wang, F., Jin, F., McMillan, O., Al-Tabbaa, A., 2016. Long-term impact of biochar on the immobilisation of nickel (II) and zinc (II) and the revegetation of a contaminated site. *Science of the Total Environment* 542, 771–776.

Siebecker, M.G., Chaney, R.L., Sparks, D.L., 2017. Nickel speciation in several serpentine (ultramafic) topsoils via bulk synchrotron-based techniques. *Geoderma* 298, 35–45.

Singh, B., Grafe, M., 2010. *Synchrotron-Based Techniques in Soils and Sediments.* Elsevier, Oxford, UK.

Sparks, D.L., 1998. Kinetics and mechanisms of chemical reactions at the soil mineral/water interface. In: Sparks, D.L. (Ed.). *Soil Physical Chemistry.* CRC Press, Boca Raton, pp. 135–192.

Sposito, G., 2004. *The Surface Chemistry of Natural Particles.* Oxford University Press, New York.

Strathmann, T.J., Myneni, S.C.B., 2004. Speciation of aqueous Ni(II)-carboxylate and Ni(II)-fulvic acid solutions: Combined ATR-FTIR and XAFS analysis. *Geochimica et Cosmochimica Acta* 68, 3441–3458.

Strathmann, T.J., Myneni, S.C.B., 2005. Effect of soil fulvic acid on nickel(II) sorption and bonding at the aqueous-boehmite (γ-AlOOH) interface. *Environmental Science & Technology* 39, 4027–4034.

Tack, F.M.G., Verloo, M.G., 1995. Chemical speciation and fractionation in soil and sediment heavy metal analysis: A review. *International Journal of Environmental Analytical Chemistry* 59, 225–238.

Tamura, H., Furuichi, R., 1997. Adsorption affinity of divalent heavy metal ions for metal oxides evaluated by modeling with the Frumkin isotherm. *Journal of Colloid and Interface Science* 195, 241–249.

Templeton, D.M., Ariese, F., Cornelis, R., Danielsson, L., Muntau, H., van Leeuwen, H.P., Lobinski, R., 2000. Guidelines for terms related to chemical speciation and fractionation of elements. Definitions, structural aspects, and methodological approaches. *Pure and Applied Chemistry* 72, 1453–1470.

Tessier, A., Campbell, P.G.C., Bisson, M., 1979. Sequential extraction procedure for the speciation of particular trace metals. *Analytical Chemistry* 51, 844–851.

Tewari, G., Tewari, L., Srivastava, P.C., Ram, B., 2010. Nickel chemical transformation in polluted soils as affected by metal source and moisture regime. *Chemical Speciation and Bioavailability* 22, 141–155.

Thomasi, S.S., Fernandes, R.B.A., Frontes, R.L.F., Jordao, C.P., 2015. Sequential extraction of copper, nickel, zinc, lead and cadmium from Brazilian Oxysols: Metal leaching and metal distribution in soil fractions. *International Journal of Environmental Studies* 72, 41–55.

Trescases, J.J., 2012. The lateritic nickel-ore deposits. In: Paquet, H., Clauer, N. (Eds.). *Soils and Sediments: Mineralogy and Geochemistry.* Springer Science & Business Media, Berline, pp. 125–138.

Trivedi, P., Axe, L., 2001. Predicting divalent metal sorption to hydrous Al, Fe, and Mn oxides. *Environmental Science & Technology* 35, 1779–1784.

Trivedi, P., Axe, L., Tyson, T.A., 2001. XAS studies of Ni and Zn sorbed to hydrous manganese oxide. *Environmental Science & Technology* 35, 4515–4521.

Ure, A.M., Davidson, C.M., 2002. *Chemical Speciation in the Environment, Second Edition.* Wiley.

Uren, N.C., 1992. Forms, reaction, and availability of nickel in soils. *Advances in Agronomy* 48, 141–203.

Veith, J.A., Sposito, G., 1977. Reactions of aluminosilicates, aluminum hydrous oxides, and aluminum oxide with o-phosphate: The formation of X-ray amorphous analogs of variscite and montebrasite. *Soil Science Society of America Journal* 41, 870–876.

Vijayaraghavan, K., Raja, F.D., 2015. Interaction of vermiculite with Pb(II), Cd, Cu(II) and Ni(II) ions in single and quaternary mixtures. *Clean-Soil Air Water* 43, 1174–1180.

Violante, A., Cozzolino, V., Perelomov, L., Caporale, A.G., Pigna, M., 2010. Mobility and bioavailability of heavy metals and metalloids in soil environments. *Journal of Soil Science and Plant Nutrition* 10, 268–292.

Vithanage, M., Rajapaksha, A.U., Oze, C., Rajakaruna, N., Dissanayake, C.B., 2014. Metal release from serpentine soils in Sri Lanka. *Environmental Monitoring and Assessment* 186, 3415–3429.

Vodyanitskii, Y.N., 2006. Methods of sequential extraction of heavy metals from soils: New approaches and the mineralogical control (A review). *Eurasian Soil Science* 39, 1074–1083.

Wells, M.A., Ramanaidou, E.R., Verrall, M., Tessarolo, C., 2009. Mineralogy and crystal chemistry of "garnierites" in the Goro lateritic nickel deposit, New Caledonia. *European Journal of Mineralogy* 21, 467–483.

Xu, Y., Axe, L., Boonfueng, T., Tyson, T.A., Trivedi, P., Pandya, K., 2007. Ni(II) complexation to amorphous hydrous ferric oxide: An x-ray absorption spectroscopy study. *Journal of Colloid and Interface Science* 314, 10–17.

Yamaguchi, N.U., Scheinost, A.C., Sparks, D.L., 2001. Surface-induced nickel hydroxide precipitation in the presence of citrate and salicylate. *Soil Science Society of America Journal* 65, 729–736.

Zachara, J.M., Cowan, C.E., Resch, C.T., 1991. Sorption of divalent metals on calcite. *Geochimica et Cosmochimica Acta* 55, 1549–1562.

Zhao, X., Qiang, S., Wu, H., Yang, Y., Shao, D., Fang, L., Liang, J., Li, P., Fan, Q., 2017. Exploring the sorption mechanism of Ni(II) on illite: Batch sorption, modeling, EXAFS and extraction investigations. *Scientific Report* 7, 8495.

Zimmerman, A.J., Weindorf, D.C., 2010. Heavy metal and trace metal analysis in soil by sequential extraction: A review of procedure. *International Journal of Analytical Chemistry*, Article ID 387803.

11

Nickel Adsorption–Desorption and Mobility in Soils: Evidence of Kinetics

Lixia Liao and H. Magdi Selim

CONTENTS

11.1 Introduction

Nickel (Ni) is a trace element that is carcinogenic to humans. Nickel enters the environment through application of fertilizer, sewage sludge, lime, and industrial waste. The average concentration of nickel in soils is 40 mg/kg. A large number of mineral and chemical compounds containing nickel are formed in soils (McIlveen and Negusanti 1994). Atmospheric sources of nickel include windblown dust, derived from the weathering of rocks and soils, volcanic emissions, forest fires, and vegetation (Cempel and Nikel 2006). The bioavailability and mobility of Ni influence the amounts taken up by plants and microorganisms and thus the toxicity of the environment. Moreover, Ni is among five trace elements (Cu, Zn, Pb, Ni, and Cd) selected by the United States Environmental Protection Agency (USEPA) according to Code of Federal Regulation CFR 40 Part 503 for land-applied sewage sludges (USEPA 1993, 1995).

Several investigations suggested that Ni may be considered as weakly sorbed by most soils in comparison to others such as Pb, Cu, and Mg (Atanassova 1999; Tiller et al. 1984). Ni was observed to have low affinity for reactive phases in acidic soils and was thus considered mobile and susceptible to transport. Under neutral to alkaline conditions, Ni was found to form multinuclear complexes on several mineral phases (Scheidegger et al. 1998),

illite (Elzinga and Sparks 2001), and kaonilite (Erick et al. 2001). These findings were based on X-ray absorption fine structure (XAFS) spectroscopy.
However, due to sorbent heterogeneity possessing a broad array of sorption
sites and each processing a unique spectroscopic signature, limited studies
were carried out on clay-sized isolates (Businelli et al. 2004; Roberts et al.
1999) and soil (Voegelin and Kretzschmar 2005).

11.2 Sorption Isotherms

Sorption of Ni on minerals, clay fractions, and soils has been frequently
measured using the traditional batch equilibration method. The relationship between the equilibrium concentrations in the aquatic solution and the
amount adsorbed on the solid surface, that is, the partition distribution coefficient, is commonly described with adsorption isotherms. Linear and nonlinear forms are usually employed to describe Ni adsorption on mineral and
soil surfaces. The Freundlich isotherm method is one of the simplest ways to
quantify the behavior of retention of reactive solute with the matrix surfaces.
It is certainly one of the oldest nonlinear sorption equations and was based
on quantifying the gas adsorbed by unit mass of solid with pressure, which
was described as

$$S = K_f C^b, \tag{11.1}$$

where S is the amount of solute retained by the soil in µg/g or mg/kg, C is
the solute concentration in solution in mg/L or µg/ml, K_f is the partitioning
coefficient in L/kg or ml/g, and b is often considered a dimensionless parameter. A major disadvantage of the Freundlich approach is that it is incapable
of describing sorption at or near saturation or when adsorption maximum
is reached. The parameter K_f represents the partitioning of a solute species
between solid and liquid phases over the concentration range of interest and
is analogous to the equilibrium constant for a chemical reaction. For b equals
unity, the Freundlich equation takes on the form of the linear equation.

$$S = K_d C, \tag{11.2}$$

where the parameter K_d is the distribution coefficient in L/kg or ml/g. As
K_d is used, this implies a linear, zero-intercept relationship between sorbed
and solution concentration, which is a convenient assumption but certainly
not universally true. This linear model is often referred to as a constant partition model and the parameter K_d is a universally accepted environment
parameter and it reflects the affinity of matrix surfaces to solute species.

The K_d parameter provides an estimate for the potential for the adsorption of dissolved contaminants in contact with soil and typically used in fate and contaminant transport calculations. According to USEPA (1999), K_d is defined as the ratio of the contaminant concentration associated with the solid to the contaminant concentration in the surrounding aqueous solution when the system is at equilibrium.

Although somewhat simplistic, the K_d approach is easy to integrate into various chemical models and allows estimations of metal dissolved in soil solution and predictions of mobility as well as potential leaching losses (Mellis et al. 2004). Covelo et al. (2004) employed the linear form to estimate the competitive sorption and desorption of heavy metals in mine soil and correlated K_d values to principal soil components such as organic matter, Fe oxides, and CEC. Modeling metal sorption using a single-valued K_d approach presumes that the sorption capacity of a material is relatively independent of soil physicochemical properties. However, due to the heterogeneity of the soil matrix, models of Freundlich (Antoniadis and Tsadilas 2007) and Langmiur (Papini et al. 2004; Srivastava et al. 2006) equations are more commonly used to model batch data. Both types are nonlinear and indicative of high-affinity chemical adsorption.

An example of adsorption isotherms for Ni is presented in Figure 11.1 for three different soils (Liao and Selim 2010a). Physical and chemical properties of the three soils are given in Table 11.1. These Ni sorption isotherms exhibit strong nonlinear retention behavior for all soils, which was well described using the Freundlich equation. Comparison of sorption capacity among the three soils indicated that Webster soil has the highest sorption for Ni, whereas Windsor exhibits the lowest sorption (see Table 11.1).

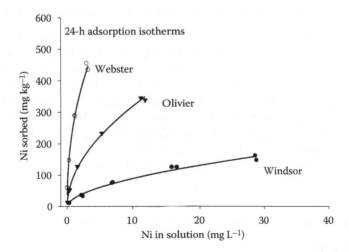

FIGURE 11.1
Nickel adsorption isotherms for Webster, Olivier, and Windsor soils after 24 h of reaction. Solid curves depict results of curve fitting using Freundlich Equation 11.1.

TABLE 11.1

Selected Physical and Chemical Properties of the Soils Studied

Soil		Olivier Loam	Webster Loam	Windsor Sand
Taxonomic Classification		Fine-Silty, Mixed, Thermic Aquic Fragiudalf	Fine-Loamy, Missed, Mesic Typic Haplaquoll	Mixed, Mesic Typic Udipsamment
pH		5.80	6.92	6.11
TOC[a]	%	0.83	4.02	2.03
CEC[b]	cmol/kg	8.6	27.0	2.0
$CaCO_3$	%	–	3.7	–
Sand[c]	%	5	39	77
Silt	%	89	39	20
Clay	%	6	22	3
Clay mineralogical composition (fraction <2 μm)[d]	%	Kaolinite (31%), illite (30%), smectite (28%), quartz(11%)	Smectite (73%), quartz (11%), kaolinite (9%), illite (7%),	Illite (33%), kaolinite (29%), chlorite (15%), smectite (12%), quartz (10%)
Selective extraction by Ammonium oxalate (pH 3.0)				
Fe	g/kg	0.32	0.98	0.36
Al	g/kg	0.08	0.89	0.69
Citrate–bicarbonate–dithionite (CBD)				
Fe	g/kg	4.09	4.42	3.68
Al	g/kg	1.29	0.77	3.65

[a] TOC, total organic carbon.
[b] CEC, cation exchange capacity.
[c] Grain size distribution: sand (0.05–2.00 mm), silt (0.05–0.002 mm), and clay (<0.002 mm).

The 24-h K_f values for Windsor, Olivier, and Webster soils were 25.34, 96.35, and 268.53 L/kg, respectively. Consistent with observations by other researchers (Gomes et al. 2001; Papini et al. 2004), soils with high CEC exhibited strong affinity for Ni sorption. As expected, Webster soil showed the highest sorption for Ni because of its high organic matter content and the fact that its clay type is dominated by smectites. On the other hand, Olivier and Windsor soils exhibited the lowest Ni affinity because of their relatively low organic matter content and the fact that their clay was dominated by kaolinite and illite, and less smectites (see Table 11.1).

11.3 Kinetics

Kinetic or time-dependent sorption of Ni has been frequently observed and often related to soil heterogeneity, slow diffusion to reaction sites within the soil matrix, rate-limited reactions, and precipitation at mineral surfaces (Erick et al. 2001; Jeon et al. 2003; Pang and Close 1999; Voegelin and Kretzschmar 2005). Jeon et al. (2003) reported that Ni and Cd kinetic sorption by hematite was instantaneous, followed by a relatively slow stage that continued for several days. Also, they found that the extent of retention of metal ions by hematite was Ni > Cd. Under neutral or somewhat alkaline conditions, Ni was found to form multinuclear complexes on several mineral phases, including pyrophyllite, montmorillonite, gibbsite, illite, and kaonilite, and even on natural soils using extended X-ray fine structure spectroscopy (Eric et al. 2001; Scheidegger et al. 1996). Moreover, this slow buildup of multinuclear complex was highly time-dependent and irreversible or weakly reversible. Such a slow release or no release was observed following leaching by dilute HNO_3 at pH 3 when competitive ions were absent.

The example shown in Figure 11.2 illustrates the effect of time of reaction on Ni sorption on the three soils of Figure 11.1. Time-dependent retention is clearly indicated by increased Ni sorption with time. Furthermore, the set of Ni sorption isotherms shown exhibited strong nonlinear Ni retention behavior and was well described using the Freundlich model. Estimates for the Freundlich parameters N and K_f along with R^2 values are presented for the different reaction times in Table 11.2. Time dependency of Ni sorption is demonstrated by increasing values of the Freundlich parameter K_f with reaction time (Figure 11.3, top). The nonlinearity of Ni isotherms is indicated by the small values of the Freundlich N (less than 1) for all reaction times (see Figure 11.3, bottom). The parameter N did not change after reaction of 24 h with average values of 052, 0.55, and 0.50 for Windsor, Olivier, and Webster soil, respectively. The small decrease in N before 24 h may be due to slow diffusion through intra-particle microspores, which also account for kinetic behavior of Ni (Strawn and Sparks 1999). Moreover, It is important to emphasize that N represents the energy distribution or the heterogeneity of the sorption site, where the highest-energy sites are preferentially sorbed at low concentrations, and as the concentration increases, successively lower-energy sites become occupied (Sheindorf et al. 1981). A major implication of the nonlinearity of the isotherms is that the magnitude of N directly affects the mobility of Ni in the soil profile. At low concentrations, solute affinity is highest, which results in limited mobility of Ni in the soil. As concentrations in solution increases, solute affinity decreases and the Zn becomes increasingly more mobile.

The time dependency of Ni sorption and release for the different input (initial) concentrations is given in Figure 11.4. The amount of Ni sorbed with time was dependent on the input Ni concentration. For example, the kinetics for Webster were characterized by an extremely rapid initial step with nearly 91% of Ni sorbed in 2 h, followed by a much slower sorption region where some 95% of the Ni was sorbed within 672 h. Such two-stage reaction is characteristic of several heavy metal sorption on clays, oxide surfaces, and soils as suggested by Eick et al. (2001) and Jeon et al. (2003), among others.

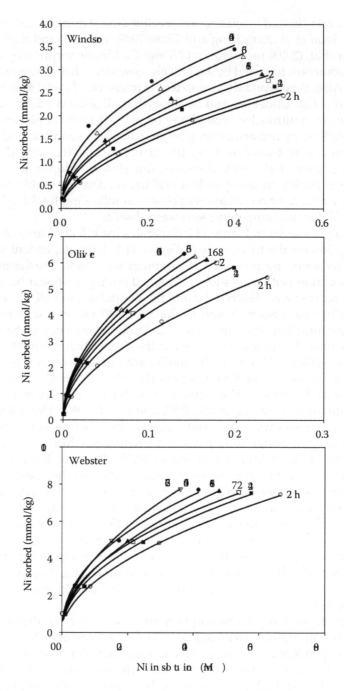

FIGURE 11.2
Adsorption isotherms of Ni by three soils at different reaction times. Symbols are for different reaction times of 2, 24, 72, 168, 336, and 504 h for Windsor and Olivier and 2, 24, 72, 168, 504, and 672 h for Webster.

TABLE 11.2

Estimated Freundlich Parameters with Stand Errors for Nickel Adsorption and Desorption at Different Reaction Times for Three Soils

Time (h)	Windsor			Olivier			Webster		
	K_f (L/kg)	b	R^2	K_f (L/kg)	b	R^2	K_f (L/kg)	b	R^2
2	3.680 ± 0.153	0.575 ± 0.033	0.998	12.332 ± 0.458	0.558 ± 0.019	0.999	28.338 ± 3.897	0.454 ± 0.039	0.999
8	3.654 ± 0.126	0.556 ± 0.027	0.998	12.740 ± 0.347	0.542 ± 0.013	0.999	29.879 ± 4.036	0.486 ± 0.040	0.997
12	3.724 ± 0.204	0.542 ± 0.042	0.996	13.301 ± 0.489	0.540 ± 0.018	0.999	30.083 ± 3.684	0.462 ± 0.063	0.997
24	4.068 ± 0.271	0.556 ± 0.027	0.994	13.716 ± 0.532	0.521 ± 0.018	0.999	30.404 ± 1.761	0.501 ± 0.018	0.999
72	4.203 ± 0.299	0.501 ± 0.050	0.996	14.753 ± 0.738	0.513 ± 0.022	0.999	31.385 ± 3.513	0.447 ± 0.047	0.986
168	4.405 ± 0.263	0.492 ± 0.041	0.996	15.131 ± 0.757	0.498 ± 0.021	0.998	31.388 ± 3.212	0.458 ± 0.088	0.995
336	5.146 ± 0.273	0.486 ± 0.034	0.997	16.246 ± 1.036	0.503 ± 0.026	0.998	35.509 ± 1.021	0.505 ± 0.008	0.999
504	5.363 ± 0.452	0.477 ± 0.050	0.992	17.053 ± 1.114	0.499 ± 0.026	0.997	42.998 ± 1.100	0.515 ± 0.007	0.993

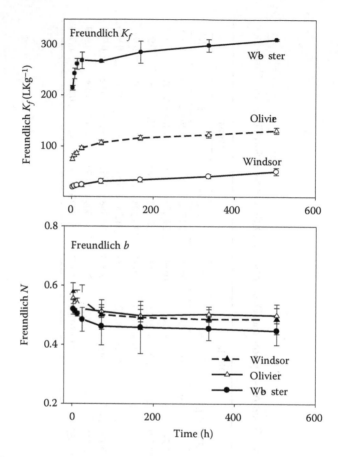

FIGURE 11.3
Freundlich parameters K_f and b versus retention time for Ni sorption for Windsor, Olivier, and Webster soils. Vertical bars represent one standard deviation.

A number of sorption mechanisms have been proposed over the past decades, including the formation of surface-induced precipitates of Ni sorption in neutral nonacidic soils. For acidic soils, cation exchange is perhaps the major mechanism for Ni sorption (Echeverría 1998; Gomes et al. 2001; Papini et al. 2004). Antoniadis and Tsadilas (2007) reported that as the pH increases, Ni sorption was related to hydrolysis of divalent ions capable of forming inner-sphere complexes with clay lattice edges. In addition, Schulthess and Huang (1990) showed that Ni adsorption by clays was strongly influenced by pH as well as silicon and aluminum oxide surface ratios. These investigators suggested that, at high pH, increased Ni sorption was due to precipitates on mineral surfaces that were characterized as a time-dependent mechanism. Our results shown in Figures 11.2 and 11.4 suggest that Ni sorption in the neutral Webster soil was much higher than the other two (acidic) soils.

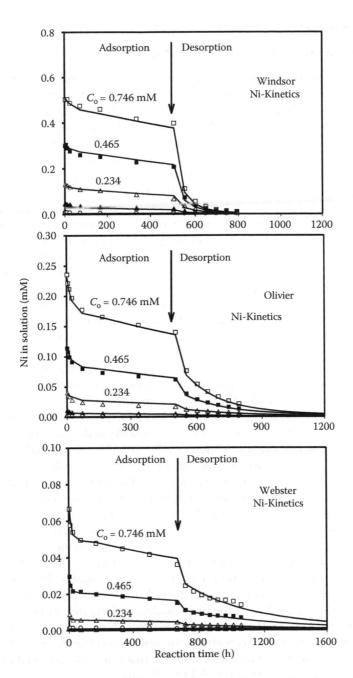

FIGURE 11.4
Nickel concentration in solution versus reaction time for three soils. Symbols are for different initial concentrations: 0.023, 0.093, 0.234, 0.465, and 0.746 mM (from bottom to top, respectively). Solid lines are MRM simulations by utilizing parameters optimized from experimental adsorption data sets.

11.4 Desorption Hysteresis and Release

Results of Ni desorption are presented as isotherms in the traditional manner in Figure 11.5. This family of desorption isotherms represents the amount of Ni sorbed during release or desorption for different initial Ni concentrations. These results indicate extensive hysteresis as depicted by the discrepancies of the adsorption from the desorption isotherms. Lack of equilibrium and irreversible sorption are perhaps responsible for the observed hysteresis (Selim 2015; Strawn and Sparks 1999). This was not surprising in view of the observed kinetic retention behavior of Ni sorption by these soils and various minerals

Based on the results presented in Figure 11.5, the amount of Ni desorbed as a percentage of that sorbed varied among soils. At the early stages of desorption, a number of processes may be responsible for the observed hysteresis, including transformation of weakly bound Ni to more strongly bound phases, for example, change in the type of surface complex, diffusion into micropores, or intraparticle spaces (Eick et al. 2001). For Webster soil with neutral pH, nuclear precipitation on the mineral phase may be responsible for the limited desorption or partial reversibility. Moreover, desorption results in Figure 11.5 indicate that at low Ni surface coverage, only a small portion of Ni was desorbed, indicating high sorption affinity of Ni by the soil matrix. In contrast, at high Ni input concentrations, the percentage of desorption increased, indicating lower Ni affinity.

At the termination of desorption, sequential extractions were carried out. Results of the sequential fractions (exchangeable, carbonate, Fe/Al, organic matter, and residual Ni) are shown in Figure 11.6. Sequential extractions were carried out to provide an insight into the understanding of the chemical binding of Ni in soil (Tessier et al. 1979). The exchangeable fraction was considered as weakly sorbed and nonspecific, and sorption on the remaining three fractions is of high binding strength and considered as specific. Metal cations were spontaneously sorbed on the exchangeable fractions, of which 90% were completed in 1 min and equilibrium was obtained within 30 min, but sorption on specific fractions continued with time (Tsang and Lo 2006). For all three soils, Ni bound with Fe/Al and organic matter fractions ranged from 19% to 60%. Singh et al. (1992) found that major proportions of Ni in soils were concentrated with the iron oxides, and the dissolution kinetics of these elements indicate that some may be present in the structure of the iron oxides, which is partially irreversible or slow reversible. Such findings provide evidence that iron oxides may be responsible for the observed hysteretic or partially reversible sorption of Ni for all soils.

For Webster soil, considerable irreversible fractions were observed for all input concentrations. This was expected since Webster soil is a fine loamy Haplaquoll with 3.7% $CaCO_3$. Businelli et al. (2004) found that calcium carbonate contributes to Ni retention through the formation of a strong

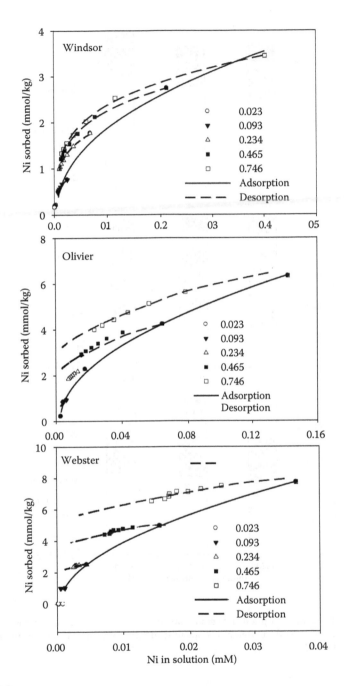

FIGURE 11.5
Traditional desorption isotherms of nickel by three soils. The solid curve is the adsorption isotherms of 504 h. Symbols are for different initial concentrations: 0.023, 0.093, 0.234, 0.465, and 0.746 mM (from bottom to top, respectively). The dashed lines are MRM simulations by utilizing parameters optimized from experimental adsorption data sets.

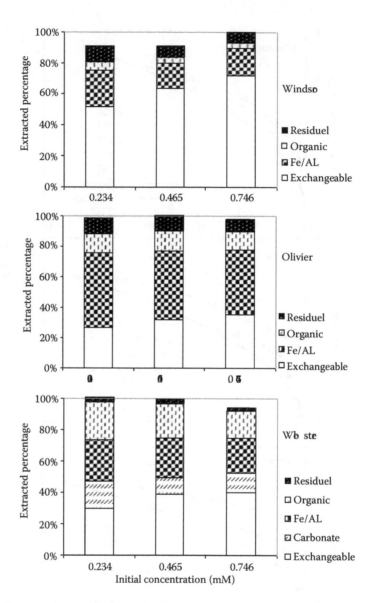

FIGURE 11.6
Recoveries of Ni from desorption and sequential extractions as percentages of total adsorption amounts for different soils. Different groups indicate different initial concentrations of 0.234, 0.465, and 0.746 mM.

complex via coprecipitation that involves Ni/Ca carbonate double salt or mixed Ni/Al hydroxides and carbonate formation. Ni/Al layered double hydroxide was observed at pH 6.5 or higher (Scheidegger et al. 1998) and builds up with time. Moreover, Ni/Al layered double hydroxides are highly stable and irreversible, and resistant to dissolution in dilute HNO_3 (Scheckel and Sparks 2001).

11.5 Multireaction Modeling

The multireaction model (MRM) schematically illustrated in Figure 11.7 accounts for several interactions of heavy metals with soil matrix surfaces. Based on soil heterogeneity and observed kinetics of sorption–desorption, the MRM accounts for nonlinear and kinetic reactivities of heavy metals in the soil environment (Selim 2015). Basic to the multireaction or multisite approach is that the soil solid phase is made up of different constituents (soil minerals, organic matter, iron, and aluminum oxides) and that a solute species is likely to react with various constituents (sites) by different mechanisms. The uniqueness of this model is that its aim is to describe the reactivity of solutes with natural systems versus time. Moreover, MRM provides a comprehensive accounting of adsorption–desorption processes, where a

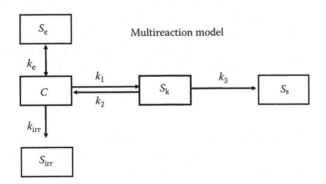

FIGURE 11.7
A schematic diagram of the MRM. Here, C is concentration in solution, S_e, S_k, S_s, and S_{irr} are the amounts sorbed on equilibrium, kinetic, consecutive, and concurrent irreversible sites, respectively, where k_e, k_1, k_2, k_3, and k_{irr} are the respective rates of reactions.

single set of parameters is sought that is applicable for an entire data set and for a wide range of initial (or input) concentrations. In the model, S_e represents the amount retained on equilibrium sites (mM/kg), S_k and S_s represent the amount retained on reversible kinetic sites (mM/kg), S_{irr} and S_s represent the amount irreversibly retained (mg/kg). The retention reactions associated with MRM can be expressed as

$$S_e = K_e \left(\frac{\theta}{\rho}\right) C^n \tag{11.3}$$

$$\frac{\partial S_k}{\partial t} = k_1 \left(\frac{\theta}{\rho}\right) C^n - [k_2 + k_3]S_k \tag{11.4}$$

$$\frac{\partial S_s}{\partial t} = k_3 S_k \tag{11.5}$$

$$\frac{\partial S_{irr}}{\partial t} = k_{irr} \left(\frac{\theta}{\rho}\right) C, \tag{11.6}$$

where C is concentration in solution (mg/L), ρ is soil bulk density (g/cm³), θ is soil water content (cm³/cm³), n is a dimensionless reaction order, and t is reaction time (h). The parameter K_e is an equilibrium constant (dimensionless) and k_1, k_2, k_3, and k_{irr} (h⁻¹) are the associated rates of reactions illustrated in Figure 11.7. The total amount sorbed S (mM/kg) is defined as

$$S = S_e + S_k + S_s + S_{irr}. \tag{11.7}$$

As illustrated by the simulations shown in Figures 11.4 and 11.5, the time-dependent behavior of Ni sorption and desorption by all soils was well described by the MRM. Parameter estimates for the simulation shown and the goodness of fit for the different MRM formulations are given in Liao and Selim (2010a). It should also be emphasized that for other models (e.g., linear, Freundlich, Langmiur, dual domain reactivity models, and treble domain reactivity models), two distinct sets of parameter are often obtained: one for adsorption and one for desorption. The uniqueness of the MRM is that it provides one set of parameters applicable for both adsorption and desorption, that is, an entire data set.

The capability of MRM to describe Ni desorption versus time was tested by Liao and Selim (2010a). They tested the hypothesis that desorption results can be predicted using MRM where the necessary model parameters were derived solely from adsorption data sets. In other words, the model was tested to predict Ni desorption or release with reliance on adsorption results only. Liao and Selim (2010a) found that there were no significant differences between model parameter estimates based on adsorption data sets or the entire (adsorption–desorption) data sets. In fact, the simulations shown in Figures 11.3 and 11.4 were obtained using adsorption parameters only. Therefore, Ni desorption can be predicted based on parameter from adsorption data alone.

Since different versions of the MRM represent different reactions from which one can deduce Ni retention mechanisms, several model versions were examined. Based on model predictions of Ni sorption with time for all soils, there were little distinguishable differences among several model variations. Moreover, for all three soils, it was not possible to determine whether best predictions were obtained when either S_{irr} or S_s was used in the model to account for irreversible reactions. Nevertheless, for all soils, model versions that accounted for kinetic reactions (S_k) are essential to provide better model predictions of measured Ni retention compared to model versions with S_e and S_{irr}. For Webster soil, incorporation of the equilibrium retention phase (S_e) was necessary to describe the initially rapid retention of Ni in this soil. A poor fit of the model to measured results was obtained when S_e was not incorporated (results not shown). This is consistent with measured results where 90% of retention was observed in the first 2 h of reactions. Therefore, for Webster soil, the full model formulation was the best to describe the highly irreversible and low desorption behavior of Ni.

11.6 Transport

Several approaches have been proposed over the past four decades to describe the transport of reactive solutes in soils. The earliest approaches focus on linear-type retention of solutes where local equilibrium is assumed dominant. For example, Liu et al. (2006) used linear adsorption with the convection–dispersion transport equation to quantify retardation factors and dispersion coefficients for Cd, Ni, and Zn transport in an acidic soil. Their assumptions are acceptable for transport under equilibrium conditions or in a homogenous porous medium. Pang and Close (1999) used the CXTFIT program, which incorporates two regions/sites with nonequilibrium reaction, to estimate transport parameters of Cd in an alluvial gravel column. Their results showed that a two-region (nonequilibrium) approach provided improved description of measured Cd data. Another model combining cation exchange/specific sorption (Voegelin et al. 2001) was used to describe the Ni, Cd, and Zn transport in an acidic soil. This model provided good prediction when heavy metal adsorption was shown to be reversible and kinetic effects were negligible.

Transport under conditions where nonequilibrium is dominant is often characterized by late arrival or extensive retardation and asymmetrical breakthrough curves (BTCs) (Goltz and Roberts 1986; Pang and Close 1999; Selim et al. 1989; van den Brink and Zaadnoordijk 1997). Recently, asymmetrical Ni BTCs were observed by Antoniadis et al. (2007) in a clay soil using a modified centrifuge infiltration column. Such asymmetry of Ni BTC displays a relatively slow breakthrough front as well as prolonged tailing during leaching. We should emphasize here that most Ni transport studies were often carried out using continuous application of Ni pulse, which results in a plateau of concentration over time (Liu et al. 2006; Michel et al. 2007). This method is not recommended since the release of applied Ni and mobility in soils cannot be accurately estimated.

In this section, we present experimental measurements and modeling efforts of the transport of Ni in the soils discussed earlier. The measurements were based on miscible displacement designed to quantify Ni interaction and mobility in soil columns. Such measurements provide means for quantifying retardation kinetics and observed extensive release during leaching. To achieve this, we first incorporated the multireaction reaction described earlier into the convection–dispersion equation (CDE)

$$\frac{\partial C}{\partial t} + \frac{\rho}{\theta}\frac{\partial S}{\partial t} = D\frac{\partial^2 C}{\partial z^2} + v\frac{\partial C}{\partial z}. \tag{11.8}$$

Here, the reaction mechanisms are presented by the term $(\partial S/\partial t)$, where S is the total amount of solute retained expressed in Equation 11.1, D is the hydrodynamic dispersion coefficient (cm^2/h), v $(= q/\theta)$ is the pore water velocity (cm/h), q is Darcy's velocity (cm/h^1), and z is distance (cm) in the soil. Equation 11.8 accounts for the above reaction mechanisms during transport (CDE) where steady-state water flow conditions are maintained.

Breakthrough results (BTCs) for Ni in all three soils are shown in Figure 11.8 (Liao and Selim 2010b). All BTCs appear retarded relative to the transport and peaks show a significant shift to the right as indicated by the late arrival of Ni in the effluent solution. Such strong retardation is indicative of the extent of Ni retention during transport and suggests kinetic (reversible and irreversible) retention in soils. The extent of retardation varied among the soils. BTCs for Windsor and Olivier soils indicated a sharp rise of effluent concentration where higher peak maxima were associated with the second pulse application of Ni in the miscible displacement columns. These peaks were accompanied by moderate tailing during desorption (right side of BTCs). This observed tailing may be due to the nonlinear and kinetic adsorption behavior, which is consistent with our batch experimental results. Moreover, measured Ni BTC for Webster soil shown in Figure 11.8 illustrates a gradual

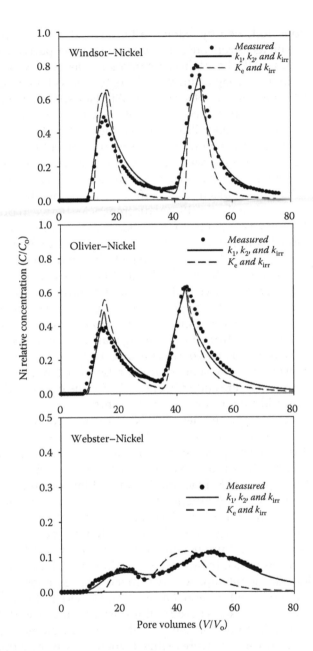

FIGURE 11.8
Nickel BTCs result from two applied Ni pulses for Windsor, Olivier, and Webster soils. Solid and dashed curves are multireaction transport mode predictions where model parameters were those from nonlinear inverse modeling.

(or diffuse) front, followed by extensive tailing and a retardation of Ni in the effluent as well as lower concentration maxima compared to Olivier and Windsor soils. Overall, the highest retardation was observed for Webster soil, which exhibited strong affinity (with highest K_f value). In contrast, the lowest retardation for Ni was observed for Windsor soil, which is consistent with sorption affinities based on our kinetic batch experiments, where the affinity followed the sequence Windsor < Olivier < Webster.

In an attempt to describe Ni transport from the soil columns, simulations were obtained when MRTM was utilized in a simulation mode (Liao and Selim 2010b). In these simulations, the necessary model parameters were derived from the kinetic batch rate coefficient. The simulated BTCs are not presented here. For all soils, significant discrepancies between predicted and experimental results were evident for all model variations used. Overall, the use of batch model parameters resulted in underestimated Ni retention during transport in all soil columns. As a result, the model overestimated the potential mobility of Ni in all soils. Such MRTM failure was also observed by Kohler et al. (1996) and Barnett et al. (2000) for uranium (VI) and Zhang and Selim (2006) for arsenic (As) transport through soils. Zhang and Selim (2006) argued that different retention capacities determined from batch and column experiments might result from the following reasons: difference between the sorption time used for batch experiments and the hydrologic retention time of column experiments, low solid/solution ratio of batch experiments, and the fact that chemicals were added in one spike for the batch study compared with continuous addition in column experiments.

In subsequent attempts to describe Ni transport in the soil columns, MRTM was utilized in an inverse mode to test its capability for describing Ni BTCs without reliance on parameter estimates from the batch experiments. Excellent fit of the data was achieved in describing Ni mobility for all soils, as illustrated in Figure 11.8. Liao and Selim (2010b) concluded that, for all three soils, incorporation of kinetic phase (S_k) was necessary to obtain the predictions shown in Figure 11.8. This is consistent with results from batch experiments where Ni reaction exhibited highly time-dependent behavior. For Windsor soil, the best predictions were obtained when S_k was the only sorbed phase in the MRM. In fact, incorporation of S_{irr} (or S_s) along with S_k did not provide additional improvements in model prediction. Incorporation of S_{irr} most likely resulted in model overfitting as evident by the large standard error associated of k_{irr}. These findings are consistent with those from the adsorption batch study. For Olivier soil, similar findings were obtained where limited improvements were realized when S_{irr} was incorporated in the model. Moreover, for all model versions, incorporation of equilibrium phase (S_e) resulted in poor predictions. Therefore, incorporation of S_e reaction was not recommended in describing Ni BTCs as illustrated in Figure 11.8.

For Webster soil, several model versions produced good model predictions of the Ni BTCs. MRM variation that accounted for equilibrium and concurrent irreversible reactions (S_e and S_{irr}) provided a poor description of measured Ni BTCs. All other model variations provided equally good predictions as long as the kinetic phase (S_k) and irreversible reactions (S_{irr}) were incorporated into the model. However, incorporation of the equilibrium phase (S_e) and/or irreversible reactions (S_s) did not provide additional improvements in model predictions. This finding is contrary to that based on Ni adsorption results (see Table 11.2) for Webster soil and illustrates the differences between results from kinetic (batch) experiments and those of BTC from column studies as previously discussed.

Differences in Ni transport BTCs from the soil columns are due to the different properties of the soils investigated by Liao and Selim (2010a,b). Windsor soil has a high content of sand and kaolinite, a highly weathered mineral. Sparks (1995) argued that in soils where kaolinite and illite minerals are dominant, cations tend to form outer-sphere complexes due to ion exchange (electrostatic). Sorbed Ni as outer-sphere complexes is considered mobile, whereas when smectitic-type clay is dominant, stronger affinities are expected between heavy metals and soils due to inner-sphere complexes as is the case for Webster soil (Serrano et al. 2005). We can associate time-dependent (specific) sorption phase as S_k, whereas S_{irr} may be associated with irreversible or slowly irreversible sorption. Tsang and Lo (2005) suggested that heavy metal sorption with organic matter and iron oxides forms specific sorption complexes that are time dependent. For Olivier and Windsor soil columns, time-dependent sorption of Ni is likely responsible for the significant S_k phase in MRM description.

Strong irreversible Ni retention by Webster soil is evident by the low recovery of Ni in the effluent. At the termination of the miscible displacement experiment, 81.1% of applied Ni was retained by the Webster soil column. This was expected since Webster soil is a fine loamy Haplaquoll with 3.7% $CaCO_3$. Businelli et al. (2004) found that calcium carbonate contributes to Ni retention through the formation of a strong complex via coprecipitation that involves Ni/Ca carbonate double salt or mixed Ni/Al hydroxides and carbonate formation. Ni/Al layered double hydroxides were observed at pH 6.5 or higher (Scheidegger et al. 1998). Moreover, the formation of Ni/Al layered double hydroxides is time dependent, which continues for several days or months. This form was highly stable and irreversible, and resistant to dissolution in diluted HNO_3 (Scheidegger et al. 1998). On the other hand, contrary to our findings based on transport data for Webster soil, batch results indicated that the use of reversible reactions of the equilibrium type (S_e) was recommended for describing Ni sorption data. It is conceivable that fundamental differences between batch and column experiments are perhaps responsible for the fact that different model variations are necessary for describing batch results separate from those describing transport BTCs. Therefore, additional research is needed to distinguish mechanisms necessary to provide an understanding of adsorption reaction associated with Ni retention during transport in soils. Such knowledge is necessary in order to provide improvements of the MRTM model formulation presented here.

References

Antoniadis, V. and C. D. Tsadilas. 2007. Sorption of cadmium, nickel and zinc in mono- and multimetal systems. *Appl. Geochem.* 22:2375–2380.

Atanassova, I. 1999. Competitive effect of copper, zinc, cadmium and nickel on ion adsorption and desorption by soil clays. *Water Air Soil Pollut.* 113:115–125.

Barnett, M. O., P. M. Jardine, S. C. Brooks, and H. M. Selim. 2000. Adsorption and transport of uranium(VI) in subsurface media. *Soil Sci. Soc. Am. J.* 64:908–917.

Bussnelli, D., F. Casciari, and G. Gigliotti. 2004. Sorption mechanisms determining Ni(II) retention by a calcareous soil. *Soil Sci.* 169:355–362.

Cempel, M., and G. Nikel. 2006. Nickel: A review of its sources and environmental toxicology. *Polish J. Environ. Stud.* (Roma, 1904) 15:375–382.

Covelo, E. F., Couce, M. L. A., and F. A. Vega. 2004. Competitive adsorption and Desorption of cadmium, chromium, copper, nickel, lead, and zinc by humic umbrisols. *Communications in Soil Science and Plant Analysis* 35:2709–2729.

Echeverría, J.C., M.T. Morera, C. Mazkiarán, and J.J. Garrido. 1998. Competitive sorption of heavy metal by soils: Isotherms and fractional factorial experiments. *Environ. Pollut.* 101:275–284.

Eick M. J., B. R. Naprstek, and P. V. Brady. 2001. Kinetics of Ni(II) sorption and desorption on kaolinite: Residence time effects. *Soil Sci.* 166:11–17.

Elzinga, E. J., and D. J. Sparks. 2001. Reaction condition effects on Nickel sorption mechanisms in illite–water suspensions. *Soil Sci. Soc. Am. J.* 65:94–101.

Goltz, M. N., and P. V. Roberts. 1986. Interpreting organic solute transport data from a field experiment using physical non-equilibrium models. *J. Contam. Hydrol.* 1:77–93.

Gomes, P. C., M. P. F. Fontes, and A. G. da Silva. 2001. Selectivity sequence and competitive adsorption of heavy metals by Brazilian soils. *Soil Sci. Soc. Am. J.* 65:1115–1121.

Jeon, B., B. A. Dempsey, and W. D. Burgos. 2003. Sorption kinetics of Fe(II), Zn(II), Co(II), Ni(II), Cd(II), and Fe(II)/Me(II) onto hematite. *Water Res.* 37:4135–4142.

Kohler, M., G. P. Curtis, D. B. Kent, and J. A. Davis. 1996. Experimental investigation and modeling of uranium(VI) transport under variable chemical conditions. *Water Resour. Res.* 32:3539–3551.

Liao, L. and H. M. Selim. 2010a. Reactivity of nickel in soils: Evidence of retention kinetics. *J. Environ. Qual.* 39:1290–1297.

Liao, L. and H. M. Selim. 2010b. Transport of nickel in different soils: Column experiments and kinetic modeling. *Soil Sci. Soc. Am. J.* 74:1945–1955.

McIlveen, W. D., and J. J. Negusanti. 1994. Nickel in the terrestrial environment. *Sci. Total Environ.* 148:109–138.

Mellis, E. V., M. C. P. Cruz, and J. C. Casagrande. 2004. Nickel adsorption by soils in relation to pH, organic matter, and iron oxides. *Sci. Agric.* 61:190–195.

Michel, K., M. Roose, and B. Ludwig. 2007. Comparison of different approaches for modeling heavy metal transport in acidic soils. *Geoderma.* 140:207–214.

Pang, L. and M. E. Close. 1999. Non-equilibrium transport of Cd in alluvial gravels. *J. Contam. Hydrol.* 36: 185–206.

Papini, M. P., T. Saurini, and A. Bianchi. 2004. Modeling the competitive adsorption of Pb, Cu, Cd and Ni onto a natural heterogeneous sorbent material (Italian "Red Soil"). *Ind. Eng. Chen. Res.* 43:5032–5041.

Roberts, D., A. M. Scheidegger, and D. L. Sparks. 1999. Kinetics of mixed Ni- Al precipitation formation on a soil clay fraction. *Environ. Sci. Technol.* 33:3749–3754.

Scheckel, K. G., and D. L. Sparks. 2001. Dissolution kinetics of nickel surface precipitates on clay mineral and oxide surfaces. *Soil Sci. Soc. Am. J.* 65:685–694.

Scheidegger, A. M., D. G. Strawn, G. M. Lamble, and D. L. Sparks. 1998. The kinetics of mixed Ni-Al hydroxide formation on clay and aluminum oxide minerals: A time-resolved XAFS study. *Geochim. Cosmochim. Acta.* 62(13):2233–2245.

Scheidegger, A. M., G. M. Lamble, and D. L. Sparks. 1996. Investigation of Ni sorption on pyrophyllite: An XAFS study. *Environ. Sci. Technol.* 30:548–554.

Schulthess, C. P., and C. P. Huang. 1990. Adsorption of heavy-metals by silicon and aluminum-oxide surfaces on clay-minerals. *Soil Sci. Soc. Am. J.* 54:679–688.

Sheindorf, C., M. Rebhun, and M. Sheintuch. 1981. A Freundlich-type multicomponent isotherm. *J. Colloid Interface Sci.* 79:136–142.

Selim, H. M. 1989. Modeling the kinetics of heavy metals reactivities in soil, pp. 91–106. In: H. M. Selim and K. I. Iskandar (eds.), *Fate and Transport of Heavy Metals in the Vadose Zone.* CRC Press, Boca Raton, FL.

Selim, H. M. 2015. *Chemical Transport and Fate in Soils: Principles and Applications.* CRC/ Taylor and Francis, Boca Raton, FL (352 pp.).

Serrano, S., F. Garrido, C. G. Campbell, and M. T. Garcia-Gonzalez. 2005. Competitive sorption of cadmium and lead in acid soils of central Spain. *Geoderma* 124:91–104.

Singh, B., and R. J. Gilkes. 1992. Properties and distribution of iron oxides and their association with minor elements in the soils of south-western Australia. *J. Soil Sci.* 43:77–98.

Sparks, D. L. 1995. Environmental soil chemistry. Academic Press, San Diego.

Srivastava, V. C., Mall, I. D., and I. M. Mishra. 2006. Equilibrium modelling of single and binary adsorption of cadmium and nickel onto bagasse fly ash. *Chemical Engineering Journal* 117:79–91.

Strawn, D. G. and D. L. Sparks. 1999. Sorption kinetics of trace metals in soils and soil materials, pp. 1–28. In: H. M. Selim and K. I. Iskandar (eds.), *Fate and Transport of Heavy Metals in the Vadose Zone.* CRC Press, Boca Raton, FL.

Tessier, A., P. Campbell, and M. Bisson. 1979. Sequential extraction procedure for the speciation of particulate trace metals. *Anal. Chem.* 51:844–851.

Tiller, K. G., J. Gerth, and G. Brümmer. 1984. The sorption of Cd, Zn and Ni by soil clay fractions: Procedures for partition of bound forms and their interpretation. *Geoderma* 34:1–16.

Tsang, D. C. W. and I. M. C. Lo. 2006. Competitive Cu and Cd and transport in soils: A combined batch kinetics, column, and sequential extraction study. *Sci. Total Environ.* 40:6655–6661.

USEPA. 1993. 40 CFR-Part 257 and 503, Standards for the disposal of sewage sludge; final rule. *Fed. Regist.* 58:9248–9415. USEPA, Washington, DC.

USEPA. 1995. Land application of sewage sludge and domestic seepage: Process design manual. EPA-625-R-95-001. USEPA, Washington, DC.

USEPA. 1999. Understanding variation in partition coefficient, K_d, values. Report EPA 402-R-99-004A.

264

van den Brink, C., and W. J. Zaadnoordijk. 1997. Non-equilibrium transport and sorption of organic chemicals during aquifer remediation. *Hydrol. Sci. J.* 42:245–264.

Voegelin, A., and R. Kretzschmar. 2005. Formation and dissolution of single and mixed Zn and Ni precipitates in soil: Evidence from column experiments and extended x-ray absorption fi ne structure spectroscopy. *Environ. Sci. Technol.* 39:5311–5318.

Voegelin, A., V. M. Vulava, and R. Kretzschmar. 2001. Reaction-based model describing competitive sorption and transport of Cd, Zn, and Ni in an acidic soil. *Environ. Sci. Technol.* 35:1651–1657.

Zhang, H. and H. M. Selim. 2006. Modeling the transport and retention of arsenic (V) in soils. *Soil Sci. Soc. Am. J.* 70:1677–1687.

12

Nickel Mobilization/Immobilization and Phytoavailability in Soils as Affected by Organic and Inorganic Amendments

Sabry M. Shaheen, Svetlana Antić-Mladenović,
Shan-Li Wang, Nabeel Khan Niazi, Christos D. Tsadilas,
Yong Sik Ok, and Jörg Rinklebe

CONTENTS

12.1 Introduction

Nickel (Ni) is an essential element for animals and plants (Kabata-Pendias 2011). Nickel contamination of soils can occur from weathering of parent material and anthropogenic activities (Rinklebe and Shaheen 2017a,b). However, as with other potentially toxic elements (PTEs), Ni can be mobilized, thus leading to soil and groundwater contamination, which increase the possibility of entering the human food chain and, as such, cause potential health risk (Antoniadis et al. 2017; Rinklebe et al. 2017). Elevated Ni concentrations in soils could have negative impact on the plants, microorganisms, and animals (Thakali et al. 2006; Yusuf et al. 2011). High soil Ni concentration can also cause adverse impacts on soil functions and considerable environmental problems regarding the mobility and thus soil–plant transfer and consequently Ni transfer into the food chain (Antoniadis et al. 2017). In addition, the toxicity of Ni in plants has become a worldwide problem threatening sustainable agriculture as well (Ma and Hooda 2010). Therefore, soil contamination with Ni could be a global environmental and health concern (Rinklebe and Shaheen 2017a).

Remediation of Ni-contaminated soils has received increasing attention and is important for adequate environmental management (Shaheen et al. 2015a). Over the past 5 years, the term "green remediation" has been extensively used (Shaheen and Rinklebe 2015a; Tomasevic et al. 2013). In situ immobilization of PTEs, including Ni, is a remediation approach based on adding easily available amendments to polluted soils aiming to reduce the solubility of PTEs in soil without altering their total concentrations (Lee et al. 2013; Shaheen et al. 2015a, 2017a; Tica et al. 2011; Varrault and Bermond 2011; Rinklebe et al. 2017).

In situ immobilization of PTEs including Ni depends on the soil characteristics, concentrations, and forms of Ni in soils, as well as mobilization and bioavailability of soil Ni. In this chapter, we will include and discuss the following: (1) total content of Ni in different soils, (2) geochemical fractions and mobilization of Ni in soils, (3) the impact of different soil amendments on geochemical fractions and mobilization/immobilization of Ni, and (4) the impact of soil amendments on phytoavailability and plant tissue concentrations of Ni in different soils. We will gather the scattered knowledge from papers in this chapter and compare these results, which are crucial for soil scientists and readers to understand the geochemical behavior of Ni in soil; this will in turn help in understanding Ni mobilization, hazard, and ecotoxicity. Additionally, detailed knowledge about the impact of soil amendments on mobilization and phytoavailability of Ni in soils is required, in order to know suitable mobilizing/immobilizing agents for Ni-contaminated soils.

12.2 Total Contents of Ni in Soils

The Ni content in soils is highly dependent on its contents in the parent material. The mean concentration of Ni in various types of rocks (mg kg^{-1}) are 2000 in ultramafic igneous, 140 in basaltic igneous, 68 in shales and clays, 50 in black shales, 20 in limestone, 8 in granitic igneous, and 2 in sandstone. Nickel is also found in association with mafic and ultramafic rock formations (Iyaka 2011; Ma and Hooda 2010). In addition to geogenic origin, Ni in soils can be derived from anthropogenic deposits. Therefore, the concentration of Ni in surface soils reflects the combined impact of both soil-forming processes and anthropogenic activities. Soils throughout the world contain Ni in the very broad range (Rinklebe and Shaheen 2017a).

The Ni content in most natural soils varies widely (0.2 to 450 mg kg^{-1}). However, Ni mean content as reported for various countries is within the range from 13 to 40 mg kg^{-1} (Iyaka 2011; Ma and Hooda 2010). The content of Ni in many arable soils seldom exceeds 50 mg kg^{-1}, but with ultramafic bedrock such as serpentine or peridotite (which are naturally enriched with Ni), it can reach more than 10,000 mg kg^{-1} (Kabata-Pendias 2011). The highest natural Ni contents given for soils of various countries and different pedoclimatic environment are as follows: 450 mg kg^{-1} in China (Quiping et al. 1984), 550 mg kg^{-1} in Serbia (Antić-Mladenović et al. 2011), 675 mg kg^{-1} in Italy (Abollino et al. 2002), 660 mg kg^{-1} in Japan and 3240 mg kg^{-1} in Italy (Bini et al. 1988), 2600 mg kg^{-1} in France and 12,000 mg kg^{-1} in New Caledonia (Massoura et al. 2006), 758 mg kg^{-1} in Austria and 2080 mg kg^{-1} in Taiwan (Chang et al. 2013; Hseu et al. 2017), 6776 mg kg^{-1} in Sri Lanka (Rajapaksha et al. 2012), and 3700 mg kg^{-1} along the ultramafic toposequence in Albania (Bani et al. 2014). Soils derived from serpentine rocks of the Coast Range and Sierra Nevada of California contained 1300–3900 mg Ni kg^{-1} (Morrison et al. 2009). Soils of the taiga zone of Western Siberia contain Ni within the range of 20–100 mg kg^{-1} (Niechayeva 2002). In surface horizons of Russian chernozems, mean Ni contents range from 28 to 34 mg kg^{-1} (Protasova and Kopayeva 1985). The median Ni content of sandy soils of Lithuania is 9 mg kg^{-1}, and that of loamy clay soils is 17 mg kg^{-1} (Kadunas et al. 1999). The total content of Ni in soils of Finland is 60 mg kg^{-1} at the 90th percentile (Koljonen 1992). Soils of arid and semiarid regions are likely to have a high Ni content. Roca et al. (2008) observed, in Argentina, the relationship between soil types and Ni content, with Calcisols being the lowest (10.6 mg kg^{-1}) and Fluvisols being the highest (21.0 mg kg^{-1}). Rinklebe and Shaheen (2017b) studied total concentration of Ni in soil profiles of Fluvisols, Luvisols, Gleysols, and Calcisols originating from Germany and Egypt and found that the maximum total concentration of Ni (105 mg kg^{-1}) was recorded in German Eutric Fluvisols, while they observed the minimum Ni concentration (5 mg kg^{-1}) in Tidalic Fluvisols.

One of the best documented cases of high soil pollution with Ni is in the area of the Ni–Cu smelter at Sudbury, Canada, with up to 26,000 mg Ni kg^{-1} reported for topsoils near the smelter (Cox and Hutchinson 1981). A well-known PTE-emiting smelter in Europe is the "Severonickel" Ni smelting complex, located in the central part of Kola Peninsula, south of Monchegorsk city, Russia. Viventsova et al. (2005) found 2000–4600 mg Ni kg^{-1} in topsoil at a 2-km distance from the "Severonickel" smelter with a median value of 4200 mg Ni kg^{-1}. In Albania, soils around the Burrel chromium smelter contain Ni concentrations of up to 1243 mg kg^{-1} (Shtiza et al. 2005). More details about the concentrations of Ni in different soils and sediments are provided in Rinklebe and Shaheen (2017a).

Burning of fuel and residual oils is the largest diffuse anthropogenic source of Ni in Ni-polluted soils, with an estimated global emission of 27 kt Ni year^{-1} (Kabata-Pendias and Mukherjee 2007). Combustion of coal is the next most important emission followed by Ni mining and smelting, which is likely the largest point source with severe local impact on soil and vegetation. In addition, application of various biosolids (e.g., livestock manures, composts, and industrial and municipal sewage sludge) to agricultural land either as fertilizers or soil conditioners inadvertently leads to accumulation of many metals, including Ni in soil (Bhogal et al. 2003; Malinowska 2017).

12.3 Geochemical Fractions and Potential Mobilization of Ni in Soils

Total content of Ni has been used to assess the risk of soil pollution by comparing it with background or guideline values and creating a pollution index such as enrichment factor, geo-accumulation index, contamination factor, and complex quality index (Rinklebe and Shaheen 2014). However, total contents of PTEs, like Ni in soils, are usually a poor indicator of their availability to plants. This is mainly because the total PTE content in soil represents the sum of the contents in various geochemical fractions (soluble + exchangeable, carbonate-associated, Fe-Mn oxides associated, organic matter associated, and residual fractions) of different reactivity (Pueyo et al. 2008). Sequential extraction procedures were developed in order to extract geochemical fractions of Ni in soils to predict retention, mobility, and bioavailability of Ni to plants and groundwater. Consequently, sequential extraction procedures enable an appropriate environmental risk assessment of soil pollution by PTEs including Ni (Fijalkowski et al. 2017; Rinklebe and Shaheen 2014).

Distribution of PTEs in the fractions reflects the dynamic relationship between ions in solution and solid phases, which is governed by many factors, such as soil pH, redox potential, total and dissolved organic matter content and quality, mineralogical composition of soil, cation and anion exchange capacity, and soil amendments (Rodríguez et al. 2009). The potential mobile fractions (PMFs) (carbonate-, Fe-Mn oxides associated, and OM-bound fractions) of Ni may become bioavailable if the pH and redox potential of the soil changed (Shaheen and Rinklebe 2014). Therefore, determining the geochemical fractions of Ni is appropriate to assess its potential mobilization and pollution status in soils and it is a key issue in many environmental studies. More important, determining the geochemical fractions of Ni in soils can be helpful in understanding the dynamics of shift from one pool to another, especially before and after remediation of Ni-contaminated soils (Kabata-Pendias and Mukherjee 2007).

Rinklebe and Shaheen (2017b) studied the geochemical distribution of Ni in soil profiles of Fluvisols, Luvisols, Gleysols, and Calcisols originating from Germany and Egypt (Figure 12.1). They found found that the reducible fraction of Ni was the dominant nonresidual fraction (as precentages of total) in all Egyptian soils and in German Eutric Fluvisols. The PMF (PMF = F1 + F2 + F3) of Ni accounted for 2%–73% in all soils. The high PMF of Ni in the German Eutric Fluvisols, Haplic Gleysols, and Sodic Fluvisols might reflect the anthropogenic origin of Ni in these soils. However, dominance of

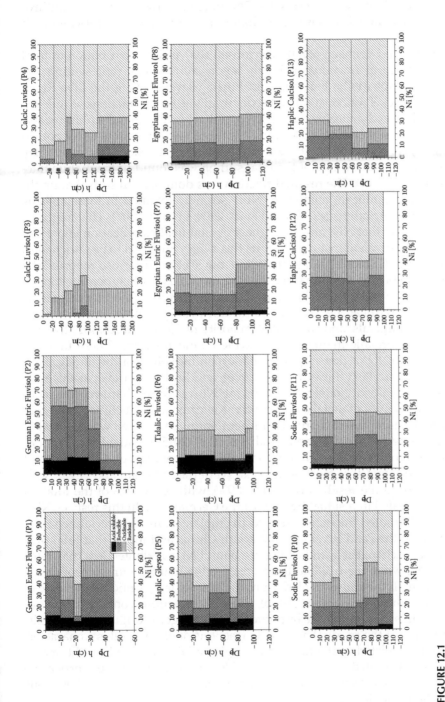

FIGURE 12.1
Sequentially extracted fractions of Ni in the studied soil profiles. (Reproduced from Rinklebe and Shaheen, 2017b. *Geoderma* 307, 122–138, with permission from the publisher.)

Ni in the residual fraction in Egyptian soils supports the assumption of the geogenic origin of Ni in these soils. The potential risk of Ni in the anthro-pogenically polluted soils is higher than in the soils naturally enriched with this element.

Rinklebe and Shaheen (2014) determined the vertical distribution of total and geochemical fractions of Ni in seven soil profiles representing the two soil groups Mollic Fluvisols and Eutric Gleysols along the Elbe River, Germany. Nickel was fractionated sequentially to seven fractions as follows: F1: soluble + exchangeable, F2: easily mobilizable, F3: bound to Mn oxides, F4: bound to soil organic matter, F5: bound to low crystalline (amorphous) Fe oxides, F6: bound by crystalline Fe oxides, and F7: residual fraction. They found that only up to 13% of the total soil Ni was found in the mobile fraction (MF = F1 + F2). The residual fraction was dominant in soils that implicated that a significant amount of Ni is strongly retained in these soils as also reported by others (e.g., Jordán et al. 2009; Li et al. 2007). This might be due to the fact that Ni^{2+} has the highest crystal field stabilization energy of the common divalent metals, and thus has a high potential to be enriched in clay minerals (Osakwe 2013). Antić-Mladenović et al. (2011) also reported the residual fraction as a dominating binding form of Ni in a serpentine topsoil in Serbia. Similarly, Rinklebe et al. (2016a) determined the binding forms of Ni in a Fluvisol at the River Velika Morava valley (Serbia), enriched with serpentine minerals, and found that majority of Ni was in residual fraction followed by organic matter and Fe oxide bound fractions. These findings coincide with Ni Fractions reported for serpentine soils worldwide [e.g., Italy (Abollino et al. 2002), Taiwan (Hseu 2006), and Sri Lanka (Rajapaksha et al. 2012)]. Thus, majority of serpentine-borne Ni was found hard-reactive concerning dynamics of metals. Differently, Kierczak et al. (2008) reported that Fe/Mn oxides represent a primary mineral form to which Ni was associated in serpentine soil from Poland. Shaheen et al. (2015b) studied the fractionation and mobilization of Ni in floodplain soils from Egypt, Germany, and Greece and found that most of the total Ni in the Egyptian and Greek soils was present in the residual fraction (>89%), while only 45% of the total Ni was fractionated in residual fraction in the German soil. In this respect, Xiao et al. (2012) studied fractionation of Ni in cropland soils from reclaimed tidal wetlands in the Pearl River estuary, south China, and found that total proportion of exchangeable and carbonate bound Ni contributed less than 6.6% of total Ni, while the residual form was predominant, having a proportion of >68% in all the soils sampled. In many studies (e.g., Antić-Mladenović et al. 2017; Shaheen et al. 2014a, 2015b), these two fractions contained ~83% of total Ni, thus confirming the hypothesis of its geogenic origin in these soils. Domination of the residual and oxide fractions of Ni in a soil is a characteristic pattern related to the lithogenic origin of Ni in this soil. Additionally, major Ni associated to the relative stable soil solid phases in serpentine soils is in good agreement with the mineralogical composition of these soils (Antić-Mladenović et al. 2017).

Among the nonresidual fractions, Fe oxide fraction was the dominant Ni-binding solid phase form as reported in many studies (e.g., Antić-Mladenović et al. 2011; Rinklebe and Shaheen 2014, 2017b; Shaheen et al. 2015a,b). The role of Fe-Mn (hydr)oxides as scavengers for soil Ni was pointed out in many studies of both Ni-polluted and unpolluted soils (Bani et al. 2014; Kierczak et al. 2008; Knox et al. 2006). In addition, Rinklebe and Shaheen (2014) showed that among the oxide fractions, Ni bound by crystalline Fe oxides (F6) was dominant followed by Ni bound to low crystalline (amorphous) Fe oxides (F5) and Mn oxides (F3). However, Ni fractionation in serpentine Fluvisol from the alluvial plain of the Velika Morava River valley indicated higher Ni association with amorphous Fe oxides compared to crystalline Fe oxides and less Ni was bound to Mn oxides compared to Fe oxides (Antić-Mladenović et al. 2017; Rinklebe et al. 2016a). Accordingly, Ni associated to Fe oxide is likely to be released if these minerals dissolve because Fe/Mn oxides are thermodynamically unstable under anoxic conditions (Rinklebe and Shaheen 2017a).

Rinklebe and Shaheen (2014) also found a substantial amount of Ni bound to the soil organic matter (SOM), which was in line with previous reports of Ni partitioning in arable soils (e.g., Barman et al. 2015; Shaheen et al. 2015a,b) and in soils along the serpentine toposequence (Antić-Mladenović et al. 2011, 2017; Hseu 2006; Rinklebe et al. 2016a).

In relation to the high content of total Ni in the soil, the smallest proportion of Ni is usually found in F1 (e.g., Antić-Mladenović et al. 2017; Hseu 2006; Kierczak et al. 2008), which implies a low risk under native conditions. For example, Rennert and Rinklebe (2010) found that only 2% to 7% of total Ni was bound into the first two fractions of the topsoil of a Mollic Fluvisol. However, the solubility of Ni might increase under different reducing/oxidizing conditions as affected by the changes of soil E_H, pH, and the chemistry of Fe-Mn oxides and dissolved organic carbon (Rinklebe and Shaheen 2017a).

12.4 Effect of Amendments on Mobilization/ Immobilization of Ni in Soils

High contents of Ni in soils (75–150 mg kg^{-1}; trigger action values; Kabata-Pendias 2011, p. 24) have a negative impact on plants and microorganisms and simultaneously pose long-term risks to humans and ecosystems (Thakali et al. 2006; Yusuf et al. 2011). As demonstrated in many studies, Ni can be mobilized under different conditions [acidic/oxic (Frohne et al. 2014); neutral to alkaline, anoxic (Antić-Mladenović et al. 2011, 2017; Rinklebe et al. 2016a,b; Shaheen et al. 2014a,b)], leading to soil and groundwater contamination, which increases the possibility of Ni entering the food chain via vegetation, especially if these soils are arable and used for growing vegetables, citrus, and cereals.

Thus, remediation of Ni-contaminated soils has received significant attention (Ahmad et al. 2013). Immobilization of Ni in soil via additives is a very promising technique because of its simplicity and high effectiveness, in situ applicability, and low cost (Bolan et al. 2014; Shaheen et al. 2015a,b, 2017a). Recently, the term "green remediation" has become established within this context. It refers to remedial processes, products, and activities that have little or no impact on the environment. This applies to the finished products of remediation and remedial processes, which should have minimal environmental impact (Tomasevic et al. 2013). However, applying fertilizers and amendments to multi-metal contaminated soils may have contradictory effects on the speciation, mobility, and bioavailability of Ni depending on the type of soil and amendment (Shaheen et al. 2017a). Therefore, we will present here the impact of various soil amendments on the mobilization/immobilization and phytoavailability of Ni.

12.4.1 Impact of Amendments on Solubility and Mobilization of Ni

Concentration of Ni in natural solutions of surface horizons of different soils vary from 3 to 150 μg L^{-1}, depending on soil types, total Ni concentration in soil, and techniques used for obtaining solution (Uren 1992). However, serpentine-derived Ni-rich soils may contain soil solution Ni concentrations ranging from 130 to 3250 mg L^{-1} (Johnston and Proctor 1981). Nickel in solution phase could be rapidly sorbed to any of the surfaces proffered by soil solid phase, which is affected by soil properties and competition with other metals (Antoniadis and Tsadilas 2007; Shaheen et al. 2017b). Thus, application of organic and/or inorganic soil amendments might affect the solubility of Ni by precipitation, sorption, or complexation (Shaheen et al. 2017b). However, application of soil amendments might cause contradictory effects on the dissolved and mobile Ni and thus affect its bioavailability (Shaheen et al. 2017a), as will be discussed in succeeding studies.

Shaheen et al. (2017a) assessed the impact of application of several low-cost amendments and environmental wastes on the (im)mobilization and potential availability of Ni in a long-term sewage effluent irrigated sandy soil collected from Egypt (Figure 12.2). They found that the acidic amendments, like sulfur and triple superphosphate, increased both the mobility and availability of Ni significantly compared to control (Figure 12.2). On the other hand, the alkaline and liming materials such as sugar beet factory lime (SBFL) and cement bypass kiln dust decreased the mobilization and availability of Ni significantly compared to control, which might be caused by their high alkalinity, high content of calcium, and total calcium carbonates (Lee et al. 2013; Shaheen et al. 2015a). This indicates that the alkaline materials can be used to immobilize Ni in contaminated soils. Also, they found that although activated charcoal and potassium humate decreased the mobile Ni by 38%

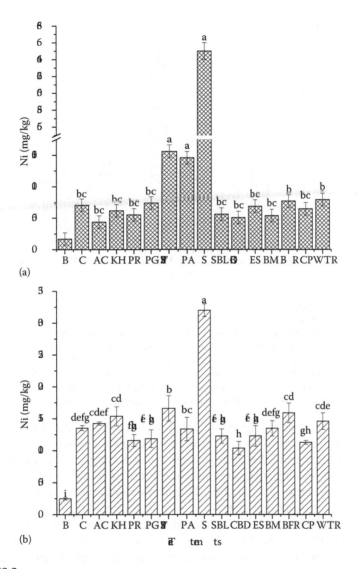

FIGURE 12.2

Effect of soil amendments on the NH₄NO₃- and AB-DTPA-extractable Ni, and in the plant tissue concentrations of Ni. AC, activated charcoal; KH, potassium humate; PR, phosphate rock; PG, phosphogypsum; TSP, triple superphosphate; PA, phosphoric acid; S, sulfur; SBFL, sugar beet factory lime; CBD, cement bypass kiln dust; ES, eggshell; BM, bone mill; BFR, brick factory residual; CP, ceramic powder; DWTR, drinking water treatment residual. B: Blank: Water-irrigated soil. C: Control: Untreated sewage effluent irrigated soil (a) Mobile Ni, (b) Available Ni, (c) Plant-Ni. (Reproduced from Shaheen et al., 2017a. *Ecotoxicology Environmental Safety* 142, 375–387.)   (*Continued*)

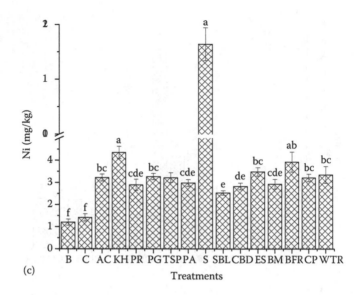

(c)

FIGURE 12.2 (CONTINUED)
Effect of soil amendments on the NH_4NO_3- and AB-DTPA-extractable Ni, and in the plant tissue concentrations of Ni. AC, activated charcoal; KH, potassium humate; PR, phosphate rock; PG, phosphogypsum; TSP, triple superphosphate; PA, phosphoric acid; S, sulfur; SBFL, sugar beet factory lime; CBD, cement bypass kiln dust; ES, eggshell; BM, bone mill; BFR, brick factory residual; CP, ceramic powder; DWTR, drinking water treatment residual. B: Blank: Water-irrigated soil. C: Control: Untreated sewage effluent irrigated soil (a) Mobile Ni, (b) Available Ni, (c) Plant-Ni. (Reproduced from Shaheen et al., 2017a. *Ecotoxicology Environmental Safety* 142, 375–387.)

and 13%, respectively (Figure 12.2a), which indicate the ability of these organic amendments to sorb Ni from soil solution and reduce its mobility. In addition, they indicated that phosphate rock decreased both the mobile and available Ni concentrations as compared to control by 22% and 14%, respectively (Figure 12.2a and b). The abovementioned results of Shaheen et al (2017a) conclude that that the acidic phosphates like triple superphosphate and phosphoric acid increased the mobilization of Ni; however, neutral phosphates like phosphate rock decreased the mobilization of Ni. We can assume that the phosphate compounds might be able to form complexes with Ni; nevertheless, these complexes are stable under neutral and alkaline conditions but dissolve under acidic conditions and the associated Ni release to soil solution as reported in Rochayati et al. (2011), Shaheen and Tsadilas (2015), and Sanderson et al. (2014). Application of phosphates either as fertilizer or soil amendment to soils might affect (im)mobilization of Ni and its downward movement and uptake by plants. Several studies have provided conclusive evidence for the potential value of both water-soluble [e.g., diammonium phosphate (DAP)] and water-insoluble [e.g., hydroxyapatite (HA) and phosphate rock (PR)] P compounds to immobilize metal(loid)s in soil, thereby reducing their bioavailability to plant (e.g., Bolan et al. 2013; Chen et al. 2007; Sanderson et al. 2014; Shaheen and Tsadilas 2015). It is, however, important to recognize that application of these amendments can cause either mobilization or immobilization of Ni in soils depending on the nature of P compounds (Shaheen and Tsadilas 2015; Wang et al. 2001). Also, phosphate compounds might form complexes with Ni in soil; these complexes can be stable under neutral and alkaline conditions, but can be dissolved under acidic conditions, thus releasing Ni into soil solution. Sanderson et al. (2014) found that commercial phosphate amendment reduced exchangeable Ni, while soft rock phosphate mobilized Ni in shooting range soils. In another study, Seaman et al. (2001) found that hydroxyapatite was effective in reducing the solubility of Ni.

Shaheen et al. (2015a) investigated the impact of different organic and inorganic soil amendments (e.g., activated carbon, bentonite, biochar, cement bypass kiln dust, chitosan, coal fly ash, limestone, nano-hydroxyapatite, organo-clay, SBFL, and zeolite) on the water-soluble and exchangeable Ni contents in a contaminated floodplain soil (Figure 12.3). They reported that application of the amendments (except the organo-clay) significantly decreased water-soluble Ni. The SBFL, cement bypass kiln dust, limestone, bentonite, activated carbon, and biochar were the most effective amendments, resulting in a 58%–99% decrease of water-soluble Ni. In contrast, organo-clay increased water-soluble Ni by 138% compared to control. The soluble + exchangeable Ni was also significantly decreased by amendments with the exception of organo-clay. The decreasing rate in soluble + exchangeable Ni pool followed the order of 74%, 73.5%, and 68% for SBFL, cement bypass kiln dust, and limestone, respectively (Figure 12.3). We can extract from the results of Figure 12.3 that the alkaline and carbonate-rich materials can be used to treat Ni-contaminated soil, reducing its solubility due to their alkalinity and high contents of calcium and carbonates, which increase soil pH and precipitation/sorption of metals in treated soil (Shaheen et al. 2015a).

Application of organic amendments affects solubility and mobilization of Ni in soils. Shaheen et al. (2015a) found that among the organic amendments, AC showed the highest efficiency in reducing water-soluble and soluble + exchangeable pool concentrations of Ni compared to control followed by BI and CH (Figure 12.3). Carbonaceous amendments such as AC have high affinity for sorbing metals including Ni. The addition of organic amendments such as AC can reduce bioavailability of Ni by forming stable complexes with humic substances (Shaheen et al. 2015a, 2017a). Biochar might significantly reduce solubility of Ni despite its low pH (Figure 12.3). Shaheen et al. (2015a) observed a strong reduction of soluble Ni concentration, indicating a strong tendency of biochar to retain Ni on its surface. Shen et al. (2016) examined the long-term effect of biochar on immobilization of Ni to remediate Ni-contaminated soil. Rinklebe et al. (2016b) studied the impact of biochar on the redox-induced dissolved concentration of Ni in a contaminated floodplain soil. More details about the impact of biochars on Ni in soils are provided in a chapter ("Potential of Biochar to Immobilize Nickel in Contaminated Soils") in the same book authored by El-Naggar et al. (2018).

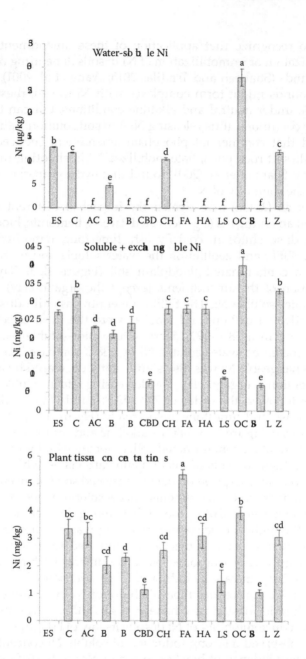

FIGURE 12.3
Impact of soil amendments on the water-soluble and mobile concentrations and on the plant
tissue concentrations of Ni. ES, experimental soil; C, control; AC, activated carbon; BE, bentonite; BI, biochar; CBD, cement bypass kiln dust; CH, chitosan; FA, coal fly ash; HA, nano-
hydroxyapatite; LS, limestone; OC, organo-clay; SBFL, sugar beet factory lime; Z, zeolite.
Values accompanied by different letters are significantly different within columns at the level
($P < 0.05$). (Reproduced from Shaheen et al., 2015a. *Int. J. Environ. Sci. Technol.* 12, 2765–2776,
with permission from the publisher.)

Shaheen et al. (2017c) investigated the impact of compost and sulfur in two rates (1.25% and 2.5%) on the solubility and mobilization of Ni in a fluvial soil spiked with 200 mg Ni/kg soil and cultivated by sorghum (dry soil) and barnyard grass (wet soil). They found that compost decreased solubility and mobilization of Ni in both soils, while sulfur increased solubility of Ni by 4.6% in the wet soil and by 8.7% in the dry soil (Figure 12.4). The organic matter content in compost can act as an important factor in evaluating the efficiency of amending compost to Ni-polluted soils in decreasing the solubility of Ni in the soils. Many other researchers have confirmed that SOM is a significant factor to control Ni solubility in soil (Al Mamun et al. 2016). On the other hand, the high solubility of Ni in sulfur treatment can be explained by the decrease in soil pH after sulfur treatment (pH 4.2–5.8) compared to the untreated soil (pH 8.3–8.7) (Shaheen et al. 2017c). These results indicate that the mobile fraction of Ni is greatly influenced by soil pH and generally increases as soil pH decreases (Kayser et al. 2000). In this respect, Wang et al. (2008) also found that application of S decreased soil pH by about 3 units and thus solubility of Ni was significantly increased after 64 days of incubation. The use of sulfur has therefore been opted for soil acidification, which can increase solubility of Ni in contaminated soils (Kaplan et al. 2005; Salati et al. 2010; Sirguey et al. 2006).

Shaheen et al. (2017c) also studied the impact of compost and S on the potential availability of Ni as extracted using ammonium bicarbonate (1 M NH_4HCO_3)–diethylene triamine pentaacetic acid (0.005 M DTPA) (AB-DTPA). Application of compost increased Ni availability in soil (26%–159%) under sorghum cultivation, but decreased it in grass soil (14%–29%) (Figure 12.4). The impact of the higher application rate of compost was stronger than that of the lower rate. These results indicate that SOM in compost treatments might act as an important carrier in determining availability of Ni under dry and wet conditions (Shaheen et al. 2014b).

On the other hand, increasing Ni availability under dry conditions compared to wet conditions might be explained by the hypothesis that SOM can increase solubility of Ni in dry soils, but decrease its solubility under wet conditions. Stevenson (1994) indicated that soluble humic substances such as humic acids can increase solubility of metals such as Ni under neutral and alkaline conditions. In the study of Shaheen et al. (2017c), sulfur application affected availability and the impact of sulfur differed based on the application rate; the low application rate increased Ni availability in dry soil (300%) but decreased it in wet soil (50%), and the high rate of S increased availability of Ni in both soils (Figure 12.4). These results indicate that the impact of sulfur on Ni availability depends on the application rate, which affects soil pH. The use of sulfur has therefore been opted for soil acidification, which can increase mobility and availability of Ni in contaminated soils and thus might be used for enhancing Ni phytoextraction (Cui et al. 2004; Kayser et al. 2000; Wang et al. 2008).

Shaheen and Rinklebe (2017) verified the scientific hypothesis that SBFL is able to immobilize Ni under different redox potentials (E_H) in a contaminated

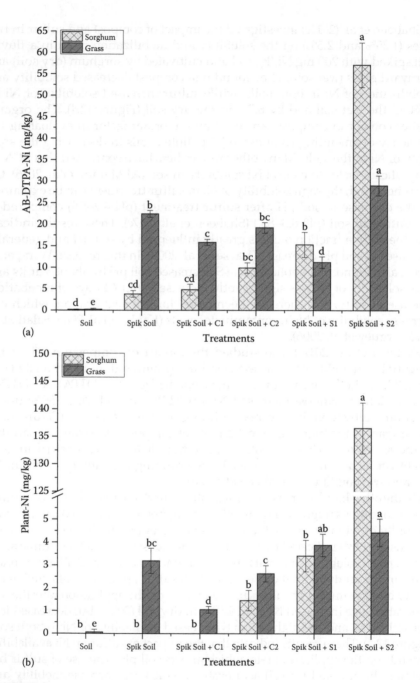

FIGURE 12.4
Impact of application of sulfur and compost on the AB-DTPA-extractable Ni (a) and in the plant tissue concentrations of Ni (b) under sorghum and barnyard grass cultivation. S, sorghum; G, barnyard grass. (Reproduced from Shaheen et al., 2017c. *Environmental Geochemistry and Health* 39(6), 1305–1324.)

floodplain soil. For this purpose, the nontreated contaminated soil (CS) and the same soil treated with SBFL (CS + SBFL) were flooded in the laboratory using a highly sophisticated automated biogeochemical microcosm apparatus. The mobilization of Ni was higher in CS + SBFL than in CS under reducing/neutral conditions. However, an opposite behavior was observed under oxic/acidic conditions and Ni concentrations were lower in CS + SBFL than in CS. They concluded that SBFL immobilized Ni under oxic acidic conditions, while mobilizing Ni under reducing neutral conditions in the studied soil (Figure 12.5)

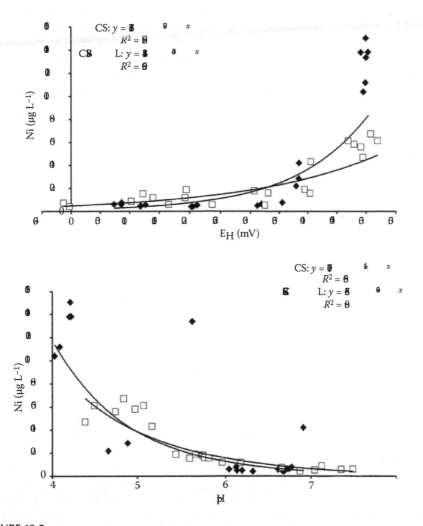

FIGURE 12.5

Relationships between dissolved Ni with E_H and pH in the soil suspension as affected by application of sugar beet factory lime. CS and in CS + SBFL ($n = 21$). (Reproduced from Shaheen and Rinklebe, 2017. *J. Environ. Manage.* 186, 253–260.)

12.4.2 Impact of Soil Amendments on Geochemical Fractions of Ni

Soil amendments affect the mobile form of Ni in soils through their impact on redistribution of Ni among its different geochemical fractions. In this respect, Shaheen et al. (2015a) found that, in comparison with control, sorbed and carbonate fractions of Ni in amended soils decreased significantly (7%–54%) with all amendments studied except for cement bypass kiln dust, which increased the sorbed and carbonate fraction of Ni by about 28% compared to control (Figure 12.6). These results indicate that amendments with high content of $CaCO_3$ (i.e., SBFL, CBD, and LS) increase the carbonate bound Ni in soils. In most cases, especially with the liming materials, a decrease in water-soluble and exchangeable Ni in the amended soils resulted in an increase of carbonate bound Ni fraction. Also, the amendments significantly decreased the Fe/Mn oxide fraction of Ni compared to control (Figure 12.6). The limestone showed the highest decreasing rate of Fe/Mn oxide fraction of Ni by 67.5%. Variations in the organic fraction of Ni after adding amendments were

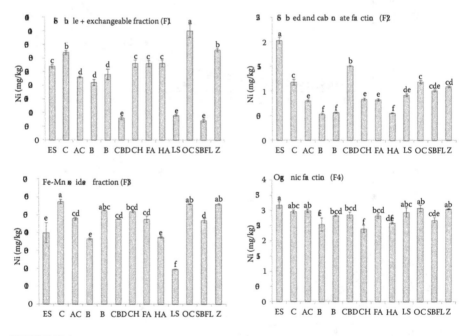

FIGURE 12.6
Impact of soil amendments on the water-soluble and geochemical fractions of Ni in the soil and on the plant tissue concentrations of Ni in the grown rapeseed. ES, experiment soil; C, control; AC, activated carbon; BE, bentonite; BI, biochar; CBD, cement bypass kiln dust; CH, chitosan; FA, coal fly ash; HA, nano-hydroxyapatite; LS, limestone; OC, organo-clay; SBFL, sugar beet factory lime; Z, zeolite. Low detection limit = 0.005 mg L^{-1}. Values accompanied by different letters are significantly different within columns at the level ($P < 0.05$). Please notice the different scales. (Reproduced from Shaheen et al., 2015a. *Int. J. Environ. Sci. Technol.* 12, 2765–2776, with permission from the publisher.)

narrow and not significant in most cases in the study by Shaheen et al. (2015a) (Figure 12.6). These results imply that the impact of the studied amendments on the organic bound fraction of Ni is relatively small compared to soluble, exchangeable, and oxide fractions.

Shaheen et al. (2017c) found that application of compost and sulfur altered distribution of Ni among the fractions and affected its mobilization in dry and wet soil significantly. However, the impact of sulfur in fractions and mobilization of Ni was stronger than that of compost in both dry and wet soils (Figure 12.7). Compost application decreased the mobile fraction of Ni (F1 + F2) (Figure 12.7). Also, the application of compost increased the organic fraction of Ni under wet conditions, which might be explained by the relatively high added organic matter with compost and higher soil pH in compost-amended soil compared to untreated soil (Shaheen et al. 2017c). In this subject, Alburquerque et al. (2011) found that an increase in the soil pH by the application of compost was responsible for changes in metal fractionation in soil. On the other hand, the application of sulfur significantly increased Ni solubility (F1) (Figure 12.7). Sulfur application increased the potential mobilization (PMF = F1 + F2 + F3 + F4) of Ni from 14.7% to 39% of

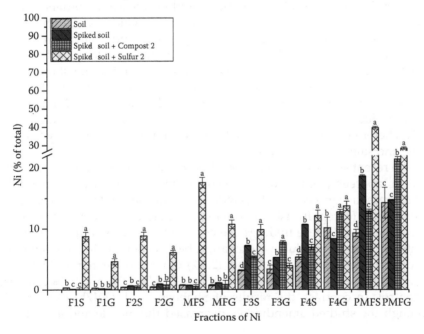

FIGURE 12.7
(See color insert.) Impact of application of sulfur and compost on the fractions and potential mobilization of Ni under sorghum and barnyard grass cultivation. S, sorghum; G, barnyard grass; F1 = soluble + exchangeable fraction; F2 = sorbed and carbonate fraction; F3 = Fe/Mn oxide fraction; F4 = organic fraction; MF = mobile fraction (=$\sum$F1 – F2); PMF = potential mobile fraction (=$\sum$F1 – F4). (Reproduced from Shaheen et al., 2017c. *Environmental Geochemistry and Health* 39(6), 1305–1324.)

total Ni in S-treated soil (Figure 12.7). The results of the Ni fractions in the study of Shaheen et al. (2017c) show that a small part of the spiked Ni (9%–14%) was distributed in the PMF. Compost application decreased mobilization of Ni in both soils. Sulfur application significantly increased solubility of Ni (4.6%–8.7%). Sulfur application altered residual fraction to nonresidual fraction (PMF) for Ni.

12.4.3 Effect of Amendments on Ni Concentration in Plant Tissue

Nickel uptake by plants from soil depends on the plant-available fraction of Ni in soil. Plant uptake of Ni provides important insight into the effect of amendments added to soil on the bioavailability of these elements. Although some amendments might decrease mobile and available forms of PTEs, these amendments can increase plant tissue contents of these elements depending on the amendments and element (Mahar et al. 2016; Rinklebe and Shaheen 2015; Shaheen and Rinklebe 2015). High Ni toxicity in crop plants is a widespread problem (He et al. 2012). Like most other PTEs, Ni is an essential micronutrient and required for normal growth and development of plants. However, Ni toxicity leads to a variety of physiological disorders in plants (Guo et al. 2010; Kamran et al. 2016). For example, Guo et al. (2010) reported that excess Ni accumulation decreased dry weight and grain yield of maize grown in Ni-contaminated soil. Nickel toxicity linearly decreased seed germination, shoot and root lengths, and biomass in Indian mustard, but the response varied among the studied genotypes (Ansari et al. 2015). Similarly, Ni toxicity decreased plant height and rice yield (Nazir et al. 2016) and biomass production and chlorophyll contents in *Eruca sativa* (Kamran et al. 2016). Therefore, it is imperative to decrease Ni uptake and toxicity to ensure stable and safe food production.

Some recent studies (e.g., Shaheen et al. 2015a) revealed that Ni concentrations in rapeseed plants grown in a contaminated weakly acidic floodplain soil decreased significantly (23%–68%) after addition of the various studied amendments (except fly ash and organo-clay; Figure 12.3) into the soil compared to control. These findings are consistent with the impact of amendments in decreasing soluble + exchangeable Ni contents in a contaminated floodplain (see Sections 12.4.1 and 12.4.2) On the other hand, fly ash and organo-clay increased the concentrations of Ni by 60% and 18% compared to control. In contrast, the results of Shaheen et al. (2017b) reported that although the studied amendments decreased the mobile and available Ni, they increased, particularly sulfur, the plant tissue concentrations of Ni in sorghum plants grown in sewage effluent irrigated sandy soil (Figure 12.2). The increased Ni uptake by sorghum with the alkaline and carbonate-rich amendments may be attributed to the formation of Ni carbonate fraction during the immobilization process, which was considered to easily mobilize in the acid rhizosphere zone (Mahar et al. 2016; Shaheen and Rinklebe 2015a,b). Also, it is assumed that competition between the high contents of Ca, Al, and Fe in fly ash and the sorbed Ni might increase release and desorption of Ni to the soil solution in an available form for plant uptake. Shaheen et al. (2013) reported that competition among various element species for available sorption sites on soil matrix surfaces can enhance the mobility of contaminants in the soil environment.

Kabata-Pendias (2011) reported that Ni phytoavailability can be reduced in soils amended with OM and alkaline materials due to sorption, complexation, coprecipitation, or a combination of the three mechanisms. Sanderson et al. (2014) found that application of soft rock phosphate, lime, commercial phosphate amendment, red mud, and magnesium oxide to shooting range soil reduced uptake of Ni significantly—that is, from 1 to 0.25 mg/g DW.

Shaheen et al. (2017c) investigated the impact of compost and sulfur in two rates (1.25% and 2.5%) on the plant uptake of Ni to sorghum (dry soil) and barnyard grass (wet soil) in a fluvial soil spiked with 200 mg Ni/kg soil. Sulfur increased Ni in both plants and the increasing rate of Ni was higher in sorghum than in grass (Figure 12.4). These results suggest that sulfur (with rate 2.5%) can be used for enhancing the phyto-extraction of Cd and Ni from contaminated alkaline soils.

Rehman et al. (2016) concluded that organic amendments like biochar (BC), farm manure (FM) from cattle farm, and compost (Cmp) are capable of alleviating Ni toxicity through decreasing Ni uptake by maize seedlings and stabilizing it in the soil. Overall, FM was most effective in reducing Ni toxicity in maize (Figure 12.8). Similarly, several studies showed that Cmp, FM, and BC application decreased metal concentrations in maize (Sabir et al. 2015; Zhao et al. 2014). The reduction in Ni content in maize might be due to the immobilization of Ni in the soil (Shen et al. 2016) and/or due to higher nutrient contents in the soil, which increased the competition at the root surface. The decrease in Ni content might also be due to the amendment-mediated improvement in soil physicochemical properties and variation in dissolved organic matter contents in the soil (Rizwan et al. 2016; Yousaf et al. 2016).

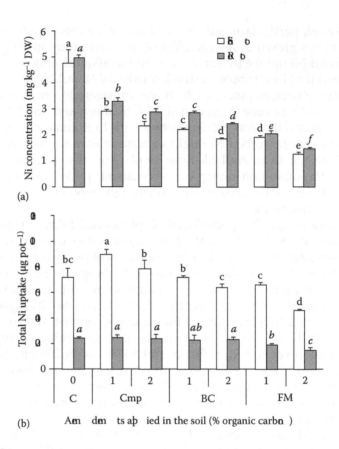

(a)

(b)   Am   dm   ts ap   ied in the soil (% organic carbn)

FIGURE 12.8
Nickel concentrations (a) and uptake (b) in shoots and roots of maize seedlings grown in Ni-contaminated soil treated with different amendments. C, control (without any amendments); Cmp, compost at 1% and 2% organic carbon; BC, biochar at 1% and 2% organic carbon; FM, farm manure at 1% and 2% organic carbon. Bars represent SD of three replicates. Different letters indicate significant differences among the treatments. (Reproduced from Rehman et al., 2016. *Ecotoxicol. Environ. Safety* 133, 218–225.)

12.5 Conclusions

In this chapter, we discussed total content and distribution of Ni among various fractions in different soils and the impact of different organic and inorganic amendments on solubility, potential mobilization, and phytoavailability of Ni in different soils. Overall, we demonstrate the high potential of many inorganic and organic amendments (e.g., active carbon, cement bypass kiln dust, limestone, SBFL, bentonite, activated carbon, biochar, farmyard manure, compost, sulfur, potassium humate, phosphate rock, phosphogypsum, triple superphosphate, phosphoric acid, eggshell, bone mill, brick factory residual, ceramic powder, and drinking water treatment residual) for remediation of soils contaminated with Ni—either by immobilizing Ni and reducing its uptake by plants and potential toxicity to plants and humans, or by mobilizing Ni and thus enhancing its phytoextraction using nonedible plants.

An understanding of the effects of the soil amendments on Ni geochemistry, dynamics, and plant uptake might contribute to the development of novel remediation strategies and management options for Ni-contaminated sites as well as to shortening the remediation period. However, many of the experiments were conducted at a laboratory scale; thus, future research should include verification of the effects of amendments on Ni at a field level. Also, spectroscopic techniques should be combined with sequential extraction in the future to provide a deeper understanding of the mechanisms of Ni (im)mobilization as affected by the soil amendments.

Acknowledgments

We thank the German Alexander von Humboldt Foundation (Ref. 3.4-EGY-1185373-GF-E) for financial support of the experienced researcher's fellowship of Prof. Dr. S.M. Shaheen at the University of Wuppertal, Germany.

References

Abollino, O., Aceto, M., Malandrino, M., Mentasti, E., Sarzanini, C., Petrella, F., 2002. Heavy metals in agricultural soils Piedmont, Italy. Distribution, speciation and chemometric data treatment. *Chemosphere* 49, 545–557.

Ahmad, M., Rajapaksha, A.U., Lim, J.E., Zhang, M., Bolan, N., Mohan, D., Vithanage, M., Lee, S.S., Ok, Y.S., 2013. Biochar as a sorbent for contaminant management in soil and water: A review. *Chemosphere* 99, 19–33.

Alburquerque, J.A., de la Fuente, C., Bernal, M.P., 2011. Improvement of soil quality after "alperujo" compost application to two contaminated soils characterised by differing heavy metal solubility. *J. Environ. Manage.* 92, 733–741.

Al Mamun, S., Chanson, G., Muliadi, Benyas, E., Aktar, M., Lehto, N., McDowell, R., Cavanagh, J., Kellermann, L., Clucas, L., Robinson, B.H., 2016. Municipal composts reduce the transfer of Cd from soil to vegetables. *Environ. Pollut.* 213, 8–15.

Ansari, M.K.A., Ahmad, A., Umar, S., Zia, M.H., Iqbal, M., Owens, G., 2015. Genotypic variation in phytoremediation potential of Indian mustard exposed to nickel stress: A hydroponic study. *Int. J. Phytorem.* 17, 135–144.

Antić-Mladenović, S., Frohne, T., Kresović, M., Staerk, H.-J., Tomić, Z., Ličina, V., Rinklebe, J., 2017. Biogeochemistry of Ni and Pb in a periodically flooded arable soil: Fractionation and redox-induced (im)mobilization. *J. Environ. Manage.* 186, 141–150.

Antić-Mladenović, S., Rinklebe, J., Frohne, T., Staerk, H.-J., Wennrich, R., Tomić, Z., Ličina, V., 2011. Impact of controlled redox conditions on nickel in a serpentine soil. *J. Soils Sed.* 11, 406–415.

Antoniadis, V., Levizou, E., Shaheen, S.M., Ok, Y.S., Sebastian, A., Baum, C., Prasad, M.N.V., Wenzel, W.W., Rinklebe, J., 2017. Trace elements in the soil-plant interface: Phytoavailability, translocation, and phytoremediation—A review. *Earth-Sci. Rev.* 171, 621–645.

Antoniadis, V., Tsadilas C.D., 2007. Sorption of cadmium, nickel, and zinc in mono- and multimetal systems. *Appl. Geochem.* 22, 2375–2380.

Bani, A., Echevarria, G., Montarges-Pelletier, E., Gjoka, F., Sulce, S., Morel, J.L., 2014. Pedogenesis and nickel biogeochemistry in a typical Albanian ultramafic toposequence. *Environ. Monit. Assess.* 186, 4431–4442.

Barman, M., Datta, S.P., Rattan, R.K., Meena, M.C., 2015. Chemical fractions and bioavailability of nickel in alluvial soils. *Plant Soil Environ.* 61, 17–22.

Bhogal, A., Nicholson, F.A., Chambers, B.J., Shepherd, M.A., 2003. Effects of past sewage sludge additions on heavy metal availability in light textured soils: Implications for crop yields and metal uptakes. *Environ. Pollut.* 121, 413–423.

Bini, C., Dall'Aglio, M., Ferretti, O., and Gragnani, R., 1988. Background levels of microelements in soils of Italy. *Environ. Geochem. Health* 10, 35.

Bolan, N., Kunhikrishnan, A., Thangarajan, R., Kumpiene, J., Park, J., Makino, T., Kirkham, M.B., Scheckel, K., 2014. Remediation of heavy metal(loid)s contaminated soils—To mobilize or to immobilize? (Review). *J. Hazard. Mater.* 266, 141–166.

Bolan, N., Mahimairaja, S., Kunhikrishnan, A., Choppala, G., 2013. Phosphorus–arsenic interactions in variable-charge soils in relation to arsenic mobility and bioavailability. *Sci. Total Environ.* 463–464, 1154–1162.

Chang, Y.T., Hseu, Z.Y., Iizuka, Y., Yu, C.D., 2013. Morphology, geochemistry, and mineralogy of serpentine soils under a tropical forest in Southeastern Taiwan. *Taiwan J. For. Sci.* 28(4), 185–201.

Chen, S., Xu, M., Ma, Y., Yang, J., 2007. Evaluation of different phosphate amendments on availability of metals in contaminated soil. *Ecotoxicol. Environ. Safety* 67, 278–285.

Cox, R.M., Hutchinson, T.C., 1981. Environmental factors influencing the rate of spread of the grass *Deschampsia caespitosa* invading areas around the Sudbury Nickel-Copper Smelter. *Water Air Soil Pollut.* 16, 83.

Cui, Y., Dong, Y., Li, H., Wang, Q., 2004. Effect of elemental sulphur on solubility of soil heavy metals and their uptake by maize. *Environ. Int.* 30, 323–328.

El-Naggar, A., Rajapaksha, A.U., Shaheen, S.M., Rinklebe et al., Ok, Y.S., 2018. Potential of biochar to immobilize nickel in contaminated soils. In: *Nickel and Soils and Plants*. Chapter 13.

Fijalkowski, K., Rorat, A., Grobelak, A., Kacprzak, M.J., 2017. The presence of contaminations in sewage sludge—The current situation. *J. Environ. Manage.* 203, 1126–1136.

Frohne, T., Rinklebe, J., Diaz-Bone, R.A., 2014. Contamination of floodplain soils along the Wupper River, Germany, with As, Co, Cu, Ni, Sb, and Zn and the impact of pre-definite redox variations on the mobility of these elements. *Soil Sediment Contam.* 23, 779–799.

Guo, X.Y., Zuo, Y.B., Wang, B.R., Li, J.M., Ma, Y.B., 2010. Toxicity and accumulation of copper and nickel in maize plants cropped on calcareous and acidic field soils. *Plant Soil* 333, 365–373.

He, S., He, Z., Yang, X., Baligar, V.C., 2012. Mechanisms of nickel uptake and hyper-accumulation by plants and implications for soil remediation. In: D.L. Sparks (Ed.). *Advances in Agronomy* 117, pp. 117–189.

Hseu, Z.-Y., 2006. Concentration and distribution of chromium and nickel fractions along a serpentinitic toposequence. *Soil Sci.* 171, 341–353.

Hseu, Z.-Y., Su, Y.-C., Zehetner, F., His, H.-C., 2017. Leaching potential of geogenic nickel in serpentine soils from Taiwan and Austria. *J. Environ. Manage.* 186(151–157), 141–150.

Iyaka, Y.A., 2011. Nickel in soils: A review of its distribution and impacts. *Sci. Res. Essays* 6(33), 6774–6777.

Johnston, W.R., Proctor, J., 1981. Growth of serpentine and non-serpentine races of *Festuca rubra* in solutions simulating the chemical conditions in a toxic serpentine soil. *J. Ecol.* 1981, 69, 855–869.

Jordán, M.M., Montero, M.A., Pina, S., García-Sánchez, E., 2009. Mineralogy and distribution of Cd, Ni, Cr, and Pb in biosolids amended soils from Castellon Province (NE, Spain). *Soil Sci.* 174, 14–20.

Kabata-Pendias, A., 2011. *Trace Elements in Soils and Plants*, 4th Ed. Boca Raton: CRC Press.

Kabata-Pendias, A., Mukherjee, A.B., 2007. *Trace Elements from Soil to Human*. Berlin Heidelberg New York: Springer.

Kadunas, V., Budavičius, R., Gregorauskiene, V., Katinas, V., Kliuugiene, E., Radzevicius, A., Taraskievicius, R. 1999. *Geochemical Atlas of Lithuania*, Geological Inst., Vilnius, 162, 1999.

Kamran, M.A., Eqani, S.A.M.A.S., Bibi, S., Xu, R.K., Monis, M.F.H., Katsoyiannis, A., Bokhari, H., Chaudhary, H.J., 2016. Bioaccumulation of nickel by E. sativa and role of plant growth promoting rhizobacteria (PGPRs) under nickel stress. *Ecotoxicol. Environ. Saf.* 26, 256–263.

Kayser, A., Wenger, K., Keller, A., Attinger, W., Felix, H.R., Gupta, S.K., Schulin, R., 2000. Enhancement of phytoextraction of Zn, Cd, and Cu from calcareous soil: The use of NTA and sulfur amendments. *Environ. Sci. Technol.* 34, 1778–1783.

Kierczak, J., Neel, C., Aleksander-Kwaterczak, U., Helios-Rybicka, E., Bril, H., Puziewicz, J., 2008. Solid speciation and mobility of potentially toxic elements from natural and contaminated soils: A combined approach. *Chemosphere* 73, 776–784.

Knox, A., S., Paller, M.H., Nelson, E.A., Specht, W.L., Halverson, N.V., Gladden, J.B., 2006. Metal distribution and stability in constructed wetland sediment. *J. Environ. Qual.* 35, 1948–1959.

Koljonen, T., 1992. *The Geochemical Atlas Finland, Part 2: Till*. Espoo: Geol. Survey of Finland.

Lee, S.S., Lim, J.E., Abd El-Azeem, S.M., Choi, B., Oh, S., Moon, D.H., Ok, Y.S., 2013. Heavy metal immobilization in soil near abandoned mines using eggshell waste and rapeseed residue. *Environ. Sci. Pollut. R.* 20, 1719–1726.

Li, Q.S., Wu, Z.F., Chu, B., Zhang, N., Cai, S.S., Fang, J.H., 2007. Heavy metals in coastal wetland sediments of the Pearl River Estuary, China. *Environ. Pollut.* 149, 158–164.

Ma, Y., Hooda, P.S., 2010. Chromium, nickel and cobalt. In: Hooda, P.S. (Ed.). Trace Elements in Soils. 1st Ed. UK: John Wiley & Sons, pp. 461–480.

Mahar, A., Wang, P., Ali, A., Guo, Z., Awasthi, M.K., Lahori, A.H., Wang, Q., Shen, F., Li, R., Zhang, Z., 2016. Impact of CaO, fly ash, sulfur and Na_2S on the (im)mobilization and phytoavailability of Cd, Cu and Pb in contaminated soil. *Ecotoxicol. Environ. Safety* 134, 116–123.

Malinowska, E., 2017. The effect of liming and sewage sludge application on heavy metal speciation in soil. *Bull. Environ. Contam. Toxicol.* 98, 105–112.

Massoura, S.T., Echevarria, G., Becquer, T., Ghanbaja, J., Leclerc-Cessac, E., Morel, J.-L., 2006. Control of nickel availability by nickel bearing minerals in natural and anthropogenic soils. *Geoderma* 136, 28–37.

Morrison, J.M., Goldhaber, M.B., Lee, L., Holloway, J.M., Wanty, R.B., Wolf, R.E., Ranville, J.F., 2009. A regional-scale study of chromium and nickel in soils of northern California, USA. *Appl. Geochem.* 24, 1500–1511.

Nazir, H., Asghar, H.N., Zahir, Z.A., Akhtar, M.J., Saleem, M., 2016. Judicious use of kinetin to improve growth and yield of rice in nickel contaminated soil. *Int. J. Phytorem.* 18(7): 651–655.

Niechayeva E.G., 2002. Heavy metals in taiga zone of Western Siberia. 2 Int. Conf. Heavy Metals, Radionuclides and Elements—Biofills in the Environment, 115–121, Semipalatynsk, Kazakhstan (in Russian).

Osakwe, S.A., 2013. Chemical partitioning of iron, cadmium, nickel and chromium in contaminated soils of south-eastern Nigeria. *Chem. Speciation Bioavail.* 25, 71–78.

Protasova, N.A., Kopayeva, M.T., 1985. Trace and dispersed elements in soils of Russian Plateau. *Pochvovedenie* 1, 29 (Ru).

Pueyo, M., Mateu, J., Rigol, A., Vidal, M., López-Sánchez, J.F., Rauret, G., 2008. Use of the modified BCR three-step sequential extraction procedure for the study of trace element dynamics in contaminated soils. *Environ. Pollut.* 152(2), 330–341.

Quiping, Z., Chuliang, Y., Lihua, T., Junxiang, X., 1984. Content and distribution of trace elements in limestone soils of China, *Acta Pedologica Sinica*, 21, 58.

Rajapaksha, A.U., Vithanage, M., Oze, C., Bandara, W.M.A.T., Weerasooriya, R., 2012. Nickel and manganese release in serpentine soil from the Ussangoda Ultramafic Complex, Sri Lanka. *Geoderma* 189–190, 1–9.

Rehman, M.Z.U., Rizwan, M., Ali, S., Fatima, N., Yousaf, B., Naeem, A., Sabir, M., Ahmad, H.R., Ok, Y.S., 2016. Contrasting effects of biochar, compost and farm manure on alleviation of nickel toxicity in maize (*Zea mays* L.) in relation to plant growth, photosynthesis and metal uptake. *Ecotoxicol. Environ. Safety* 133, 218–225.

Rennert, T., Rinklebe, J., 2010. Release of Ni and Zn from contaminated floodplain soils under saturated flow conditions. *Water Air Soil Pollut.* 205, 93–105.

Rinklebe, J., Antić-Mladenović, S., Frohne, T., Staerk, H.-J., Tomić, Z., Ličina, V., Staerk, H.-J., Tomić, Z., Ličina, V., 2016a. Nickel in a serpentine-enriched Fluvisol: Redox affected dynamics and binding forms. *Geoderma* 263, 203–214.

Rinklebe, J., Knox, A.S., Paller, M., 2016b. *Trace Elements in Waterlogged Soils and Sediments*. New York: CRC Press/Taylor & Francis Group.

Rinklebe, J., Kumpiene, J., Du Laing, G., Ok, Y.S., 2017. Biogeochemistry of trace elements in the environment—Editorial to the special issue. *J. Environ. Manage.* 186, 127–130.

Rinklebe, J., Shaheen, S.M., 2014. Assessing the mobilization of cadmium, lead, and nickel using a seven-step sequential extraction technique in contaminated floodplain soil profiles along the Central Elbe River, Germany. *Water Air Soil Pollut.* 225(8), 2039; DOI: 10.1007/s11270-014-2039-1.

Rinklebe, J., Shaheen, S.M., 2017a. Redox chemistry of nickel in soils and sediments: A review. *Chemosphere* 179, 265–278.

Rinklebe, J., Shaheen, S.M., 2017b. Geochemical distribution of Co, Cu, Ni, and Zn in soil profiles of Fluvisols, Luvisols, Gleysols, and Calcisols originating from Germany and Egypt. *Geoderma* 307, 122–138.

Rinklebe, J., Shaheen, S.M., Frohne, T., 2016b. Amendment of biochar reduces the release of toxic elements under dynamic redox conditions in a contaminated floodplain soil. *Chemosphere* 142, 41–47.

Rizwan, M., Ali, S., Qayyum, M.F., Ibrahim, M., Rehman, M.Z., Abbas, T., Ok, Y.S., 2016. Mechanisms of biochar-mediated alleviation of toxicity of trace elements in plants: A critical review. *Environ. Sci. Pollut. Res.* 23, 2230–2248.

Roca, N., Pazos, M.S., Bech, J., 2008. The relationship between WRB soil units and heavy metals content in soils of Catamarca (Argentyna). *J. Geochem. Explor.* 96, 77–85.

Rochayati, S., Du Laing, G., Rinklebe, J., Meissner, R., Verloo, M., 2011. Use of reactive phosphate rocks as fertilizer on Indonesian acid upland soils: Accumulation of cadmium and zinc in soils and shoots of maize plants. *J. Plant Nutr. Soil Sci.* 174, 186–194.

Rodríguez, L., Ruiz, E., Alonso-Azcárate, J., Rincón, J., 2009. Heavy metal distribution and chemical speciation in tailings and soils around a Pb–Zn mine in Spain. *J. Environ. Manage.* 90, 1106–1116.

Sabir, M., Ali, A., Rehman, M.Z., Hakeem, K.R., 2015. Contrasting effects of Farmyard Manure (FYM) and compost for remediation of metal contaminated soil. *Int. J. Phytorem.* 17, 613–621.

Sanderson, P., Naidu, R., Bolan, N., 2014. Ecotoxicity of chemically stabilized metal(loid)s in shooting range soils. *Ecotoxicol. Environ. Safety* 100, 201–208.

Seaman, J.C., Arey, J.S., Bertsch, P.M., 2001. Immobilization of nickel and other metals in contaminated sediments by hydroxyapatite addition. *J. Environ. Qual.* 30, 460–469.

Shaheen, S.M., Antoniadis, V., Biswas, J., Wang, H., Ok, Y.-S., Rinklebe, J., 2017b. Biosolids application affects the competitive sorption and lability of cadmium, copper, nickel, lead, and zinc in fluvial and calcareous soils. *Environ. Geochem. Health* 39(6):1365–1379.

Shaheen, S.M., Balbaa, A.A., Khatab, A.M., Rinklebe, J., 2017c. Compost and sulfur affect the mobilization and phytoavailability of Cd and Ni to sorghum and barnyard grass in a spiked fluvial soil. *Environ. Geochem. Health* 39(6), 1305–1324.

Shaheen, S.M., Rinklebe, J., 2014. Geochemical fractions of chromium, copper, and zinc and their vertical distribution in floodplain soil profiles along the Central Elbe River, Germany. *Geoderma* 228–229, 142–159.

Shaheen, S.M., Rinklebe, J., 2015a. Impact of emerging and low cost alternative amendments on the (im)mobilization and phytoavailability of Cd and Pb in a contaminated floodplain soil. *Ecol. Eng.* 74, 319–326.

Shaheen, S.M., Rinklebe, J., 2015b. Phytoextraction of potentially toxic elements from a contaminated floodplain soil using Indian mustard, rapeseed, and sunflower. *Environ. Geochem. Health* 37, 953–967.

Shaheen, S.M., Rinklebe, J., 2017. Sugar beet factory lime affects the mobilization of Cd, Co, Cr, Cu, Mo, Ni, Pb, and Zn under dynamic redox conditions in a contaminated floodplain soil. *J. Environ. Manage.* 186, 253–260.

Shaheen, S.M., Rinklebe, J., Frohne, T., White, J.R., DeLaune, R.D., 2014a. Biogeochemical factors governing cobalt, nickel, selenium, and vanadium dynamics in periodically flooded Egyptian North Nile Delta rice soils. *Soil Sci. Soc. Am. J.* 78, 1065–1078.

Shaheen, S.M., Rinklebe, J., Rupp, H., Meissner, R., 2014b. Temporal dynamics of pore water concentrations of Cd, Co, Cu, Ni, and Zn and their controlling factors in a contaminated floodplain soil assessed by undisturbed groundwater lysimeters. *Environ. Pollut.* 191, 223–231.

Shaheen, S.M., Rinklebe, J., Selim, H.M., 2015a. Impact of various amendments on the bioavailability and immobilization of Ni and Zn in a contaminated floodplain soil. *Int. J. Environ. Sci. Technol.* 12, 2765–2776.

Shaheen, S.M., Rinklebe, J., Tsadilas, C.D., 2015b. Fractionation and mobilization of toxic elements in floodplain soils from Egypt, Germany and Greece; a comparison study. *Eurasian Soil Sci.* 1317–1328.

Shaheen, S.M., Shams, M.S., Khalifa, M.R., El-Daly, M.A., Rinklebe, J., 2017a. Various soil amendments and wastes affect the (im)mobilization and phytoavailability of potentially toxic elements in a sewage effluent irrigated sandy soil. *Ecotoxicol. Environ. Safety* 142, 375–387.

Shaheen, S.M., Tsadilas, C.D., 2015. Influence of phosphates on fractionation, mobility, and bioavailability of soil metal(loid)s. In: Selim, H.M. (Ed). *Phosphate in Soils: Interaction with Micronutrients, Radionuclides and Heavy Metals.* New York: CRC Press/Taylor & Francis Group. Chapter 7, pp. 169–201.

Shaheen, S.M., Tsadilas, C.D., Rinklebe, J., 2013. A review of the distribution coefficient of trace elements in soils: Influence of sorption system, element characteristics, and soil colloidal properties. *Adv. Colloid Interf. Sci.* 201–202, 43–56.

Shen, Z., Som, A., Wang, F., Jin, F., McMillan, O., Al-Tabbaa, A., 2016. Long-term impact of biochar on the immobilisation of nickel (II) and zinc (II) and the revegetation of a contaminated site. *Sci. Total Environ.* 542, 771–776.

Shtiza, A., Swennen, R., Tashko, A., 2005. Chromium and nickel distribution in soils, active river, overbank sediments and dust around the Burrel chromium smelter (Albania). *J. Geochem. Explor.* 87, 92–108.

Sirguey, C., Schwartz, C., Morel, J.L., 2006. Response of Thlaspi caerulescens to nitrogen, phosphorus and sulfur fertilisation. *Int. J. Phytoremediation.* 8, 149–161.

Thakali, S., Herbert, E.A., Di Toro, D.M. et al., 2006. A terrestrial biotic ligand model. 2. Application to Ni and Cu toxicities to plants, invertebrates, and microbes in soil. *Environ. Sci. Technol.* 2006, 40, 7094–7100.

Tica, D., Udovic, M., Lestan, D., 2011. Immobilization of potentially toxic metals using different soil amendments. *Chemosphere* 85, 577–583.

Tomasevic, D.D., Dalmacija, M.B., Prica, M., Dalmacija, B.D., Kerkez, D.V., Becelic-Tomin, M.R., Roncevic, S.D., 2013. Use of fly ash for remediation of metals polluted sediment—Green remediation. *Chemosphere* 92, 1490–1497.

Uren, N.C., 1992. Forms, reaction, and availability of nickel in soils. *Adv. Agron.* 1992, 48, 141–203.

Varrault, G., Bermond, A., 2011. Kinetics as a tool to assess the immobilization of soil trace metals by binding phase amendments for in situ remediation purposes. *J. Hazard. Mater.* 192, 808–812.

Viventsova (Ruth), E., Kumpiene, J., Gunneriusson, L., Holmgren, A., 2005. Changes in soil organic matter composition and quantity with distance to a nickel smelter—A case study on the Kola Peninsula, NW Russia. *Geoderma* 127, 216–226.

Wang, Y., Li, Q., Hui, W., Shi, J., Lin, Q., Chen, X., Chen, Y., 2008. Effect of sulphur on soil Cu/Zn availability and microbial community composition. *J. Hazard. Mater.* 159, 385–389.

Wang, Y.M., Chen, T.C., Yeh, K.J., Shue, M.F., 2001. Stabilization of an elevated heavy metal contaminated site. *J. Hazard. Mater.* B88, 63–74.

Xiao, R., Bai, J., Gao, H., Huang, L., Huang, C., Liu, P., 2002. Heavy metals (Cr and Ni) distribution and fractionation in cropland soils from reclaimed tidal wetlands in Pearl River estuary, South China. *Proc. Environ. Sci.* 13, 1684–1687.

Yousaf, B., Liu, G., Wang, R., Rehman, M.Z., Rizwan, M.S., Imtiaz, M., Murtaza, G., Shakoor, A., 2016. Investigating the potential influence of biochar and traditional organic amendments on the bioavailability and transfer of Cd in the soil–plant system. *Environ. Earth Sci.* 75, 1–10.

Yusuf, M., Fariduddin, Q., Hayat, S., Ahmad, A., 2011. Nickel: An overview of uptake, essentiality and toxicity in plants. *Bull. Environ. Contam. Toxicol.* 86, 1–17.

Zhao, Y., Yan, Z., Qin, J., Xiao, Z., 2014. Effects of long-term cattle manure application on soil properties and soil heavy metals in corn seed production in Northwest China. *Environ. Sci. Pollut. Res.* 21, 7586–7595.

13

Potential of Biochar to Immobilize Nickel in Contaminated Soils

Ali El-Naggar, Anushka Upamali Rajapaksha,
Sabry M. Shaheen, Jörg Rinklebe, and Yong Sik Ok

CONTENTS

13.1 Introduction

Nickel (Ni) is formed in soil through anthropogenic (e.g., mining, waste disposal) and non-anthropogenic (e.g., volcanic eruptions, rocks weathering) resources (Adrees et al. 2015; Khan et al. 2016). Nickel is an essential micronutrient, which is needed for the healthy growth of plants (Rehman et al. 2016). However, the high concentration of Ni in soils could lead to adverse consequences in plants and soil (Kabata-Pendias 2011). Moreover, the transformation of Ni to groundwater by leaching is a critical issue especially in the acidic and the coarse-textured soils (Alloway 2013; Lockwood et al. 2015;

Zhang et al. 2015). Soil contamination by Ni is a globally serious concern due to its high environmental and human health risk (El-Naggar et al. 2018a; Li et al. 2018). Thus, sustainable management strategies for remediating Ni-contaminated soils are required to achieve soil security and food safety (Tandy et al. 2009).

Nickel is presented in soil among different fractions such as mobile, easily mobile, bound to manganese oxide, complex with soil organic matter, bound to sulfides, bound to amorphous iron oxide, occluded in crystalline iron oxide, and residual fractions (Kabata-Pendias 2011; Rinklebe and Shaheen 2017). However, the main portion of total Ni concentration in soils is concentrated in the residual and bound to amorphous and crystalline Fe oxides (Kabata-Pendias 2011). Similar to the other potentially toxic elements (PTEs), the toxicity risk by Ni exists when elevated concentrations of the element are presented in the mobile or potentially mobile fractions in soils (Abbas et al. 2017; Shaheen et al. 2017). Therefore, the stabilization/immobilization of PTEs, including the Ni, in soils has been proposed as a cost-effective strategy for remediating contaminated soils (Zhao et al. 2014). In the last decade, there has been an increasing interest in the application of biochar as a soil amendment for the restoration of contaminated soils.

In this chapter, the main focus is on the biochar potential for immobilization of Ni in soils. The possible mechanisms of interaction between biochar and Ni in soils and the cutting edge of biochar production and development for better performance of Ni immobilization are also discussed.

13.2 Common Technologies for Immobilization of Ni in Soils

Different management strategies have been employed for the remediation of contaminated soils with PTEs including Ni. Conventional soil remediation strategies, including solidification, soil replacement, soil washing, and electrokinetic extraction, have been used for such purpose (Karer et al. 2015; Sruthy and Jayalekshmi 2014). However, those technologies are usually expensive and substantial, and frequently lead to the destruction of the structure and fertility of soils (Karer et al. 2015). To avoid these drawbacks, considerable interest has been expressed in the use of "gentle remediation options (GROs)" (Kumpiene et al. 2014). The GRO includes the utilization of plants as phytoremediation techniques and the approach of in situ metal immobilization using various soil amendments (Shaheen et al. 2015a,b). Immobilization of Ni in soil via amendments is a promising and cost-effective strategy, aiming to decrease the mobility and phytoavailability of Ni in soil without changing their total concentrations (Ahmad et al. 2012a; Shaheen et al. 2015b).

Numerous soil amendments have been used to reduce the mobility of PTEs in soil including inorganic substrates, such as carbonates, phosphates, iron oxide substrates, or clay mineral-based amendments, or organic materials, such as biosolids, agricultural residues, composts, and manures (Almaroai et al. 2014; Shaheen et al. 2015a; Usman et al. 2013). Compared to the inorganic amendments, the organic amendments are more beneficial to the soil properties and have higher biodegradability (Rizwan et al. 2016). Recently, biochar has been recommended for the immobilization of PTEs including Ni in soils (Shaheen et al. 2015a). Biochar showed a promising role on the immobilization of Ni in contaminated soils as compared to other organic/inorganic soil amendments (Shaheen et al. 2015a).

13.3 Biochar as an Emergent Amendment for Soil Remediation

Biochar is a carbon-based soil amendment that has received growing interest recently. The biochar movement was started by the recognition of Terra Preta: tropical soils that were improved by the addition of biochar, or at least combusted waste, by native Amazonians 1–2000 years ago (Glaser et al. 2001).

13.3.1 Biochar Definition and Characteristics

The International Biochar Initiative (IBI) defined the biochar as "a solid material obtained from the thermochemical conversion of biomass in an oxygen-limited environment" (IBI 2012). More specifically, biochar is a porous carbonaceous solid by-product of the pyrolysis process of biomass such as wood, manure, or leaves (Al-Wabel et al. 2015; Shackley et al. 2012; Verheijen et al. 2010). Biochar contains volatile and aromatic organic substances (Brewer et al. 2011) and is mainly composed of C, O, H, N, and other ionic contents such as Ca^{2+}, Mg^{2+}, K^+, PO_4^{3-}, which are essentially required by plants (Spokas et al. 2012). Biochar, often, has a large inner surface area with high porosity, high contents of organic carbon, high adsorption capacity, commonly high pH (Ahmad et al. 2014; Park et al. 2015), and high cation exchange capacity (CEC) (Mohan et al. 2018; Tan et al. 2015a).

In general, pyrolysis conditions and feedstock characteristics considerably control the physicochemical properties of biochar, for example, composition, surface area, water holding capacity, pH, electrical conductivity, and particle and pore size distribution (Ahmad et al. 2012b; Rajapaksha et al. 2014; Solaiman and Anawar 2015). Thus, biochar has a wide range of characteristics (Lehmann and Joseph 2009). Furthermore, aging of biochar in soil can also alter its properties (Cheng et al. 2006; Joseph et al. 2010). Hence, the potential benefits from biochar application vary with the biochar and the soil properties (Butnan et al. 2015).

13.3.2 Biochar for Environmental Management

Biochar is recognized as a potentially feasible tool for dealing with numerous environmental issues, including soil remediation, waste management, climate change mitigation, and energy production (Ok et al. 2015). Application of biochar can enhance soil water availability (Baronti et al. 2014), water holding capacity (Abel et al. 2013; Kammann et al. 2012), soil aeration (Cayuela et al. 2013), soil organic carbon (El-Naggar et al., 2018b), soil microbial biomass and activity (Thies and Rillig 2009), enzymatic activity (Paz-Ferreiro et al. 2012), and nutrient retention and availability (El-Naggar et al. 2015; Usman et al. 2016b), which result in less fertilizer needs and reduced nutrients leaching (Laird 2008; Lehmann et al. 2003). Biochar has the potential to immobilize PTEs and improve the quality of the contaminated soil, and could reduce the bioavailability of the PTEs in soil and their uptake by plants (Li et al. 2016; Yang et al. 2017; Zhang et al. 2013). The capacity of biochar to adsorb PTEs mainly relies on the feedstock type and pyrolysis temperature (Al-Wabel et al. 2013; Fristak et al. 2015; Park et al. 2018).

13.4 Biochar Potential for Ni Immobilization in Soils

In a number of studies, biochar proved to be a suitable candidate to stabilize Ni and remedy contaminated soils (Ahmad et al. 2014; Paz-Ferreiro et al. 2014; Shen et al. 2016). Transformation of the agricultural wastes and materials to the form of biochar and its application to soil has the potential to be a cost-appropriate and eco-friendly strategy for the remediation of Ni-contaminated soils. For instance, Mendez et al. (2012) had applied both raw sewage sludge and sewage sludge-derived biochar, and they found that the biochar significantly decreased the leaching risk and bioavailability of Ni in the soil as compared to the raw sewage sludge. Bian et al. (2014) conducted a 3-year field study adding wheat straw biochar to a contaminated paddy soil. They observed a reduction of extracted Ni concentrations in the leachates 3 years after the biochar treatment. Moreover, they reported that biochar potential for Ni immobilization in this study was similar to conventional cement-based stabilization trials carried out in parallel on the same site (Wang et al. 2015), suggesting that the field performance of biochar may be comparable with technologies currently being used. It has been reported recently that biochar exhibited a better performance for Ni immobilization as compared to other soil amendments. Shaheen et al. (2015a) conducted a pot experiment for examining the efficiency of various amendments to reducing the Ni phytoavailability in soil. The applied soil amendments included biochar, activated carbon, cement bypass kiln dust, bentonite, chitosan, coal fly ash, limestone, organo-clay, sugar beet factory lime, nano-hydroxyapatite, and zeolite. They concluded that biochar followed by chitosan showed the best performance in terms of Ni immobilization in soil (Shaheen et al. 2015a). Biochar capability to immobilize Ni in soil leads to decreasing its uptake by plant as reported by Rehman et al. (2016). They found that biochar addition to soil stabilized the Ni in the soil and alleviated the Ni toxicity risk by lowering its uptake by maize seedlings. In another study, the application of biochar to Ni-contaminated soil reduced the Ni phytoavailability for sunflower seed by up to 17% as compared to biochar untreated soil (Turan et al. 2017). Table 13.1 shows data obtained from several reports on the utilization of biochar for remediating Ni-contaminated soil. However, several factors determine the biochar potential for Ni immobilization such as pyrolysis temperature, type of biochar and soil, the application rate of biochar, and the period of biochar aging in soil.

13.4.1 Factors Affecting Biochar Potential for Ni Immobilization

13.4.1.1 Pyrolysis Temperature

Production condition of biochar is a key factor in determining its characteristics and chemical composition (Ahmad et al. 2013; Al-Wabel et al. 2013). Biochar potential for Ni immobilization depends on the pyrolysis temperature of biochar (Rajapaksha et al. 2015a; Uchimiya et al. 2010). For instance, biochar produced from dried olive pomace at 700°C was more efficient to immobilize the Ni in soil as compared to the same biochar produced at 400°C (Pellera and Gidarakos 2015). It has been reported that application of paper sludge biochar (500°C) to the soil decreased the leached and bioavailable concentrations of Ni after 77 days of incubation (Mendez et al. 2014). It could be concluded, from a number of reports, that a pyrolysis temperature range of 500°C–700°C is recommended for the biochar applied to remediating Ni-contaminated soils (Uchimiya et al. 2012).

13.4.1.2 Soil and Biochar Type

Immobilization of PTEs including the Ni in soil varies with the type of biochar and soil (Tan et al. 2015b; Yang et al. 2015). Uchimiya et al. (2012) examined the performance of biochars produced from five feedstocks on the stabilization of Ni in soil. They concluded that Ni immobilization in soil varied strongly with the type of biochar (Paz-Ferreiro et al. 2014). Biochar derived from unfertilized dates was applied to a sandy loam alkaline soil contaminated with Ni and showed a substantial reduction in Ni toxicity risk in the soil (Ehsan et al. 2014). In another study, the application of 80% coniferous and 20% hardwood biochar was a good candidate to reduce the extractable Ni in soil after only 1 week (Rees et al. 2014). Such potential of biochar to immobilize Ni in soil needs to be further validated via long-term investigation.

TABLE 13.1

Biochar Utilization for Remediation of Ni-Contaminated Soil

Biochar Feedstock	Production Temperature	Application Rate/ Type of Experiment	Soil Type (Texture/ Classification)	Effect	References
Mix of shell limestone, perlite, organic substrates	NA	1% Greenhouse pot experiment	Silty soil— weakly acidic	34.4% decrease in the available fraction of Ni (soluble + exchangeable fraction) as compared to control 58.7% decrease in the water-soluble Ni in soil as compared to control	Shaheen et al. (2015a)
Cottonseed hulls	200°C–800°C	Sorption experiment	Loamy sand	Nickel sequestration by the surface functional groups of biochar	Uchimiya et al. (2011b)
Eucalyptus saligna wood	450°C	0%, 1%, 2% Pot experiment	Sandy clay loam—neutral	22% decrease of bioavailable Ni by biochar 1% and 33% decrease of bioavailable Ni by BC 2% as compared to the control Stabilization of Ni due to the rise in soil pH and the high CEC of the biochar	Rehman et al. (2016)
Woody biomass	900°C	0%, 1%, 2.5%, and 5% Pot experiment	NA	Nickel bioavailability decreased in soil by 14%, 36%, and 61% with the biochar application rates of 1%, 2.5%, and 5%, respectively, as compared to the control	Herath et al. (2015)

(Continued)

TABLE 13.1 (CONTINUED)

Biochar Utilization for Remediation of Ni-Contaminated Soil

Biochar Feedstock	Production Temperature	Application Rate/ Type of Experiment	Soil Type (Texture/ Classification)	Effect	References
Broiler litter	350°C–700°C	0%, 5%, 10%, and 20% Leaching experiment	NA	Increased biochar application rate from 0%, 5%, 10%, and 20% enhanced the immobilization of Ni in soil	Uchimiya et al. (2010)
Unfertilized dates	500°C	0%, 0.5%, 1%, and 2% (w/w) Incubation experiment	Sandy loam—alkaline	The NH_4NO_3 extractable Ni decreased by 53% as compared to the control	Ehsan et al. (2014)
British broadleaf hardwood	600°C	0%, 0.5%, 1% and 2% (w/w) Field experiment	High degree of heterogeneity in soil texture	The biochar treatment decreased the leachabilities of Ni up to 200% as compared to the untreated soil after 3 years of the application	Shen et al. (2016)
Sewage sludge	500°C	5% and 10% (w/w) Pot experiment	Silty sand—acidic	Decreased Ni bioavailability in soil due to the rise in the pH	Khan et al. (2013)

Abbreviations: NA, not applicable; BC, biochar.

The potential of biochar for Ni immobilization in soil depends on the aging of biochar in soil (Qian et al. 2015). However, only a few reports are available on the effect of the biochar aging process on Ni immobilization. A field experiment was carried out to investigate the long-term influence of biochar on the immobilization of Ni in a contaminated site in Castleford, UK (Shen et al. 2016). Results showed a promising performance of biochar, as it decreased the Ni leaching (the extracted Ni from carbonic acid leaching test) in soil treated with biochar by 90%–98% over 3 years as compared to the untreated soil (Figure 13.1). The sequential extraction indicated that biochar addition at 0.5%–2% increased the Ni concentration in the residual fraction (from 51% to 61%–66%; Figure 13.2), which led to the reduction of Ni leachability (from 0.35% to 0.12%–0.15%) in biochar-treated soil as compared to untreated soil (Figure 13.1). Conversely, the Ni portions bound to Fe/Mn oxides and organic matter (which are potential mobile fractions) were decreased in biochar-treated soils as compared to control (Figure 13.2; Shen et al. 2016). This long-term study suggested that biochar is an effective soil amendment for Ni immobilization in contaminated soils. However, other reports indicated that Ni immobilization decreased with the biochar aging process (Paz-Ferreiro et al. 2014; Qian et al. 2015).

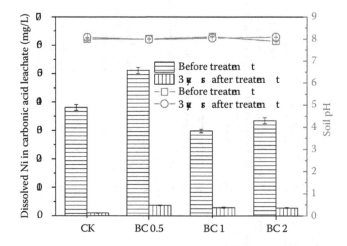

FIGURE 13.1
The dissolved concentration of Ni in carbonic acid leachates and soil pH before and 3 years after the treatments (CK: Control, BC0.5: Biochar 0.5%, BC1: Biochar 1%, and BC2: Biochar 2%). (Reproduced from Shen et al. (2016) *Sci. Total Environ.* 542:771–776, with permission from the publisher.)

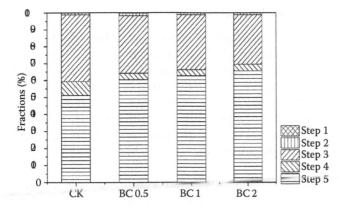

FIGURE 13.2
Geochemical fractions of Ni in soils 3 years after the treatments (CK: Control, BC0.5: Biochar 0.5%, BC1: Biochar 1%, and BC2: Biochar 2%). Step 1: Exchangeable fraction. Step 2: Bound to carbonate and phosphate fraction. Step 3: Bound to iron and manganese fraction. Step 4: Bound to organic matter fraction. Step 5: Residue fraction. (Reproduced from Shen et al. (2016) *Sci. Total Environ.* 542:771–776, with permission from the publisher.)

13.4.1.3 Application Rate

The application rate of biochar to soil has a crucial role in its performance in soil remediation, which varies based on the biochar and soil type (Ramzani et al. 2017a,b). Pellera and Gidarakos (2015) reported that an increasing rate of biochar increased the sorption of PTEs in soil and decreased their desorption. In a pot experiment, woody biochar was applied to soil at 1%, 2.5%, and 5% (w/w) to evaluate its effect as a soil amendment on the bioavailability of Ni in serpentine soil (Herath et al. 2015). The highest application rate of biochar (5%) led to the highest reduction in the exchangeable fraction of Ni as compared to the control and other application rates. The results recommended the use of woody biochar at a high rate in the serpentine soil to reduce the release of Ni in soil, thereby reducing its bioavailability (Herath et al. 2015). It has been reported that application of willow biochar at 1% and 2% rates was insufficient for the Ni immobilization in soil (Trakal et al. 2011). In another study, the application of poultry manure-derived biochar (2%) to contaminated soil increased Ni concentration in soil pore water (Marchand et al. 2016), whereas other reports suggested a 5% biochar application rate as the best optimized rate (Kumarathilaka and Vithanage 2017; Saffari et al. 2015; Younis et al. 2015). Similarly, Herath et al. (2015) reported that bioavailable Ni concentration decreased by 61% in the soil treated with biochar at 5% as compared to control (Rizwan et al. 2016).

13.4.1.4 Reduction–Oxidation (Redox) Conditions

Redox condition is an important factor controlling the behavior of Ni in soil (Rinklebe and Shaheen 2017). Nickel is presented in soil environments in a number of oxidation states: 1, 2, 3, and 4. Ma and Hooda (2010) reported that only Ni^{2+} is stable under a wide range of pH and redox conditions in soils. However, there is a lack of knowledge on Ni redox geochemistry. It has been reported that the behavior of Ni in soils is controlled by different factors including soil type, redox potential, pH, iron, manganese, dissolved organic carbon, and sulfate (Rinklebe and Shaheen 2017; Rinklebe et al. 2016a).

Recent studies showed that biochar could alter the release dynamics, potential mobility, and phytoavailability of Ni in soil under dynamic redox conditions (Rinklebe et al. 2016b). Rinklebe et al. (2016b) demonstrated that biochar was an effective amendment to immobilize the Ni in floodplain soil and to decrease its dissolved concentration as compared to control under dynamic redox conditions. Rinklebe et al. (2016a) examined the potential of biochar to reduce the dissolved Ni in soil solution of Ni-contaminated floodplain soil under dynamic redox conditions. They found that biochar decreased the dissolved Ni by 44% as compared to control, indicating the role of biochar to immobilize Ni in contaminated floodplain soil under dynamic redox conditions. However, El-Naggar et al. (2018a) found a contrasting impact of biochar on the dissolved concentration of Ni under dynamic redox conditions. We assessed the biochar effects on the release dynamics of Ni in dissolved and colloidal phases of soil under dynamic redox conditions using an automated biogeochemical microcosm apparatus (Figure 13.3). Biochar-treated soil exhibited a wider range of E_H and a lower pH than the control soil (El-Naggar et al. 2018a). Biochar increased the dissolved concentration of Ni under oxic conditions, which might be due to the lower pH values under oxic conditions in the biochar-treated soil compared to the control soil (Figure 13.3). This might create potential environmental risks with using biochar in such soil under oxic conditions, as this might increase the Ni release and transfer into the groundwater and the food chain, which should be harmful to the surrounding environment (El-Naggar et al. 2018a).

More specific future studies with different types of biochars and soils should be done to elucidate the impact of different biochars on the temporal kinetics of the Ni in soil solution and soil sediments and thus assess the potential applicability of biochars in remediation of Ni-contaminated wetland and upland soils.

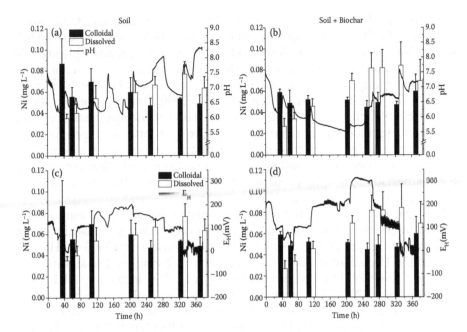

FIGURE 13.3
Development of redox potential (E_H), and pH in the soil slurry (data recorded every 10 min, averages were reported for an underling data set [$n \approx 11,000$] of four replicate samples) in the microcosms of nontreated soil (a and c), and biochar-treated soil (b and d). Columns present the distribution of Ni concentrations in the dissolved and colloidal phases in the nontreated soil and the biochar-treated soil.

13.5 Possible Mechanisms of Interaction between Biochar and Ni in Soils

Before reviewing the possible mechanisms in the interaction between biochar and Ni, it is important to note that biochars could affect only the bioavailable fraction of Ni. Thus, it can reduce the risk of Ni leachability to groundwater (Paz-Ferreiro et al. 2014).

The key properties of biochar could explain its potential to stabilize PTEs, including Ni, in soils. One of the key properties of biochar is possessing large surface areas, which afford a high capacity for PTEs complexation on their surface. This high sorption capacity of biochar can be attributed to the presence of different functional groups on the surface of biochar.

Thus, they could form complexes with PTEs on biochar surfaces, due to the exchange of PTEs with cations associated with biochar, such as Ca^{2+}, Mg^{2+}, K^+, and Na^+ (Lu et al. 2012; Uchimiya et al. 2011a). Physical adsorption could also occur (Lu et al. 2012). It has also been reported that the O-containing functional groups on biochar surfaces have a role in the stabilization of PTEs in the biochar surface (Mendez et al. 2009; Uchimiya et al. 2011a). In particular, Uchimiya et al. (2011b) reported that the application of high O-containing biochar (derived from cottonseed hull and produced at 350°C) led to a high adsorption of Ni. Sorption mechanisms are dependent on several factors such as soil type and the cations and other compounds (e.g., carbonates, phosphates, or sulfates) present in both biochar and soil (Karami et al. 2011; Park et al. 2013).

The relativistic contribution of several mechanisms to PTE immobilization by biochar is not well understood. However, it has been postulated that the biochar impact on PTE immobilization is mostly a pH effect (Houben et al. 2013). The alkaline nature of the biochar is considered to significantly influence the mobility and bioavailability of PTEs including Ni in soils (Almaroai et al. 2014). Biochar-induced raises in soil pH can be responsible for the lower concentrations of available PTEs and their precipitation in soils (Ahmad et al. 2012c). It has been reported that biochar indirect influences on soil properties (especially under dynamic redox conditions), such as its effect on pH and E_H, hence, E_H-dependent factors (e.g., dissolved organic matter, dissolved inorganic carbon, iron, manganese, sulfate), might have a crucial role in Ni immobilization in soil (Rinklebe and Shaheen 2017; Rinklebe et al. 2016b).

The possible mechanisms of biochar interaction with PTEs including Ni have been reported by Ahmad et al. (2014) (Figure 13.4). They proposed four different possible mechanisms for biochar interaction with PTEs: (i) ion exchange between PTEs and exchangeable elements in biochar, (ii) electrostatic attraction of cationic PTEs, (iii) electrostatic attraction of anionic PTEs, and (iv) precipitation of PTEs. This assumption was consistent with the postulated mechanisms for biochar interaction with Pb^{2+} by Lu et al. (2012), as they proposed that Pb^{2+} immobilization by biochar is due to electrostatic outer-sphere complexation (element exchange with cations such as K^+ and Na^+ in the biochar), coprecipitation and inner-sphere complexation of elements with organic matter and mineral oxides of biochar, surface complexation (with active -COOH and -OH functional groups of biochar). Uchimiya et al. (2010) attributed the enhanced immobilization of Ni in soil to the increased pH after the addition of basic biochar to the soil. They proposed the following mechanisms for the immobilization of Ni by biochar: (i) the precipitation as metal (hydr)oxide, carbonate, or phosphate and/or the activation of surfaces induced by the pH increase; (ii) sorptive interactions between d-electrons of Ni and aromatic π-electrons of biochar (Karer et al. 2015);

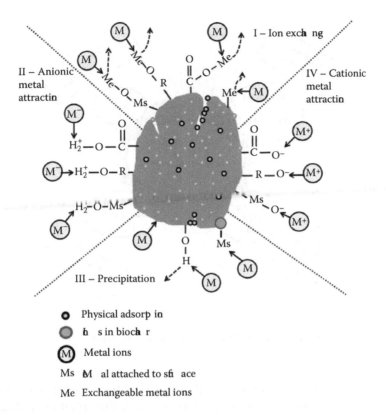

FIGURE 13.4
Mechanisms of interaction between biochar and inorganic elements in soil. (Reproduced from Ahmad et al. (2014) *Chemosphere* 99:19–33, with permission from the publisher.)

and (iii) specific metal–ligand complexation via surface functional groups of biochar (Uchimiya et al. 2010).

However, there is a lack of reports that specify the possible mechanisms of Ni (im)mobilization by biochar. As illustrated in Figure 13.5, we postulate the following mechanisms for Ni interaction with biochar:

1. Electrostatic interactions between Ni and negatively charged active sites on biochar surfaces. Electrostatic attraction between positively charged elements and negatively charged biochar was proposed to be the main mechanism for cationic elements (Ahmad et al. 2014; Hsu et al. 2009; Uchimiya et al. 2011c).

2. Physical adsorption of Ni in the pores preexisting in the biochar feedstock, which provides internal surface for adsorption.

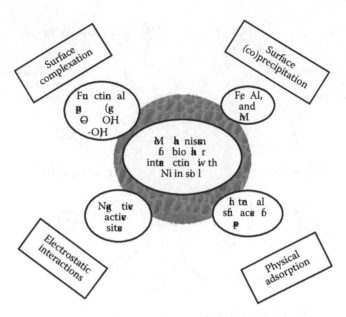

FIGURE 13.5
Postulated mechanisms of interaction between biochar and Ni in soil.

3. Surface (co)precipitation with Fe, Al, and Mn, and the formation of metal (hydr)oxide, sulfate, or phosphate. Precipitation is likely important for Ni, which has a high affinity for specific ligands (e.g., sulfate) on the surface of, or released by, biochar, as was observed in our ongoing research.

4. Biochar surface functional group (e.g., carboxyl and hydroxyl) complexation with Ni.

13.6 Designer/Modified Biochar as a Novel Approach for Immobilization of Ni in Soils

In recent years, biochars have been widely used as an adsorbent for several emerging organic and inorganic pollutants. To improve remediation efficiencies and environmental benefits of biochar, recent attention has been paid to develop designer/modified biochar (Ok et al. 2015). A variety of biochar modification methods such as chemical, physical, magnetic, clay-mineral, and biological have been widely used to alter the properties of biochars and thereby enhance their benefits (Rajapaksha et al. 2016; Figure 13.6).

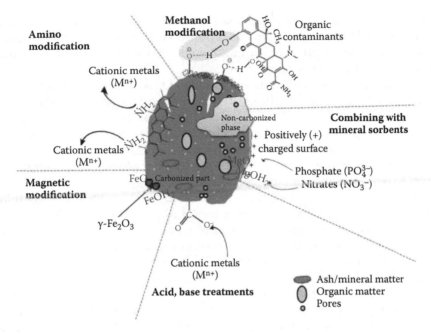

FIGURE 13.6
Schematic diagram of biochar modifications. (Reproduced from Rajapaksha et al. (2016) *Chemosphere* 148:276–291, with permission from the publisher.)

Only few studies have been carried out to understand the overall mechanism for immobilization of Ni by modified biochar in aqueous environments. To our knowledge, no prior study employed designer/modified biochar for immobilization of Ni in soils. Therefore, as a novel approach, it is now timely and essential to study the possibility of modified biochar for immobilization of Ni in soil.

Chemical modification involves mixing of chemical reagents with biochar or pretreatment of precursors before the carbonization process (Qian et al. 2013; Usman et al. 2016a). It alters the surface functional groups and/or introduces new functional groups. Alkali/acid activation and intentional surface oxidation using hydrogen peroxide (H_2O_2), potassium permanganate ($KMnO_4$), ammonium persulfate [$(NH_4)_2S_2O_8$], and ozone (O3) are the most widely used chemical modification methods (Rajapaksha et al. 2016; Xiu et al. 2017). It has been shown that the activation of biochar by alkali solutions such as NaOH and KOH enhance O-containing functional groups (Regmi et al. 2012). For example, alkali (NaOH)-modified biochar showed much higher sorption of Ni than the pristine biochar derived from hickory wood due to the high surface area and large number of oxygen-containing surface functional groups (Ding et al. 2016). The negatively charged carboxyl groups in the chemically modified biochars can increase sorption by binding with positively charged Ni metals. Xue et al. (2012) reported similar findings in which the H_2O_2-modified biochars produced from pea nut hull resulted in a remarkable sorption of Ni in aqueous media due to their increased O-containing functional groups.

The main physical modification methods of biochar modification involve steam/air activation, ball milling, and microwave heating process (Peterson 2012; Rajapaksha et al. 2015b). Physical modification increases the heavy metal adsorption capacity of modified biochar by increasing the specific surface area and creating more micropores and mesopores on biochar (Rajapaksha et al. 2016). Therefore, there is a strong tendency of physically modified biochars to remediate Ni-contaminated sites.

It has already been shown that clay minerals such as kaolinite, goethite, and gibbsite are promising sorbents for Ni (Rajapaksha et al. 2012; Yavuz et al. 2003). Therefore, impregnation of minerals with biochar, termed clay–biochar composites, would enhance Ni sorption. Biochars act as an excellent porous structure to support and host the distribution of the fine clay particles within the clay–biochar composite matrix (Yao et al. 2014). The nickel adsorption process is mainly governed by surface complexation and precipitation (Rajapaksha et al. 2012; Yavuz et al. 2003). The clay–biochar composites exhibited lower surface areas compared to unmodified biochar due to blockage of pores by clay particles (Yao et al. 2014). However, the integrated performance of biochar and clay will significantly enhance Ni immobilization in soils.

Anaerobic digestion has been suggested as a means of biological activation to make high-efficiency biochars for heavy metal removal (Inyang et al. 2011). It improved CEC and showed negative zeta potential values (Inyang et al. 2011). This would facilitate strong adsorption between Ni cation and negatively changed surfaces. Inyang et al. (2011) reported a significant increase of Ni sorption in aqueous media by anaerobically digested biochar compared with unmodified biochar. This suggests the possibility of biologically active biochars for the remediation of cations such as Ni in soil environment as well.

Generally, iron or iron oxides are used to produce magnetic biochars (Chen et al. 2011; Mohan et al. 2014). Impregnation with magnetite/maghemite significantly enhanced the sorption of Cd (Trakal et al. 2016). Magnification increased the cation release and metal complexation with carboxylic groups (Trakal et al. 2016). The CEC value of magnetically modified biochar greatly enhanced due to the presence of Fe oxides in biochars. This facilitates the heavy metal sorption.

Undoubtedly, the modified/designer biochar can have strong Ni remediation efficiencies in soil environment. Also, modified biochar application would make soil quality improvements. Very little information is available regarding the long-term investigations of modified biochar in soil. Therefore, the long-term effect of modified biochar on soil health should be further investigated.

13.7 Environmental Benefits and Impacts

The application of biochar to Ni-contaminated soils has substantial implications on the environment and the sustainable development. Biochar is considered as a "multi-win" amendment, due to its role in the recycling of waste materials, climate mitigation, energy production, and soil remediation (Park et al. 2016; Wang et al. 2017). Converting waste materials into biochar is a promising solution for the secure and profitable disposal of solid waste materials such as animal litter and sewage sludge. Moreover, all active pathogens contained in those materials will be removed during the pyrolysis process in biochar production (Ahmad et al. 2014). Additionally, syngas and bio-oil, as sources of renewable-energy, are produced during the pyrolysis process, which is an excellent contribution to solving the energy crisis worldwide.

Biochar could serve as a green environment sorbent, as it showed its capability to immobilize Ni in contaminated soils and to reduce its potential mobility and phytoavailability, and promote soil quality and restoration (Ahmad et al. 2014).

Biochar revealed its compatibility with other common sorbent materials. For instance, activated biochar has equal/greater sorption efficiency for Ni and other PTEs as compared to activated carbon. However, biochar is cost-effective as compared to activated carbon, as it produces from waste resources. The estimated prices for biochar and activated carbon are U.S. \$246 ton^{-1}, and U.S. \$1500 ton^{-1}, respectively, which means that biochar is approximately 1/6 less expensive than activated carbon (Ahmad et al. 2012c; McCarl et al. 2009).

Therefore, biochar is recognized as an environmentally friendly and low-cost amendment for immobilization of Ni in contaminated soils (Wang et al. 2017).

13.8 Future Perspective

Biochar has been proposed as a promising soil amendment to affect the (im)mobilization of Ni. However, as a novel strategy, several aspects need to be further documented to validate the potential of biochar for in situ applications.

Specifically, further investigation is required in the following sectors:

1. Standardization of the production methods and types of feedstock, which are more suited to biochar for the remediation purpose of Ni-contaminated soils. The characteristics that different types of biochar might possess make some particular feedstocks and production conditions more suitable than others to remediate Ni in

soils. Consequently, when applying biochar for soil remediation, one should pay attention not only to soil type and properties but also to biochar characteristics. Furthermore, it should also be considered that major biochar properties such as pH, surface area, ash, and carbon contents can be influenced by pre-/posttreatment of biochar and hence enhance biochars' potential to immobilize specific elements such as Ni (Lima et al. 2014; Rajapaksha et al. 2016).

2. Although short-term findings reported the feasibility of biochar application to remediate Ni-contaminated soils, information on the aging impacts of biochar in soils is lacking. This will entail in situ monitoring during long-term field experiments with different biochar feedstocks and pyrolysis conditions, and soil types. These long-term experiments should be carried out to examine the biochar efficacy on the stabilization of Ni in multi-element contaminated soils, to predict the metal stabilization mechanism of biochar on the long term.

3. The interaction between biochar and soil characteristics, changes in the biochar surface chemistry, and how these translate to improvements in soil quality and immobilization of Ni is another area demanding research. In this respect, using cutting-edge technologies such as X-ray absorption spectroscopy is essential to elucidate the specific mechanisms for biochar–element interactions.

4. The efficiency of biochar for immobilization of Ni under different redox conditions in low-fertility soils is not well understood. Further research is required on the potential of biochar to immobilize Ni under dynamic redox conditions to identify this knowledge gap.

References

Abbas, T., Rizwan, M., Ali, S., Zia-ur-Rehman, M., Qayyum, M.F., Abbas, F., Hannan, F., Rinklebe, J., Ok, Y.S., 2017. Effect of biochar on cadmium bioavailability and uptake in wheat (*Triticum aestivum* L.) grown in a soil with aged contamination. *Ecotoxicol. Environ. Safety* 140, 37–47.

Abel, S., Peters, A., Trinks, S., Schonsky, H., Facklam, M., Wessolek, G., 2013. Impact of biochar and hydrochar addition on water retention and water repellency of sandy soil. *Geoderma* 202, 183–191.

Adrees, M., Ali, S., Rizwan, M., Ibrahim, M., Abbas, F., Farid, M., Zia-ur-Rehman, M., Irshad, M.K., Bharwana, S.A., 2015. The effect of excess copper on growth and physiology of important food crops: A review. *Environ. Sci. Pollut. Res.* 22, 8148–8162.

Ahmad, M., Hashimoto, Y., Moon, D.H., Lee, S.S., Ok, Y.S., 2012a. Immobilization of lead in a Korean military shooting range soil using eggshell waste: An integrated mechanistic approach. *J. Hazard. Mater.* 209–210, 392–401.

Ahmad, M., Lee, S.S., Dou, X., Mohan, D., Sung, J.K., Yang, J.E., Ok, Y.S., 2012b. Effects of pyrolysis temperature on soybean stover- and peanut shell-derived biochar properties and TCE adsorption in water. *Bioresour. Technol.* 118, 536–544. DOI:10.1016/j.biortech.2012.05.042.

Ahmad, M., Lee, S.S., Rajapaksha, A.U., Vithanage, M., Zhang, M., Cho, J.S., Lee, S.E., Ok, Y.S., 2013. Trichloroethylene adsorption by pine needle biochars produced at various pyrolysis temperatures. *Bioresour. Technol.* 143, 615–622.

Ahmad, M., Lee, S.S., Yang, J.E., Ro, H.M., Lee, Y.H., Ok, Y.S., 2012c. Effects of soil dilution and amendments (mussel shell, cow bone, and biochar) on Pb availability and phytotoxicity in military shooting range soil. *Ecotoxicol. Environ. Safety* 79, 225–231.

Ahmad, M., Rajapaksha, A.U., Lim, J.E., Zhang, M., Bolan, N., Mohan, D., Vithanage, M., Lee, S.S., Ok, Y.S., 2014. Biochar as a sorbent for contaminant management in soil and water: A review. *Chemosphere* 99, 19–33.

Alloway, B., 2013. Heavy metals and metalloids as micronutrients for plants and animals. In: Alloway B.J. (Ed.), *Heavy Metals in Soils.* Springer, The Netherlands, pp. 195–209.

Almaroai, Y.A., Usman, A.R.A., Ahmad, M., Moon, D.H., Cho, J.S., Joo, Y.K., Jeon, C., Lee, S.S., Ok, Y.S., 2013. Effects of biochar, cow bone, and eggshell on Pb availability to maize in contaminated soil irrigated with saline water. *Environ. Earth Sci.* http://dx.doi.org/10.1007/s12665-013-2533-6.

Almaroai, Y.A., Usman, A.R., Ahmad, M., Moon, D.H., Cho, J.S., Joo, Y.K., Jeon, C., Lee, S.S., Ok, Y.S., 2014. Effects of biochar, cow bone, and eggshell on Pb availability to maize in contaminated soil irrigated with saline water. *Environ. Earth Sci.* 71(3), 1289–1296.

Al-Wabel, M.I., Al-Omran, A., El-Naggar, A.H., Nadeem, M., Usman, A.R., 2013. Pyrolysis temperature induced changes in characteristics and chemical composition of biochar produced from conocarpus wastes. *Bioresour. Technol.* 131, 374–379.

Al-Wabel, M.I., Usman, A.R., El-Naggar, A.H., Aly, A.A., Ibrahim, H.M., Elmaghraby, S., Al-Omran, A., 2015. Conocarpus biochar as a soil amendment for reducing heavy metal availability and uptake by maize plants. *Saudi J. Biol. Sci.* 22(4), 503–511.

Antić-Mladenović, S., Frohne, T., Kresovic, M., Staerk, H.-J., Tomi C.Z., Licina, V., Rinklebe, J., 2017. Biogeochemistry of Ni and Pb in a periodically flooded arable soil: Fractionation and redox-induced (im)mobilization. *J. Environ. Manag.* 186, 141–150.

Baronti, S., Vaccari, F.P., Miglietta, F., Calzolari, C., Lugato, E., Orlandini, S., Pini, R., Zulian, C., Genesio, L., 2014. Impact of biochar application on plant water relations in *Vitis vinifera* (L.). *Eur. J. Agron.* 53, 38–44.

Bian, R., Joseph, S., Cui, L., Pan, G., Li, L., Liu, X., Zhang, A., Rutlidge, H., Wong, S., Chia, C., Marjo, C., Gong, B., Munroe, P., Donne, S., 2014. A three-year experiment confirms continuous immobilization of cadmium and lead in contaminated paddy field with biochar amendment. *J. Hazard. Mater.* 272, 121–128.

Brewer, C.E., Unger, R., Schmidt-Rohr, K., Brown, R.C., 2011. Criteria to select biochars for field studies based on biochar chemical properties. *Bioenergy Res.* 4,
312–323.

Butnan, S., Deenik, J.L., Toomsan, B., Antal, M.J., Vityakon, P., 2015. Biochar characteristics and application rates affecting corn growth and properties of soils
contrasting in texture and mineralogy. *Geoderma* 237, 105–116.

Cayuela, M.L., Sánchez-Monedero, M.A., Roig, A., Hanley, K., Enders, A., Lehmann,
J., 2013. Biochar and denitrification in soils: When, how much and why doesbio-
char reduce N_2O emissions? *Sci. Rep.* 3, 1732.

Chen, B., Chen, Z., Lv, S., 2011. A novel magnetic biochar efficiently sorbs organic pollutants and phosphate. *Bioresour. Technol.* 102, 716–723.

Cheng, C.H., Lehmann, J., Thies, J.E., Burton, A.J., Engelhard, M., 2006. Oxidation
of black carbon by biotic and abiotic processes. *Organ. Geochem.* 37, 1477–1488.

Ding, Z., Hu, X., Wan, Y., Wang, S., Gao, B., 2016. Removal of lead, copper, cadmium,
zinc, and nickel from aqueous solutions by alkali-modified biochar: Batch and
column tests. *J. Ind. Eng. Chem.* 33, 239–245.

Ehsan, M., Barakat, M.A., Husein, D.Z., Ismail, S.M., 2014. Immobilization of Ni and
Cd in soil by biochar derived from unfertilized dates. *Water Air Soil Pollut.*
225(11), 2123.

El-Naggar, A., Awad, Y.M., Tang, X.Y., Liu, C., Niazi, N.K., Jien, S.H., Tsang, D.C.,
Song, H., Ok, Y.S. and Lee, S.S., 2018b. Biochar influences soil carbon pools and
facilitates interactions with soil: A field investigation. *Land Degrad Dev.* doi:
10.1002/ldr.2896.

El-Naggar, A., Shaheen, S.M., Ok, Y.S., Rinklebe, J., 2018a. Biochar affects the dissolved
and colloidal concentrations of Cd, Cu, Ni, and Zn and their phytoavailability
and potential mobility in a mining soil under dynamic redox-conditions. *Sci.
Total Environ.* doi.org/10.1016/j.scitotenv.2017.12.190.

El-Naggar, A.H., Usman, A.R., Al-Omran, A., Ok, Y.S., Ahmad, M., Al-Wabel, M.I.,
2015. Carbon mineralization and nutrient availability in calcareous sandy soils
amended with woody waste biochar. *Chemosphere* 138, 67–73.

Fristak, V., Pipiska, M., Lesny, J., Soja, G., Friesl-Hanl, W., Packová, A., 2015. Utilization
of biochar sorbents for Cd^{2+}, Zn^{2+}, and Cu^{2+} ions separation from aqueous solutions: Comparative study. *Environ. Monit. Assess.* 187, 4093.

Glaser, B., Haumaier, L., Guggenberger, G., Zech, W., 2001. The 'Terra Preta'
phenomenon: A model for sustainable agriculture in the humid tropics.
Naturwissenschaften 88, 37–41.

Glaser, B., Lehmann, J., Zech, W., 2002. Ameliorating physical and chemical properties of highly weathered soils in the tropics with charcoal—A review. *Biol.
Fertil. Soils* 35, 219–230.

Herath, I., Kumarathilaka, P., Navaratne, A., Rajakaruna, N., Vithanage, M., 2015.
Immobilization and phytotoxicity reduction of heavy metals in serpentine soil
using biochar. *J. Soils Sedim.* 15(1), 126–138.

Houben, D., Evrard, L., Sonnet, P., 2013. Mobility, bioavailability and pH-dependent
leaching of cadmium, zinc and lead in a contaminated soil amended with biochar. *Chemosphere* 92, 1450–1457.

Hsu, N.H., Wang, S.L., Liao, Y.H., Huang, S.T., Tzou, Y.M., Huang, Y.M., 2009. Removal
of hexavalent chromium from acidic aqueous solutions using rice straw-derived
carbon. *J. Hazard. Mater.* 171, 1066–1070.

IBI, 2012. Standardized product definition and product testing guidelines for biochar that is used in soil. International Biochar Initiative, April 2012.

Inyang, M., Gao, B., Ding, W., Pullammanappallil, P., Zimmerman, A.R., Cao, X., 2011. Enhanced lead sorption by biochar derived from anaerobically digested sugarcane bagasse. *Sep. Sci. Technol.* 46, 1950–1956.

Joseph, S.D., Camps-Arbestain, M., Lin, Y., Munroe, P., Chia, C.H., Hook, J., Van Zwieten, L., Kimber. S., Cowie, A., Singh, B.P., Lehmann, J., 2010. An investigation into the reactions of biochar in soil. *Austral. J. Soil Res.* 48, 501–515.

Kabata-Pendias, A., 2011. *Trace Elements in Soils and Plants*, 4th ed. CRC Press, Boca Raton.

Kammann, C., Ratering, S., Eckhard, C., Müller, C., 2012. Biochar and hydrochar effects on greenhouse gas (carbon dioxide nitrous oxide, and methane) fluxes from soils. *J. Environ. Qual.* 41, 1052.

Karami, N., Clemente, R., Moreno-Jiménez, E., Lepp, N., and Beesley, L., 2011. Efficiency of green waste compost and biochar soil amendments for reducing lead and copper mobility and uptake to ryegrass (*Lolium perenne*), *J. Hazard. Mater.* 191, 41–48.

Karer, J., Wawra, A., Zehetner, F., Dunst, G., Wagner, M., Pavel, P.B., Puschenreiter, M., Friesl-Hanl, W., Soja, G., 2015. Effects of biochars and compost mixtures and inorganic additives on immobilisation of heavy metals in contaminated soils. *Water Air Soil Pollut.* 226(10), 342.

Khan, M.U., Shahbaz, N., Waheed, S., Mahmood, A., Shinwari, Z. K., Malik, R.N., 2016. Comparative health risk surveillance of heavy metals via dietary food stuff consumption in different land-use types of Pakistan. *Hum. Ecol. Risk Assess. Int. J.* 22, 168–186.

Khan, S., Chao, C., Waqas, M., Arp, H.P.H., Zhu, Y.G., 2013. Sewage sludge biochar influence upon rice (*Oryza sativa* L.) yield, metal bioaccumulation and greenhouse gas emissions from acidic paddy soil. *Environ. Sci. Technol.* 47, 8624–8632.

Kumarathilaka, P., Vithanage, M., 2017. Influence of *Gliricidia sepium* biochar on attenuate perchlorate-induced heavy metal release in serpentine soil. *J. Chem.* 2017.

Kumpiene, J., Bert, V., Dimitriou, I., Eriksson, J., Friesl-Hanl, W., Galazka, R. et al. 2014. Selecting chemical and ecotoxicological test batteries for risk assessment of trace element contaminated soils (phyto) managed by gentle remediation options (GRO). *Sci. Total Environ.* 496, 510–522.

Laird, D.A., 2008. The charcoal vision: A win–win–win scenario for simultaneously producing bioenergy, permanently sequestering carbon, while improving soil and water quality. *Agron. J.* 100, 178–181.

Lehmann, J., Da Silva, J.P., Steiner, C., Nehls, T., Zech, W., Glaser, B., 2003. Nutrient availability and leaching in an archaeological anthrosol and a Ferralsol of the central Amazon Basin: Fertilizer, manure and charcoal amendments. *Plant Soil* 249, 343–357.

Lehmann, J., Joseph, S., 2009. Biochar for environmental management: An introduction. In: Lehmann J., Joseph S. (Eds.), *Biochar for Environmental Management Science and Technology*. Earthscans, UK, pp. 1–12.

Li, H.; Ye, X., Geng, Z., Zhou, H., Guo, X., Zhang, Y., Zhao, H., Wang, G., 2016. The influence of biochar type on long-term stabilization for Cd and Cu in contaminated paddy soils. *J. Hazard. Mater.* 304, 40–48.

Li, Y., Fang, F., Wu, M., Kuang, Y. and Wu, H., 2018. Heavy metal contamination and health risk assessment in soil-rice system near Xinqiao mine in Tongling city, Anhui province, China. *Hum. Ecol. Risk. Assess.* 24(3), 743–753.

Lima, I.M., Boykin, D.L., Klasson, K.T., Uchimiya, M., 2014. Influence of post-treatment strategies on the properties of activated chars from broiler manure, *Chemosphere* 95, 96–104.

Lockwood, C.L., Stewart, D.I., Mortimer, R.J., Mayes, W.M., Jarvis, A.P., Gruiz, K., Burke, I.T., 2015. Leaching of copper and nickel in soil-water systems contaminated by bauxite residue (red mud) from Ajka, Hungary: The importance of soil organic matter. *Environ. Sci. Pollut. Res.* 22, 10800.

Lu, H., Zhang, Y. Y., Huang, X., Wang, S., Qiu, R., 2012. Relative distribution of Pb^{2+} sorption mechanisms by sludge-derived biochar. *Water Res.* 46, 854–862.

Ma, Y., Hooda, P.S., 2010. Chromium, nickel and cobalt. In: Hooda, P.S. (Ed.), *Trace Elements in Soils*, 1st ed. John Wiley & Sons, PO19 8SQ, UK, pp. 461-480.

Marchand, L., Pelosi, C., González-Centeno, M.R., Maillard, A., Ourry, A., Galland, W., Teissedre, P.L., Bessoule, J.J., Mongrand, S., Morvan-Bertrand, A., Zhang, Q., 2016. Trace element bioavailability, yield and seed quality of rapeseed (*Brassica napus* L.) modulated by biochar incorporation into a contaminated technosol. *Chemosphere* 156, 150–162.

McCarl, B.A., Peacocke, C., Chrisman, R., Kung, C.C., Sands, R.D., 2009. Economics of biochar production, utilization and greenhouse gas offsets. In: Lehmann, J., Joseph, A.S. (Eds.), *Biochar for Environmental Management: Science and Technology.* Earthscan, London, pp. 341–358.

Mendez, A., Barriga, S., Fidalgo, J.M., Gascó, G., 2009. Adsorbent materials from paper industry waste materials and their use in Cu(II) removal from water. *J. Hazard. Mater.* 165, 736–743.

Mendez, A., Gomez, A., Paz-Ferreiro, J., Gascó, G., 2012. Effects of sewage sludge biochar on plant metal availability after application to a Mediterranean soil. *Chemosphere* 89, 1354–1359.

Mendez, A., Paz-Ferreiro, J., Araujo, F., Gasco, G., 2014. Biochar from pyrolysis of deinking paper sludge and its use in the treatment of a nickel polluted soil. *J. Anal. Appl. Pyrolysis.* 107:46–52.

Mohan, D., Abhishek, K., Sarswat, A., Patel, M., Singh, P. and Pittman, C.U., 2018. Biochar production and applications in soil fertility and carbon sequestration—A sustainable solution to crop-residue burning in India. *RSC Advances* 8(1), 508–520.

Mohan, D., Kumar, H., Sarswat, A., Alexandre-Franco, M., Pittman, C.U., 2014. Cadmium and lead remediation using magnetic oak wood and oak bark fast pyrolysis bio-chars. *Chem. Eng. J.* 236, 513–528.

Ok, Y.S., Chang, S.X., Gao, B., Chung, H.-J., 2015. SMART biochar technology—A shifting paradigm towards advanced materials and healthcare research. *Environ. Technol. Innovat.* 4, 206–209.

Park, J.H., Choppala, G.H., Lee, S.J., Bolan, N., Chung, J.W., Edraki, M., 2013. Comparative sorption of Pb and Cd by biochars and its implication for metal immobilization in soil. *Water Air Soil Pollut.* 224, 1711

Park, J.H., Lee, S.J., Lee, M.E., Chung, J.W., 2016. Comparison of heavy metal immobilization in contaminated soils amended with peat moss and peat moss-derived biochar. *Environ. Sci. Proc. Impacts* 18(4), 514–520.

Park, J.H., Ok, Y.S., Kim, S.H., Kang, S.W., Cho, J.S., Heo, J.S., Delaune, R.D., Seo, D.C., 2015. Characteristics of biochars derived from fruit tree pruning wastes and their effects on lead adsorption. *J. Kor. Soc. Appl. Biol. Chem.* 58, 751–760.

Park, J.H., Wang, J.J., Kim, S.H., Kang, S.W., Cho, J.S., Delaune, R.D., Ok, Y.S., and Seo, D.C., 2018. Lead sorption characteristics of various chicken bone part-derived chars. *Environ Geochem Hlth* 1–11.

Paz-Ferreiro J., Gascó G., Gutiérrez B., Méndez A., 2012. Soil biochemical activities and the geometric mean of enzyme activities after application of sewage sludge and sewage sludge biochar to soil. *Biol. Fertil. Soils* 48, 511–517.

Paz-Ferreiro, J., Lu, H., Fu, S., Méndez, A., and Gascó, G., 2014. Use of phytoremediation and biochar to remediate heavy metal polluted soils: A review. *Solid Earth* 5(1), 65–75.

Pellera, F.M., Gidarakos, E., 2015. Effect of dried olive pomace-derived biochar on the mobility of cadmium and nickel in soil. *J. Environ. Chem. Eng.* 3, 1163–1176.

Peterson, S.C., Jackson, M.A., Kim, S., Palmquist, D.E., 2012. Increasing biochar surface area: Optimization of ball milling parameters. *Powder Technol.* 228, 115–120.

Qian, L., Chen, M., Chen, B., 2015. Competitive adsorption of cadmium and aluminum onto fresh and oxidized biochars during aging processes. *J. Soils Sedim.* 15, 1130–1138.

Qian, W., Zhao, A.-z., Xu, R.-k., 2013. Sorption of As(V) by aluminum-modified crop straw-derived biochars. *Water Air Soil Pollut.* 224, 1–8.

Rajapaksha, A.U., Ahmad, M., Vithanage, M., Kim, K.R., Chang, J.Y., Lee, S.S., Ok, Y.S., 2015a. The role of biochar, natural iron oxides, and nanomaterials as soil amendments for immobilizing metals in shooting range soil. *Environ. Geochem. Health* 37, 931–942.

Rajapaksha, A.U., Chen, S.S., Tsang, D.C.W., Zhang, M., Vithanage, M., Mandal, S., Gao, B., Bolan, N.S., Ok, Y.S., 2016. Engineered/designer biochar for contaminant removal/immobilization from soil and water: Potential and implication of biochar modification. *Chemosphere* 148, 276–291.

Rajapaksha, A.U., Vithanage, M., Ahmad, M., Seo, D.-C., Cho, J.-S., Lee, S.-E., Lee, S.S., Ok, Y.S., 2015b. Enhanced sulfamethazine removal by steam-activated invasive plant-derived biochar. *J. Hazard. Mater.* 290, 43–50.

Rajapaksha, A.U., Vithanage, M., Weerasooriya, R., Dissanayake, C.B., 2012. Surface complexation of nickel on iron and aluminum oxides: A comparative study with single and dual site clays. *Colloids Surf. A Physicochem. Eng. Aspects* 405, 79–87.

Rajapaksha, A.U., Vithanage, M., Zhang, M., Ahmad, M., Mohan, D., Chang, S.X., Ok, Y.S., 2014. Pyrolysis condition affected sulfamethazine sorption by tea waste biochars. *Bioresour. Technol.* 166, 303–308.

Ramzani, P.M.A., Anjum, S., Abbas, F., Iqbal, M., Yasar, A., Ihsan, M.Z., Anwar, M.N., Baqar, M., Tauqeer, H.M., Virk, Z.A., Khan, S.A., 2017a. Potential of miscanthus biochar to improve sandy soil health, in situ nickel immobilization in soil and nutritional quality of spinach. *Chemosphere* 185, 1144–1156.

Ramzani, P.M.A., Khalid, M., Anjum, S., Ali, S., Hannan, F., Iqbal, M., 2017b. Cost effective enhanced iron bioavailability in rice grain grown on calcareous soil by sulfur mediation and its effect on heavy metals mineralization. *Environ. Sci. Pollut. Res.* 24, 1219–1228.

Rees, F., Simonnot, M.O., Morel, J.L., 2014. Short-term effects of biochar on soil heavy metal mobility are controlled by intra-particle diffusion and soil pH increase. *Eur. J. Soil Sci.* 65, 149–161.

Regmi, P., Garcia Moscoso, J.L., Kumar, S., Cao, X., Mao, J., Schafran, G., 2012. Removal of copper and cadmium from aqueous solution using switchgrass biochar produced via hydrothermal carbonization process. *J. Environ. Manag.* 109, 61–69.

Rehman, M.Z.U., Rizwan, M., Ali, S., Fatima, N., Yousaf, B., Naeem, A., Sabir, M., Ahmad, H.R., Ok, Y.S., 2016. Contrasting effects of biochar, compost and farm manure on alleviation of nickel toxicity in maize (*Zea mays* L.) in relation to plant growth, photosynthesis and metal uptake. *Ecotoxicol. Environ. Safety* 133, 218–225.

Rinklebe, J., Shaheen, S.M., 2017. Redox chemistry of nickel in soils and sediments: A review. *Chemosphere* 179, 265–278.

Rinklebe, J., Antić-Mladenović, S., Frohne, T., Staerk, H.-J., Tomic, Z., Licina, V., Staerk, H.-J., Tomi c, Z., Licina, V., 2016a. Nickel in a serpentine-enriched Fluvisol: Redox affected dynamics and binding forms. *Geoderma* 263, 203–214.

Rinklebe, J., Shaheen, S.M., Frohne, T., 2016b. Amendment of biochar reduces the release of toxic elements under dynamic redox conditions in a contaminated floodplain soil. *Chemosphere* 142, 41–47.

Rizwan, M., Ali, S., Qayyum, M.F., Ibrahim, M., Zia-ur-Rehman, M., Abbas, T., Ok, Y.S., 2016. Mechanisms of biochar-mediated alleviation of toxicity of trace elements in plants: A critical review. *Environ. Sci. Pollut. Res.* 23(3), 2230–2248.

Saffari, M., Karimian, N., Ronaghi, A., Yasrebi, J., Ghasemi-Fasaei, R., 2015. Stabilization of nickel in a contaminated calcareous soil amended with low-cost amendments. *J. Soil Sci. Plant Nutr.* 15, 896–913.

Shackley, S., Carter, S., Knowles, T., Middelink, E., Haefele, S., Sohi, S., Cross, A., Haszeldine, S., 2012. Sustainable gasification-biochar systems? A case-study of rice-husk gasification in Cambodia, Part 1: Context, chemical properties, environmental and health and safety issues. *Energy Policy* 42, 49–58.

Shaheen, S.M., Frohne, T., White, J.R., DeLaune, R.D., Rinklebe, J., 2017. Redox-induced mobilization of copper, selenium, and zinc in deltaic soils originating from Mississippi (USA) and Nile (Egypt) River Deltas: A better understanding of biogeochemical processes for safe environmental management. *J. Environ. Manag.* 186, 131–140.

Shaheen, S.M., Rinklebe, J., Selim, M.H., 2015a. Impact of various amendments on immobilization and phytoavailability of nickel and zinc in a contaminated floodplain soil. *Int. J. Environ. Sci. Technol.* 12(9), 2765–2776.

Shaheen, S.M., Tsadilas, C.D., Rinklebe, J., 2015b. Immobilization of soil copper using organic and inorganic amendments. *J. Plant Nutr. Soil Sci.* 178(1), 112–117.

Shen, Z., Som, A.M., Wang, F., Jin, F., McMillan, O., Al-Tabbaa, A., 2016. Long-term impact of biochar on the immobilisation of nickel (II) and zinc (II) and the revegetation of a contaminated site. *Sci. Total Environ.* 542, 771–776.

Singh, B., Singh, B.P., Cowie, A.L., 2010. Characterisation and evaluation of biochars for their application as a soil amendment. *Soil Res.* 48, 516–525.

Solaiman, Z.M., Anawar, H.M., 2015. Application of biochars for soil constraints: Challenges and solutions. *Pedosphere* 25, 631–638.

Spokas, K.A., Cantrell, K.B., Novak, J.M., Archer, D.W., Ippolito, J.A., Collins, H.P., Boateng, A.A., Lima, A.A., Lamb, M.C., McAloon, A.J., Lentz, R.D., Nichols, K.A., 2012. Biochar: A synthesis of its agronomic impact beyond carbon sequestration. *J. Environ. Qual.* 41, 973–989.

Sruthy, O.A., Jayalekshmi, S., 2014. Electrokinetic remediation of heavy metal contaminated soil. *Int. J. Struct. Civil Eng.* 3.

Tan, X., Liu, Y., Gu, Y., Zeng, G., Hu, X., Wang, X., Hu, X., Guo, Y., Zeng, X., Sun, Z., 2015b. Biochar amendment to lead contaminated soil: Effects on the fluorescein diacetate hydrolytic activity and phytotoxicity to rice. *Environ. Toxicol. Chem.* 34, 1962–1968.

Tan, X., Liu, Y., Zeng, G., Wang, X., Hu, X., Gu, Y., Yang, Z., 2015a. Application of biochar for the removal of pollutants from aqueous solutions. *Chemosphere* 125, 70–85.

Tandy, S., Healey, J.R., Nason, M.A., Williamson, J.C., Jones, D.L., 2009. Remediation of metal polluted mine soil with compost: Cocomposting versus incorporation. *Environ. Pollut.* 157, 690–697.

Thies, J.E., Rillig, M.C., 2009. Characteristics of biochar: Biological properties. In: Lehmann, J., Joseph, S. (Eds.), *Biochar for Environmental Management: Science and Technology*. Earthscan, London, pp. 85–105.

Trakal, L., Komarek, M., Szakova, J., Zemanova, V., Tlustos, P., 2011. Biochar application to metal-contaminated soil: Evaluating of Cd, Cu, Pb and Zn sorption behavior using single-and multi-element sorption experiment. *Plant Soil Environ.* 57, 372–380.

Trakal, L., Veselská, V., Šafařík, I., Vítková, M., Číhalová, S., Komárek, M., 2016. Lead and cadmium sorption mechanisms on magnetically modified biochars. *Bioresour. Technol.* 203, 318–324.

Turan, V., Ramzani, P.M.A., Ali, Q., Abbas, F., Iqbal, M., Irum, A., Khan, W.U.D., 2017. Alleviation of nickel toxicity and an improvement in zinc bioavailability in sunflower seed with chitosan and biochar application in pH adjusted nickel contaminated soil. *Arch. Agron. Soil Sci.* DOI: 10.1080/03650340.2017.1410542.

Uchimiya, M., Cantrell, K.B., Hunt, P.G., Novak, J.M., Chang, S.C., 2012. Retention of heavy metals in a Typic Kandiudult amended with different manure-based biochars. *J. Environ. Qual.* 41, 1138–1149.

Uchimiya, M., Chang, S.C., Klasson, K.T., 2011a. Screening biochars for heavy metal retention in soil: Role of oxygen functional groups, *J. Hazard. Mater.* 190, 432–444.

Uchimiya, M., Klasson, K.T., Wartelle, L.H., Lima, I.M., 2011b. Influence of soil properties on heavy metal sequestration by biochar amendment: 1. Copper sorption isotherms and the release of cations. *Chemosphere* 82, 1431–1437.

Uchimiya, M., Lima, I.M., Klasson, K.T., Chang, S., Wartelle, L.H., Rodgers, J.E., 2010. Immobilization of heavy metal ions (CuII, CdII, NiII, and PbII) by broiler litter derived biochars in water and soil. *J. Agric. Food Chem.* 58, 5538–5544.

Uchimiya, M., Wartelle, L.H., Klasson, T., Fortier, C.A., Lima, I.M., 2011c. Influence of pyrolysis temperature on biochar property and function as a heavy metal sorbent in soil. *J. Agric. Food Chem.* 59, 2501–2510.

Usman, A.R.A., Ahmad, M., El-Mahrouky, M., Al-Omran, A., Ok, Y.S., Sallam, A.S., El-Naggar, A.H., Al-Wabel, M.I., 2016a. Chemically modified biochar produced from conocarpus waste increases NO_3 removal from aqueous solutions. *Environ. Geochem. Health* 38, 511–521.

Usman A.R.A., Almaroai, Y.A., Ahmad, M., Vithanage, M., Ok, Y.S., 2013. Toxicity of synthetic chelators and metal availability in poultry manure amended Cd, Pb and As contaminated agricultural soil. *J. Hazard. Mater.* 262, 1022–1030.

Usman, A.R.A., Al-Wabel, M.I., Abdulaziz, A.H., Mahmoud, W.A., El-Naggar, A.H., Ahmad, M., Abdulelah, A.F., Abdulrasoul, A.O., 2016b. Conocarpus biochar induces changes in soil nutrient availability and tomato growth under saline irrigation. *Pedosphere* 26(1), 27–38.

Verheijen, F., Jeffery, S., Bastos, A.C., van der Velde, M., Diafas, I., 2010. Biochar application to soils. A Critical Scientific Review of Effects on Soil Properties, Processes and Functions. European Commission, Italy.

Wang, F., Wang, H., Al-tabbaa, A., 2015. Time-dependent performance of soilmix technology stabilized/solidified contaminated site soils. *J. Hazard. Mater.* 286, 503–508.

Wang, M., Zhu, Y., Cheng, L., Andserson, B., Zhao, X., Wang, D., Ding, A., 2017. Review on utilization of biochar for metal-contaminated soil and sediment remediation. *J. Environ. Sci.* 63, 156–173.

Xiu, S., Shahbazi, A., Li, R., 2017. Characterization, modification and application of biochar for energy storage and catalysis: A review. *Trends Renew. Energy* 3, 86–101.

Xue, Y., Gao, B., Yao, Y., Inyang, M., Zhang, M., Zimmerman, A.R., Ro, K.S., 2012. Hydrogen peroxide modification enhances the ability of biochar (hydrochar) produced from hydrothermal carbonization of peanut hull to remove aqueous heavy metals: Batch and column tests. *Chem. Eng. J.* 200–202, 673–680.

Yang, X., Liu, J., McGrouther, K., Huang, H., Lu, K., Guo, X., Wang, H., 2015. Effect of biochar on the extractability of heavy metals (Cd, Cu, Pb, and Zn) and enzyme activity in soil. *Environ. Sci. Pollut. Res.* 23, 974–984.

Yang, X., Lu, K., McGrouther, K., Che, L., Hu, G., Wang, Q., Liu, X., Shen, L., Huang, H., Ye, Z., Wang, H., 2017. Bioavailability of Cd and Zn in soils treated with biochars derived from tobacco stalk and dead pigs. *J. Soils Sedim.* 17, 751–762.

Yao, Y., Gao, B., Fang, J., Zhang, M., Chen, H., Zhou, Y., Creamer, A.E., Sun, Y., Yang, L., 2014. Characterization and environmental applications of clay–biochar composites. *Chem. Eng. J.* 242, 136–143.

Yavuz, Ö., Altunkaynak, Y., Güzel, F., 2003. Removal of copper, nickel, cobalt and manganese from aqueous solution by kaolinite. *Water Res.* 37, 948–952.

Younis, U., Athar, M., Malik, S., Raza Shah, M., Mahmood, S., 2015. Biochar impact on physiological and biochemical attributes of Spinach (*Spinacia oleracea* L.) in nickel contaminated soil. *Global J. Environ. Sci. Manag.* 1, 245–254.

Zhang, X., Li, J., Wei, D., Li, B., Ma, Y., 2015. Predicting soluble nickel in soils using soil properties and total nickel. *PloS one* 10, e0133920.

Zhang, X., Wang, H., He, L., Lu, K., Sarmah, A., Li, J., Bolan, N.S., Pei, J., Huang, H., 2013. Using biochar for remediation of soils contaminated with heavy metals and organic pollutants. *Environ. Sci. Pollut. Res.* 20(12), 8472–8483.

Zhao, Z., Jiang, G., Mao, R., 2014. Effects of particle sizes of rock phosphate on immobilizing heavy metals in lead zinc mine soils. *J. Soil Sci. Plant Nutr.* 14(2), 258–266.

14

Overview Scheme for Nickel Removal and Recovery from Wastes

Ahamed Ashiq, Viraj Gunarathne, and Meththika Vithanage

CONTENTS

14.1 Introduction

Trace metals can be described as elements that have atomic weights from 63.5 to 200.6 amu and specific gravity higher than 5.0 (Srivastava and Majumder 2008). Release of trace metals into the natural environment occurs through several industrial activities, including mining and metallurgical, electronic, electroplating, and metal finishing (Jadhav and Hocheng 2012), involving severe environmental pollution in the last few decades (Naaz and

Pandey 2010). Heavy metals are difficult to remove naturally by degradation
and tend to accumulate in living organisms, generating highly toxic effects
on them (Qin et al. 2012).

Most of the trace metals released into the environment through anthropogenic activities are in the form of aqueous solutions (Borbély and Nagy
2009). Therefore, the kind of wastewater generated from manufacturing processes in industries such as electronic and metal finishing typically contains
higher concentrations of metals than acceptable levels established by law
(Hunsoma et al. 2002). Other than wastewater, solid wastes including spent
petroleum catalysts, fly ash, boiler ash, and waste electrical and electronic
equipment (WEEE) contain considerable amounts of trace metals (Abdel-Aal
and Rashad 2004; Akcil et al. 2015; Park and Fray 2009).

With increasing economic and environmental issues in mind, the removal
or recovery of metals from wastes is an important concern (Kaminari et al.
2007; Qin et al. 2012).

14.1.1 Nickel as a Resource for Industry

Nickel has been recognized as an important strategic resource because of its
optical, electrical, and catalytic performance; toughness; high corrosion resistance; thermal stability; chemical passivity; and ability to make super alloys
(Farrell et al. 2010; Peng et al. 2014; Zahraei et al. 2015). It is a comparatively
widespread element in nature (Förstner 1981), and raw materials including
oxidic, silicate, sulfide, and laterite ore are the main production sources of
nickel (Moskalyk and Alfantazi 2002; Panigrahi et al. 2009). However, after
the 1960s, laterites became a major commercial source of nickel since sulfide
ores declined with a drastic increase in world nickel consumption (Guo et al.
2009). Moreover, 70% of the world's nickel deposits are present in the form of
laterites (Dalvi et al. 2004).

Nickel is extensively used in the stainless steel industry, accounting for
two-thirds of the total nickel production. Also, the nickel consumption rate
for this industry is increasing at a rate of 5%–6% annually (Anderson 1996;
Anthony and Flett 1997).

Moreover, nickel is one of the most used element for coating of different
industrial materials as it provides decorative appearances and improves
resistance to corrosion (Schario 2007). Watts' baths containing chlorides
and sulfates of nickel and boric acid are a widely used industrial method to
improve surface finishing and appearance (Benvenuti et al. 2014).

Furthermore, nickel is used as a raw material for a wide variety of manufacturing processes such as mineral processing, paint formulation, phosphate fertilizers, electro chromic films, dye-sensitized solar cell, forging,
battery manufacturing, magnets, semiconductors, gas sensors, nonferrous
metals, and special alloys (e.g., nickel steels, nickel cast irons, nickel brasses,
and bronzes) as well as for minting of coins and tinting of glasses in green
(Dizge et al. 2009; Mohammadijoo et al. 2014; Peng et al. 2014).

14.2 Sources of Nickel as Waste

Nickel-containing wastes have been released into the environment in two major forms: solid and aqueous. Regularly, nickel is discharged into water bodies through wastewater streams originating from various industries (Gupta et al. 2003). However, the rapidly growing production and consumption of nickel-based products (i.e., batteries, electronic equipment, etc.) are involved in the accumulation of nickel in the environment in the form of solid waste.

14.2.1 Nickel in Wastewaters and Solid Waste

Wastewater from industries such as nickel electroplating and textile or effluents comes from washing of nickel-contaminated soil that is used as a remediation technique, which often carries up to 1000 mg L^{-1} of nickel (Dermentzis 2010). However, according to industrial sources, nickel concentration varies from tens to thousands of milligrams per liter (Dermentzis 2010; Varma et al. 2013). Also, 60%–70% of all metals used in the plating process are not utilized effectively and removed with wastewater during rinsing of the plated article (Dermentzis 2010).

Spent rechargeable batteries (i.e., Ni-Cd, Ni-MH, and Li-ion batteries) are a major contributor to nickel-containing solid wastes. The cathode material of Ni-Cd batteries encompassed with nickel hydroxy-oxide in nickel form and anode consists of metallic cadmium embedded on a wire mesh of nickel (Sullivan and Gaines 2012). Generally, Ni-Cd batteries contain nearly 20% of nickel by weight (Gaines and Singh 1995; Rydh and Karlström 2002). However, cadmium is highly toxic in nature, and Ni-Cd batteries were eventually replaced by Ni-MH batteries. After commercializing them in 1990, Ni-MH batteries have been extensively used for a wide range of applications (Li et al. 2009). Electrodes and the active electrode material of Ni-MH batteries, which mainly consist of nickel, have a 30% total nickel content (Innocenzi and Vegliò 2012; Zhang et al. 1999). Also, in 2005, the total count of spent Ni-MH batteries reached up to 1 billion, containing 7500 tons of nickel, leading to severe environmental pollution from trace metals (Innocenzi and Vegliò 2012). At present, LiNiO$_2$ is also used as a cathode material for lithium-ion batteries (Chen et al. 2015; Nishi 2001), and their disposal may contribute to nickel release into the environment.

Also, different kinds of spent catalysts that are used in various industries may contain varied amounts of nickel, according to the nature of the industrial process and operating conditions. For instance, the number of spent catalysts from an ammonia plant that contain nickel at elevated quantities ranged from 12% to 30% (Singh 2009). Spent hydrodesulfurization catalysts from petroleum refining industries (Dufresne 2007; Szymczycha-Madeja 2011) and the Raney nickel catalyst (fine particles of Ni–Al alloy) used in the pharmaceutical industry (Lee et al. 2010) contain high amounts of nickel. In addition, several other nickel-containing catalysts are extensively used in different manufacturing processes such as hydrocracking (NiS, WS3/ $SiO_2Al_2O_3$) and hydro refining (Ni, Mo/Al_2O_3) (Thomas 1970).

Moreover, most WEEE contain components that are made of nickel. As an example, metallic constituents of CRT monitors contain about 25%–45% nickel (Robotin et al. 2011). Manufacturing and consumption of WEEE have increased drastically during the last few decades and the waste stream of the products is rapidly growing (Guo et al. 2010). Thus, trace metals including nickel are accumulated in the environment at high concentrations.

14.2.2 Fractionation of Nickel in Soil

The types of nickel (both physical and chemical forms) that exist in wastes are strongly associated with their bioavailability for organisms (Schaumlöffel 2012). Speciation distribution of nickel in wastes is varied according to the type of waste, physicochemical properties, and waste management treatment. For example, nickel in WEEE is mostly associated with residual fraction and Fe-Mn oxide bound fraction than with organic matter fraction and carbonate bound fraction. However, the products resulting from incineration of WEEE contained significantly increased nickel contents in exchangeable and residual fractions (Long et al. 2013). The majority of nickel in fly ash obtained from waste filter bags of steel plants existed in residual and acid-exchangeable fractions (Zhou et al. 2013).

14.2.3 Problems Associated with Nickel-Containing Wastes

Nickel occurs in sea water, petroleum, and coal at trace levels, and Ni^{2+} present in the environment can seep into plant and animal bodies. Minor quantities of Ni^{2+} serve as an activator of certain enzymes and have a beneficial effect on humans and other organisms (Peng et al. 2014). However, common compounds that contain elevated amounts of nickel have a toxic effect on humans and animals. Nickel ions can generate major genetic effects by binding with nucleic acids; a typically known local reaction to nickel is dermatitis (Gupta 1998). Also, exposure to Ni^{2+} at higher concentrations has wide-ranging effects, from minute symptoms such as dizziness, headache, extreme weakness, chest pain, rapid respiration, dry cough, gastrointestinal disorders, and cyanosis, to serious illnesses such as lung cancer and renal edema (Argun 2008; Pillai et al. 2009; Smith-Sivertsen et al. 1997), but acute toxicity is rarely observed (Borbély and Nagy 2009).

Moreover, nickel has been considered as a nonbiodegradable toxic trace metal present in different kinds of waste materials (Peng et al. 2014). Therefore, removal of nickel from waste has been known as a difficult task and has been considered as a scientific challenge (Dermentzis et al. 2016).

14.3 Removal and Recovery Methods for Nickel from Wastes

Methods mentioned here are subcategorized either for removal or for recovery of the Ni^{2+} ions both from aqueous media and from solid wastes. Most of the extraction techniques revolve around the divalent nature of nickel either from wastewaters or solid wastes. Commercially used methods employed for nickel removal are leaching, ion exchange, ion flotation, adsorption, and membrane filtration (Figure 14.1). The most prominent commercial methods used to recover Ni^{2+} ions are electrodialysis methods and a combination of electrochemical and ionization processes. A summary of the recovery and removal methods is provided in Table 14.1.

14.3.1 Leaching

Leaching is the extraction of a soluble component from a solid by means of a solvent. The process is utilized for the extraction of a very valuable solid material like nickel or for the removal of an insoluble solute from a contaminated solvent (Harker and Backhurst 2002).

To extract nickel from battery scraps, a systematic approach is used, such as leaching through mineral acids (e.g., sulfuric acid, hydrochloric acid, and nitric acid), followed by a filtration technique to remove the rest of the elements in the battery. Subsequently, precipitation is also carried out to concentrate nickel (Lee et al. 2010; Zhang et al. 1999); in the case of Ni-MH batteries, the electrode materials are treated with 2 M sulfuric acid and maintained at

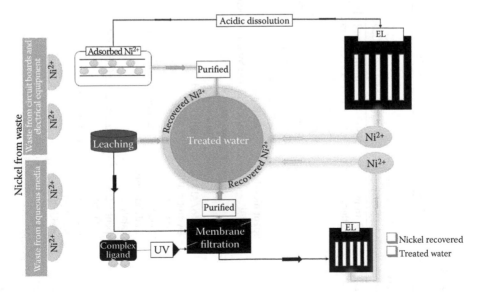

FIGURE 14.1
Nickel recovery from industrial wastewater. EL, electrochemical methods.

TABLE 14.1

Summary of Nickel Removal/Recovery Methods, Sources of Wastes, and Percentage Removal

Recovery Method	Ni²⁺ Waste Source	Materials Used	Efficiency for Recovery (%)	Adsorption Capacity (mg/g)	Dosage (g/L)	References
Leaching	Ni²⁺ from batteries	Bis (2,4,4-tri-methylpenthyl) phosphinic acid Cyanex 272 and sulfuric acid	98	–	–	Li et al. (2009)
		Sulfuric acid, citric acid, and sodium hydroxide (precipitation)	99	–	–	Innocenzi and Vegliò (2012)
		Trioctyl/decylamine (Alamine 336)	99	–	–	Fernandes et al. (2013)
	Spent nickel metal	Sulfuric acid	98	–	–	Siemens et al. (1986)
		Oxalic acid and hydrogen peroxide	65	–	–	Mulak et al. (2006)
		Tartaric acid	83	–	–	Marafi and Furimsky (2005)
		Citric acid	85	–	–	Marafi and Furimsky (2005)
	Ni Mo/Al₂O₃ hydrotreating catalyst	Sulfuric acid	98	–	–	Valverde et al. (2008)
	Nickel molybdenum catalyst	Potassium bisulfate KHSO₄	96–99	–	–	Busnardo et al. (2007)
	Spent HDS catalyst	Nitric acid, sulfuric acid	90	–	–	Lai et al. (2008)
	Ammonia leaching residue	Water leaching	91	–	–	Chen et al. (2006)

(Continued)

TABLE 14.1 (CONTINUED)

Summary of Nickel Removal/Recovery Methods, Sources of Wastes, and Percentage Removal

Recovery Method	Ni²⁺ Waste Source	Materials Used	Efficiency for Recovery (%)	Adsorption Capacity (mg/g)	Dosage (g/L)	References
Ion exchange	From wastewaters	Chelex-100	99	—	—	Leinonen et al. (1994)
		Sodium titanate	99.9	—	—	Lehto et al. (1996); Leinonen et al. (1994)
		Dowex HCR S/S	98	—	—	Alyüz and Veli (2009)
Ion flotation		Dodecyldiethylenetriamine (Ddien)	93	—	—	Liu and Doyle (2009)
		Ethylhexadecyldimethylammonium bromide (EHDABr)	88	—	—	Hoseinian et al. (2015)
Adsorption		Citrus limettioides seed carbon (CLSC)		1.5	35.54	Sudha et al. (2015)
		Parthenium hysterophorus L.		17.24	20	Lata et al. (2008)
		Carbon aerogel		2.80	12	Goel et al. (2005)
		Walnut shell carbon		15.34	30	Wang et al. (2010)
		Bentonite clay		112.69	0.5	Yang et al. (2009)
		Thermally treated attapulgite		2	10	Ren et al. (2014)
		Turkish zeolite		119.7	8	Kocaoba et al. (2007)
		Natural zeolitic tuff		16.64	10	Rajic et al. (2010)
		ZrO-montmorillonite TBA-montmorillonite		(40%)	0.05	Gupta and Bhattacharyya (2006)

pH 0.4. Rare earths (other metal parts) are recovered at first from an external solvent extraction by adding bis(2-ethylhexyl)phosphoric acid (D2EHPA) followed by precipitation with oxalic acid. The nickel in the raffinates are then extracted through bis(2,4,4-trimethylpenthyl)phosphinic acid (Cyanex 272) followed by precipitation with oxalic acid, recovering about 99.8% of the nickel (Zhang et al. 1999). Other recovery techniques employed after leaching are electrolysis and electrowinning, which will be described later.

The other solvent that is commonly used for the extraction of Ni^{2+} ions from the batteries that showed promising results is D2EHPA with 6 M nitric acid. Nickel is then made to precipitate easily from the latter using oxalic acid, leaving behind the rest of the elements (the rare metals in the case of Ni-MH batteries) (Lyman and Palmer 1993a,b).

Nickel-based catalysts have been extensively used in oil and gas processes and are disposed of as solid waste either in their pure form or as composites (Idris et al. 2010; Yang et al. 2011). Nickel found in these sources employs the leaching procedure as explained above through the use of other mineral acids, hydrogen peroxide, and potassium bisulfite (Akcil et al. 2015; Lai et al. 2008; Mulak et al. 2006).

The most common practice to recover nickel from used storage batteries (i.e., Ni-Cd batteries or Ni-MH batteries) and spent catalysts is through leaching, and it is becoming more popular because it is easy and can be carried out on a larger scale as the consumption and wastage of these batteries are exponentially increasing (Table 14.2). Majority of the hydrometallurgical processes in the literature proposed the combination of leaching and selective precipitations of the liquid–liquid phase (Innocenzi and Vegliò 2012).

TABLE 14.2

Spent Catalyst and Their Leaching Methods

Spent Catalyst	Leaching Agent Used	Nickel Recovered (%)	References
Spent nickel metal	Sulfuric acid	98	Siemens et al. (1986)
	Oxalic acid and hydrogen peroxide	65	Mulak et al. (2006)
	Tartaric acid	83	Marafi and Furimsky (2005)
	Citric acid	85	Marafi and Furimsky (2005)
Ni Mo/Al_2O_3 hydrotreating catalyst	Sulfuric acid	98	Valverde et al. (2008)
Nickel molybdenum catalyst	Potassium bisulfate $KHSO_4$	96–99	Busnardo et al. (2007)
Spent HDS catalyst	Nitric acid, sulfuric acid	90	Lai et al. (2008)
Ammonia leaching residue	Water leaching	91	Chen et al. (2006)

14.3.2 Ion Exchangers

Ion exchange is a highly selective method and clearly depends on the solution type and the desired contaminant or valuable metal to be extracted. Ion exchangers consist of solid materials that could carry cations or anions interchangeably. The ions are exchanged at a fixed stoichiometric rate with the particles in the solution. The materials with exchangeable cations are called cation exchangers and the materials that could carry anions are called anion exchangers (Helfferich 1962). Ion exchangers are specific to definite metals that have affinity toward that material. Mostly, the ions are exchanged or replaced with the desirable ion or displaced by the undesirable ion from the solution, say, Ni^{2+} from waters. The most commonly utilized exchanger types nowadays are mostly resin based with defined lattices (Barakat 2011; Daβrowski et al. 2004). Some other exchangers are coupled with electric current as in electrolysis for the recovered Ni^{2+} ions from the surfaces of the resins onto a concentrated solution, the details of which can be found in later sections.

For recovery, the important factor to take into consideration is the selectivity of ion and its medium for binding and unbinding, which is closely bound to the surface area of the exchangers used (Chitpong and Husson 2017). The ion exchangers are highly dependent on the functional group attached to the surface; carboxylate, sulfonate, amino, thiol, and amide groups as well as specialty ligands are commonly used to extract Ni^{2+} and to form a complex that can be recovered at a later stage through acidification (Chitpong and Husson 2017; Ma et al. 2013). Traditionally, zeolites were commonly used in ion exchangers and have now been replaced with synthetic resins that have been proven effective and inexpensive. The advantages were faster exchange rates, longer life cycles, higher capacity to withhold the nickel, and the ability to adsorb other metals as well (Alyüz and Veli 2009; Clifford 1999). Macroporous carboxylic cation exchangers showed positive results for the removal of nickel ions (Halle et al. 1982). Sulfonic-based cation exchangers with Styrene matrix were also suggested for commercial application for the removal of nickel from wastewaters (Alyüz and Veli 2009). Some of the commercially utilized resins specific to nickel ions are listed in Table 14.3.

TABLE 14.3

Commercially Used Ion Exchange Resins

Commercially Used Resin	Removal Efficiency for Ni^{2+}	References
Wofatit CA-20	Carboxylic acid	Halle et al. (1982)
Chelex-100	99%	Leinonen et al. (1994)
Sodium titanate	99.9%	Lehto et al. (1996); Leinonen et al. (1994)
Dowex HCR S/S	98%	Alyüz and Veli (2009)

Source: Lewinsky, A. A. 2007. *Hazardous Materials and Wastewater: Treatment, Removal and Analysis*, Nova Publishers.

The commercial application of resins utilizes the divalent ion removal resin for the removal of nickel. There are certain criteria in which the selection of a resin is looked upon (Alyüz and Veli 2009):

i. Higher charge on the ion than the resin preferred
ii. Less hydrated ions
iii. Ions with a strong affinity toward a particular functional group that contains the resins are to be considered

Ion exchangers have shown promising results when it comes to nickel recovery mainly due to its low cost and high efficiency. The major attributes notable in the literatures are high recovery, less sludge, and being able to meet the requirements of the specifications (Rengaraj et al. 2001). Among the materials used in ion exchange processes, synthetic resins are commonly preferred because they are effective and inexpensive and can be modified to certain functional groups to have higher affinity toward Ni^{2+} (Clifford 1999). Cation exchange resins generally contain sulfonic acid groups. These groups can also be carboxylic, phosphonic, or phosphinic (Alyüz and Veli 2009). The drawbacks of ion exchange include the fouling of the surfaces of the resin matrices as it cannot handle high concentrations of the solutions with organics and solid wastes and it is quite dependent on the pH of the solution (Barakat 2011).

14.3.3 Ion Flotation

The flotation method is a gravity separation method. This method follows the usage of a surfactant that is foamed with an inert gas. The foam is then made to accumulate the trace metal values using the metal–ion complex that is formed. This is usually viable when there is an appropriate surfactant utilized to separate the required cation and the foaming conditions are taken into account as well (Sebba 1962). This technique utilizes two properties of the material: surface charge and surface energy. These properties need to be controlled by precisely choosing the right surfactants, followed by flotation (Sanciolo et al. 1992). After the selective attachment of the cation with the froth or foam the solids are concentrated and transferred from the body of water to the surface. Thus, unlike settling by gravity only, it is a solid–liquid separation technique that has been utilized only after the density of the particles is made lower than that of the liquids (Deliyanni et al. 2017). This makes the recovery of Ni^{2+} plausible using separation and by further employing acid treatment and electrolysis.

Most of the metals in the periodic table are recovered by this method, especially at low concentrations, and in comparison with the other methods, flotation promises higher recovery of Ni^{2+} (Jafari et al. 2017). Table 14.4 shows a few notable surfactants available commercially and their removal percentile.

Overview Scheme for Nickel Removal and Recovery from Wastes 329

TABLE 14.4

Notable Surfactant and Their Efficiency

Surfactant Used	Ni²⁺ (Recovered)	References
Dodecyldiethylenetriamine (Ddien)	93%	Liu and Doyle (2009)
Ethylhexadecyldimethylammonium Bromide (EHDABr)	88%	Hoseinian et al. (2015)

Source: Carolin, C. F., Kumar, P. S., Saravanan, A., Joshiba, G. J., and Naushad, M. 2017. Efficient techniques for the removal of toxic heavy metals from aquatic environment: A review. *Journal of Environmental Chemical Engineering*, 5, 2782–2799.

The probability for the flotation separation to occur is highly dependent on the bubble and the particle collision. Its behavior tremendously varies at different nickel particle sizes (Dai et al. 2000). Recent studies have shown the different behaviors when the bubble–particle collision is enhanced by decreasing the bubble size or by increasing the particle size (Miettinen et al. 2010). From a nickel recovery perspective, the following are the major advantages of using this method: much lower energy requirements, lesser sludge with nickel concentrates, higher recovery, and continuous operation. This method is used for other trace metals as well (Rubio et al. 2002; Salmani et al. 2013).

14.3.4 Adsorption

This method is especially used for the removal of trace metals in aqueous media. Most of the Ni²⁺ ions are in parts-per-million levels and have several advantages over other processes. Adsorption is the most simplified and the oldest method for nickel recovery. In adsorption, nickel ions are diffused from the bulk solution onto the surface of the solid adsorbent, thereby forming a layer of adsorbed phase. The major drawback of this recovery is that the adsorbate does not hold at a certain level and thus several criteria are made to select the right adorbents for nickel recovery (Harker and Backhurst 2002). Adsorbents that have a large surface area and that are chemically stable with the adsorbate are the basic criteria. Some of the adsorbents are modified through chemical treatments to have an enhanced surface area (Ren et al. 2015). Activated carbon (AC), clays, and zeolites are the commonly used adorbents and have shown promising results for nickel removal from aqeous solutions (Raval et al. 2016).

14.3.4.1 Activated Carbon

ACs are commonly used commercial adsorbents for trace metal removals because of their large surface area and can withstand huge amounts of particles. It was found that a typical surface area for AC is 10^6 m²/kg with a pore diameter of 2 nm (Harker and Backhurst 2002). Table 14.5 summarizes the best AC and sources of AC used for nickel recovery.

TABLE 14.5

Summary of the Major Activated Carbon Used with Their Dosages Used in Ni Recovery

Source of AC	Adsorption Capacity (mg/g)	Dosage (g/L)	References
Citrus limettioides seed carbon (CLSC)	1.5	35.54	Sudha et al. (2015)
Parthenium hysterophorus L.	17.24	20	Lata et al. (2008)
Carbon aerogel	2.80	12	Goel et al. (2005)
Walnut shell carbon	15.34	30	Wang et al. (2010)

Source: Raval, N. P., Shah, P. U., and Shah, N. K. 2016. Adsorptive removal of nickel (II) ions from aqueous environment: A review. *Journal of Environmental Management*, 179, 1–20.

14.3.4.2 Clay Adsorbents and Zeolites

Both these adsorbents are known for their crystalline lattice that facilitates the adsorption of trace metals. Because of this, the positively charged trace metals are easily adsorbed and exchanged in their closely knit frameworks (Kim and Keane 2002; Qiu and Zheng 2009). These tetrahedral structures possess large channels containing negatively charged sites that could take advantage of adsorbing a cation, say, Ni^{2+} (Bailey et al. 1999).

Zeolites are aluminosilicates that have SiO_4 and Al_2O_3 linked to each other with a shared oxygen atom in a three-dimensional crystal lattice (Deer et al. 1992). They have the capability of counterbalancing with a cation because of their negatively charged sites (Malamis and Katsou 2013).

Clays are naturally occurring minerals and are hydrated aluminosilicates as well. They too possess crystal structures as in zeolites but exist in both tetrahedral and octahedral crystal lattices, which allow more affinity for adsorption of trace metals (Uddin 2017).

However, these minerals are almost completely modified to increase the cation uptake onto the adsorbents either chemically or thermally. Table 14.6 summarizes some notable and the highest adsorbent capacities with the least dosage of these materials.

14.3.5 Membrane Filtration

This method is completely driven by pressure and very selective toward the kind of membrane to be used for nickel recovery and its isolation from the aqueous solution (Carolin et al. 2017). Membrane filtration can be varied in types depending on the sizes of the particles; in nickel, the most prominent techniques used are ultrafiltration, nanofiltration, and reverse osmosis (Barakat 2011; Daβrowski et al. 2004).

Particles in the range of 10–100 nm are separated by ultrafiltration. To enhance the separation process, polymeric agents are used to clog the ions, say, Ni (II) ions. Macromolecules are formed when these polymeric agents

TABLE 14.6

Brief of Clays and Zeolites Used with Adsorption Capacity for Nickel Removal

Adsorbents	Adsorption Capacity (mg/g)	Dosage (g/L)	References
Bentonite clay	112.69	0.5	Yang et al. (2009)
Thermally treated attapulgite	2	10	Ren et al. (2014)
Turkish zeolite	119.7	8	Kocaoba et al. (2007)
Natural zeolitic tuff	16.64	10	Rajic et al. (2010)
ZrO-montmorillonite TBA-montmorillonite	(40%)	0.05	Gupta and Bhattacharyya (2006)
Gibbsite	9×10^{-5} (mol/m²)	5	Rajapaksha et al. (2012)
Goethite	2×10^{-5} (mol/m²)	-do-	Rajapaksha et al. (2012)
Laterite	4×10^{-5} (mol/m²)	-do-	Rajapaksha et al. (2012)
Natural red earth	6×10^{-6} (mol/m²)	-do-	Rajapaksha et al. (2012)

Source: Raval, N. P., Shah, P. U., and Shah, N. K. 2016. Adsorptive removal of nickel (II) ions from aqueous environment: A review. *Journal of Environmental Management*, 179, 1–20.

TABLE 14.7

Membrane and Polymer Combinations for Complexion–Ultrafiltration for Nickel Recovery

Polymer	Membrane	References
Polyethyleneimine (PEI)	Poly(ether sulfone) (PES)	Molinari et al. (2008)
Polyethyleneimine (PEI)	Carbon-zirconia composite	Cañizares et al. (2002)
Polyallylamine (PAA)	Poly(ether sulfone) (PES)	Moreno-Villoslada and Rivas (2003)
Chitosan	Ceramic	Taha et al. (1996)

combine with the metal ions and are conglomerated, thereby making the separation easier (Crini et al. 2017; Rivas et al. 2011). These macromolecules congregate containing the metal ions and are thus recovered. Table 14.7 shows the use of the membrane and the polymer for the removal of nickel.

A combination of processes (ultrafiltration and electrolysis techniques) is thus employed for the recovery of Ni^{2+} ions from industrial wastes (Chaufer and Deratani 1988). First, the ions are mixed with certain ligands to form a complex as mentioned above. A concentrated solution of nickel is obtained through ultrafiltration of the complex. Then, the Ni^{2+} ions are desorbed or decomplexed from the complex aggregate through acidification, allowing the recycle of the complex ions. Finally, through electrolysis, we can recover the Ni^{2+} ions using appropriate solutions. Mostly, the complexes used are polycarboxylate and polyethylenimine based. The decomplexation or extraction of Ni^{2+} from the complex formed is achieved by using suitable acids (Baticle et al. 2000; Chaufer and Deratani 1988).

Around 99% of nickel can be recovered by this complexation process using sodium polyacrylate (PAAS) and polyethylenimine (PEI) (Shao et al. 2013). This membrane process is based on the principle that if the polymers used have a larger molecular weight than the membrane used, then they can bind the trace metals to form the macromolecular complex, which is retained in the membrane. Further electrolysis and acid treatment are performed to recover the entire Ni^{2+} ions (Baticle et al. 2000; Shao et al. 2013).

Nanofiltration membranes are mostly made of polymers that are either positively or negatively charged depending on the ion to be isolated. There is a close relation with the electrostatic forces of the ions and the membrane and thus lesser energy or pressure is utilized unlike the other membrane techniques (Carolin et al. 2017).

The notable advantages of nanofiltration are lesser energy utilization and much lower pressure required than the other methods and operations. However, it is highly dependent on pH, pressure, temperature, configuration, and the incoming feed concentrations (Fu and Wang 2011; Tao et al. 2016). These filters are made up of synthetic polymers that have a certain positive or negative charge that enhances the trace metal removal (Mohammad et al. 2015).

Reverse osmosis is based on the particle size and the charges on the particle. It uses a semipermeable membrane for the removal of the solid present in the aqueous medium (Bilal et al. 2013). The pore sizes for the membrane range from 0.1 to 1.0 nm and are increasingly familiarized in the removal of trace metals as well. Reverse osmosis requires high energy for its operation compared to other membrane techniques (Carolin et al. 2017).

14.3.6 Electrodialysis and Ion Exchange

These recovery processes are primarily driven by electric current in the aqueous media. Electrodialysis is a membrane process that is based on the selective migration of Ni^{2+} ions through a suitable membrane using an electric current source. It is most prominent method used to recover the cations from rinse water in electroplating (Njau et al. 2000; Shao et al. 2013; Tzanetakis et al. 2003). Low concentrations of trace metals are usually removed by this method, and the effluent water is mostly recycled for use (Benvenuti et al. 2014). Electrodeionization is a combination of techniques for Ni^{2+} recovery. It includes electrolysis, electrodialysis, and ion exchange where the system to be purified is stacked with ion exchange resins, most of which are explained as above, and an external power source is applied as in electrodialysis (Alvarado and Chen 2014). Basically, a series of steps were followed to ultimately recover the Ni^{2+} and obtain 100% purity in water. Electrolysis–electrodialysis and ionization combination processes recover both high and low concentrated solutions. Around 94% purity/recovered nickel can be obtained with this method (Peng et al. 2014).

14.4 Future Outlook

In recent years, a wide range of studies have been made and have been extended for removal and recovery of trace metals before disposing them to the environment, whether they are in solid form or aqueous media; the recovery of nickel has been of primary importance due to its threat to humans and to the environment as well as its high economic value to the industry.

From the different methods of nickel extraction from wastes, membrane separation and adsorption remain the most conventionally employed method to obtain reusable water. To date, there has been not much focus on the recovery of nickel. The electricity-driven methods seem to be more promising for recovery of nickel, but the plausibility of having it in a large-scale facility to treat nickel-containing sludge is challenging, keeping in mind the energy consumed for the recovery.

Another major drawback is the clogging of the membranes at the downstream processes for filtration of the trace metals. Ongoing studies have proven that adsorption and ion exchange would be effective for the treatment of nickel from wastewaters or for any trace metal for that matter (Carolin et al. 2017; Dermentzis et al. 2016; Raval et al. 2016). Further investigation is needed to optimize the adsorption process because it gives a promising outcome and is currently plausible only at low concentrations. Improvement and modifications are needed in order to enhance the surface area of these adsorbents through the right kind of modification for the right trace metal removal, say, Ni^{2+} ion extraction.

References

Abdel-Aal, E. and Rashad, M. M. 2004. Kinetic study on the leaching of spent nickel oxide catalyst with sulfuric acid. *Hydrometallurgy*, 74, 189–194.

Akcil, A., Vegliò, F., Ferella, F., Okudan, M. D. and Tuncuk, A. 2015. A review of metal recovery from spent petroleum catalysts and ash. *Waste Management*, 45, 420–433.

Alvarado, L. and Chen, A. 2014. Electrodeionization: Principles, strategies and applications. *Electrochimica Acta*, 132, 583–597.

Alyüz, B. and Veli, S. 2009. Kinetics and equilibrium studies for the removal of nickel and zinc from aqueous solutions by ion exchange resins. *Journal of Hazardous Materials*, 167, 482–488.

Anderson, S. 1996. Nickel: Breaking new ground. *Engineering and Mining Journal (USA)*, 197, 43–44.

Anthony, M. and Flett, D. 1997. Nickel processing technology: A review. *Minerals Industry International*, 1, 26–42.

Argun, M. E. 2008. Use of clinoptilolite for the removal of nickel ions from water: Kinetics and thermodynamics. *Journal of Hazardous Materials,* 150, 587–595.

Bailey, S. E., Olin, T. J., Bricka, R. M. and Adrian, D. D. 1999. A review of potentially low-cost sorbents for heavy metals. *Water Research,* 33, 2469–2479.

Barakat, M. 2011. New trends in removing heavy metals from industrial wastewater. *Arabian Journal of Chemistry,* 4, 361–377.

Baticle, P., Kiefer, C., Lakhchaf, N., Leclerc, O., Persin, M. and Sarrazin, J. 2000. Treatment of nickel containing industrial effluents with a hybrid process comprising of polymer complexation–ultrafiltration–electrolysis. *Separation and Purification Technology,* 18, 195–207.

Benvenuti, T., Krapf, R. S., Rodrigues, M., Bernardes, A. and Zoppas-Ferreira, J. 2014. Recovery of nickel and water from nickel electroplating wastewater by electrodialysis. *Separation and Purification Technology,* 129, 106–112.

Bilal, M., Shah, J. A., Ashfaq, T., Gardazi, S. M. H., Tahir, A. A., Pervez, A., Haroon, H. and Mahmood, Q. 2013. Waste biomass adsorbents for copper removal from industrial wastewater—A review. *Journal of Hazardous Materials,* 263, 322–333.

Borbély, G. and Nagy, E. 2009. Removal of zinc and nickel ions by complexation–membrane filtration process from industrial wastewater. *Desalination,* 240, 218–226.

Busnardo, R. G., Busnardo, N. G., Salvato, G. N. and Afonso, J. C. 2007. Processing of spent NiMo and CoMo/Al_2O_3 catalysts via fusion with $KHSO_4$. *Journal of Hazardous Materials,* 139, 391–398.

Cañizares, P., Pérez, Á. and Camarillo, R. 2002. Recovery of heavy metals by means of ultrafiltration with water-soluble polymers: Calculation of design parameters. *Desalination,* 144, 279–285.

Carolin, C. F., Kumar, P. S., Saravanan, A., Joshiba, G. J. and Naushad, M. 2017. Efficient techniques for the removal of toxic heavy metals from aquatic environment: A review. *Journal of Environmental Chemical Engineering,* 5, 2782–2799.

Chaufer, B. and Deratani, A. 1988. Removal of metal ions by complexation-ultrafiltration using water-soluble macromolecules: Perspective of application to wastewater treatment. *Nuclear and Chemical Waste Management,* 8, 175–187.

Chen, X., Chen, Y., Zhou, T., Liu, D., Hu, H. and Fan, S. 2015. Hydrometallurgical recovery of metal values from sulfuric acid leaching liquor of spent lithium-ion batteries. *Waste Management,* 38, 349–356.

Chen, Y., Feng, Q., Shao, Y., Zhang, G., Ou, L. and Lu, Y. 2006. Investigations on the extraction of molybdenum and vanadium from ammonia leaching residue of spent catalyst. *International Journal of Mineral Processing,* 79, 42–48.

Chitpong, N. and Husson, S. M. 2017. High-capacity, nanofiber-based ion-exchange membranes for the selective recovery of heavy metals from impaired waters. *Separation and Purification Technology,* 179, 94–103.

Clifford, D. A. 1999. Ion exchange and inorganic adsorption. *Water Quality and Treatment,* 4, 561–564.

Crini, G., Morin-Crini, N., Fatin-Rouge, N., Deon, S. and Fievet, P. 2017. Metal removal from aqueous media by polymer-assisted ultrafiltration with chitosan. *Arabian Journal of Chemistry,* 10, S3826–S3839.

Dąbrowski, A., Hubicki, Z., Podkościelny, P. and Robens, E. 2004. Selective removal of the heavy metal ions from waters and industrial wastewaters by ion-exchange method. *Chemosphere,* 56, 91–106.

Dai, Z., Fornasiero, D. and Ralston, J. 2000. Particle–bubble collision models—A review. *Advances in Colloid and Interface Science,* 85, 231–256.

Dalvi, A. D., Bacon, W. G. and Osborne, R. C. 2004. The past and the future of nickel laterites. In: *PDAC 2004 International Convention, Trade Show & Investors Exchange.* Toronto: The Prospectors and Developers Association of Canada, 1–27.

Deer, W. A., Howie, R. A. and Zussman, J. 1992. *An Introduction to the Rock-Forming Minerals,* Longman Scientific & Technical Hong Kong.

Deliyanni, E. A., Kyzas, G. Z. and Matis, K. A. 2017. Various flotation techniques for metal ions removal. *Journal of Molecular Liquids,* 225, 260–264.

Dermentzis, K. 2010. Removal of nickel from electroplating rinse waters using electrostatic shielding electrodialysis/electrodeionization. *Journal of Hazardous Materials,* 173, 647–652.

Dermentzis, K., Marmanis, D., Christoforidis, A., Kokkinos, N. and Stergiopoylos, E. 2016. Recovery of metallic nickel from waste sludge produced by electrocoagulation of nickel bearing electroplating effluents. In: *4th International Conference on Sustainable Solid Waste Management,* 23–25.

Dizge, N., Keskinler, B. and Barlas, H. 2009. Sorption of Ni (II) ions from aqueous solution by Lewatit cation-exchange resin. *Journal of Hazardous Materials,* 167, 915–926.

Dufresne, P. 2007. Hydroprocessing catalysts regeneration and recycling. *Applied Catalysis A: General,* 322, 67–75.

Farrell, M., Perkins, W. T., Hobbs, P. J., Griffith, G. W. and Jones, D. L. 2010. Migration of heavy metals in soil as influenced by compost amendments. *Environmental Pollution,* 158, 55–64.

Fernandes, A., Afonso, J. C. and Dutra, A. J. B. 2013. Separation of nickel (II), cobalt (II) and lanthanides from spent Ni-MH batteries by hydrochloric acid leaching, solvent extraction and precipitation. *Hydrometallurgy,* 133, 37–43.

Förstner, U. 1981. Metal transfer between solid and aqueous phases. *Metal Pollution in the Aquatic Environment.* Springer.

Fu, F. and Wang, Q. 2011. Removal of heavy metal ions from wastewaters: A review. *Journal of Environmental Management,* 92, 407–418.

Gaines, L. and Singh, M. 1995. Energy and environmental impacts of electric vehicle battery production and recycling. SAE Technical Paper.

Goel, J., Kadirvelu, K., Rajagopal, C. and Garg, V. K. 2005. Investigation of adsorption of lead, mercury and nickel from aqueous solutions onto carbon aerogel. *Journal of Chemical Technology and Biotechnology,* 80, 469–476.

Guo, X., Li, D., Park, K.-H., Tian, Q. and Wu, Z. 2009. Leaching behavior of metals from a limonitic nickel laterite using a sulfation–roasting–leaching process. *Hydrometallurgy,* 99, 144–150.

Guo, Y., Huo, X., Li, Y., Wu, K., Liu, J., Huang, J., Zheng, G., Xiao, Q., Yang, H. and Wang, Y. 2010. Monitoring of lead, cadmium, chromium and nickel in placenta from an e-waste recycling town in China. *Science of the Total Environment,* 408, 3113–3117.

Gupta, S. S. and Bhattacharyya, K. G. 2006. Adsorption of Ni (II) on clays. *Journal of Colloid and Interface Science,* 295, 21–32.

Gupta, V. K. 1998. Equilibrium uptake, sorption dynamics, process development, and column operations for the removal of copper and nickel from aqueous solution and wastewater using activated slag, a low-cost adsorbent. *Industrial & Engineering Chemistry Research,* 37, 192–202.

Gupta, V. K., Jain, C., Ali, I., Sharma, M. and Saini, V. 2003. Removal of cadmium and nickel from wastewater using bagasse fly ash—A sugar industry waste. *Water Research,* 37, 4038–4044.

Halle, K., Fischwasser, K. and Fenk, B. 1982. Recovery of metals from electroplating wastes. *Technologie Umweltschutz*, 25, 120–132.

Harker, J. H. and Backhurst, J. R. 2002. *Coulson and Richardson's Chemical Engineering: Particle Technology and Separation Processes*. Butterworth-Heinemann.

Helfferich, F. G. 1962. *Ion Exchange*. Courier Corporation.

Hoseinian, F. S., Irannajad, M. and Nooshabadi, A. J. 2015. Ion flotation for removal of Ni (II) and Zn (II) ions from wastewaters. *International Journal of Mineral Processing*, 143, 131–137.

Hunsoma, M., Vergnesb, H., Duverneuilb, P., Pruksathorna, K. and Damronglerda, S. 2002. Recovering of copper from synthetic solution in 3PE reactor. *Science Asia*, 28, 153–159.

Idris, J., Musa, M., Yin, C.-Y. and Hamid, K. H. K. 2010. Recovery of nickel from spent catalyst from palm oil hydrogenation process using acidic solutions. *Journal of Industrial and Engineering Chemistry*, 16, 251–255.

Innocenzi, V. and Vegliò, F. 2012. Recovery of rare earths and base metals from spent nickel-metal hydride batteries by sequential sulphuric acid leaching and selective precipitations. *Journal of Power Sources*, 211, 184–191.

Jadhav, U. and Hocheng, H. 2012. A review of recovery of metals from industrial waste. *Journal of Achievements in Materials and Manufacturing Engineering*, 54, 159–167.

Jafari, M., Abdollahzadeh, A. A. and Aghababaei, F. 2017. Copper ion recovery from mine water by ion flotation. *Mine Water and the Environment*, 36, 323–327.

Kaminari, N., Schultz, D., Ponte, M., Ponte, H., Marino, C. and Neto, A. 2007. Heavy metals recovery from industrial wastewater using Taguchi method. *Chemical Engineering Journal*, 126, 139–146.

Kim, J. S. and Keane, M. A. 2002. The removal of iron and cobalt from aqueous solutions by ion exchange with Na-Y zeolite: Batch, semi-batch and continuous operation. *Journal of Chemical Technology and Biotechnology*, 77, 633–640.

Kocaoba, S., Orhan, Y. and Akyüz, T. 2007. Kinetics and equilibrium studies of heavy metal ions removalby use of natural zeolite. *Desalination*, 214, 1–10.

Lai, Y.-C., Lee, W.-J., Huang, K.-L. and Wu, C.-M. 2008. Metal recovery from spent hydrodesulfurization catalysts using a combined acid-leaching and electrolysis process. *Journal of Hazardous Materials*, 154, 588–594.

Lata, H., Garg, V. and Gupta, R. 2008. Sequestration of nickel from aqueous solution onto activated carbon prepared from *Parthenium hysterophorus* L. *Journal of Hazardous Materials*, 157, 503–509.

Lee, J. Y., Rao, S. V., Kumar, B. N., Kang, D. J. and Reddy, B. R. 2010. Nickel recovery from spent Raneynickel catalyst through dilute sulfuric acid leaching and soda ash precipitation. *Journal of Hazardous Materials*, 176, 1122–1125.

Lehto, J., Harjula, R., Leinonen, H., Paajanen, A., Laurila, T., Mononen, K. and Saarinen, L. 1996. Advanced separation of harmful metals from industrial waste effluents by ion exchange. *Journal of Radioanalytical and Nuclear Chemistry*, 208, 435–443.

Leinonen, H., Lehto, J. and Mäkelä, A. 1994. Purification of nickel and zinc from waste waters of metal-plating plants by ion exchange. *Reactive Polymers*, 23, 221–228.

Lewinsky, A. A. 2007. *Hazardous Materials and Wastewater: Treatment, Removal and Analysis*, Nova Publishers.

Li, L., Xu, S., Ju, Z. and Wu, F. 2009. Recovery of Ni, Co and rare earths from spent Ni–metal hydride batteries and preparation of spherical Ni(OH)2. *Hydrometallurgy*, 100, 41–46.

Liu, Z. and Doyle, F. M. 2009. Ion flotation of Co^{2+}, Ni^{2+}, and Cu^{2+} using dodecyldiethylenetriamine (Ddien). *Langmuir*, 25, 8927–8934.

Long, Y.-Y., Feng, Y.-J., Cai, S.-S., Ding, W.-X. and Shen, D.-S. 2013. Flow analysis of heavy metals in a pilot-scale incinerator for residues from waste electrical and electronic equipment dismantling. *Journal of Hazardous Materials*, 261, 427–434.

Lyman, J. and Palmer, G. 1993a. Recycling of rare earths and iron from NdFeB magnet scrap. *High Temperature Materials and Processes*, 11, 175–188.

Lyman, J. W. and Palmer, G. R. 1993b. Investigating the recycling of nickel hydride battery scrap. *JOM Journal of the Minerals, Metals and Materials Society*, 45, 32–35.

Ma, H., Hsiao, B. S. and Chu, B. 2013. Electrospun nanofibrous membrane for heavy metal ion adsorption. *Current Organic Chemistry*, 17, 1361–1370.

Malamis, S. and Katsou, E. 2013. A review on zinc and nickel adsorption on natural and modified zeolite, bentonite and vermiculite: Examination of process parameters, kinetics and isotherms. *Journal of Hazardous Materials*, 252, 428–461.

Marafi, M. and Furimsky, E. 2005. Selection of organic agents for reclamation of metals from spent hydroprocessing catalysts. *Erdoel, Erdgas, Kohle*, 121, 93–96.

Miettinen, T., Ralston, J. and Fornasiero, D. 2010. The limits of fine particle flotation. *Minerals Engineering*, 23, 420–437.

Mohammad, A. W., Teow, Y., Ang, W., Chung, Y., Oatley-Radcliffe, D. and Hilal, N. 2015. Nanofiltration membranes review: Recent advances and future prospects. *Desalination*, 356, 226–254.

Mohammadijoo, M., Khorshidi, Z. N., Sadrnezhaad, S. and Mazinani, V. 2014. Synthesis and characterization of nickel oxide nanoparticle with wide band gap energy prepared via thermochemical processing. *Journal of Nanoscience and Nanotechnology*, 4, 6–9.

Molinari, R., Poerio, T. and Argurio, P. 2008. Selective separation of copper (II) and nickel (II) from aqueous media using the complexation–ultrafiltration process. *Chemosphere*, 70, 341–348.

Moreno-Villoslada, I. and Rivas, B. L. 2003. Retention of metal ions in ultrafiltration of mixtures of divalent metal ions and water-soluble polymers at constant ionic strength based on Freundlich and Langmuir isotherms. *Journal of Membrane Science*, 215, 195–202.

Moskalyk, R. and Alfantazi, A. 2002. Nickel laterite processing and electrowinning practice. *Minerals Engineering*, 15, 593–605.

Mulak, W., Szymczycha, A., Lesniewicz, A. and Zyrnicki, W. 2006. Preliminary results of metals leaching from a spent hydrodesulphurization (HDS) catalyst. *Physicochemical Problems of Mineral Processing*, 40, 69–76.

Naaz, S. and Pandey, S. 2010. Effects of industrial waste water on heavy metal accumulation, growth and biochemical responses of lettuce (*Lactuca sativa* L.). *Journal of Environmental Biology*, 31, 273–276.

Nishi, Y. 2001. Lithium ion secondary batteries; past 10 years and the future. *Journal of Power Sources*, 100, 101–106.

Njau, K., Vd Woude, M., Visser, G. and Janssen, L. 2000. Electrochemical removal of nickel ions from industrial wastewater. *Chemical Engineering Journal*, 79, 187–195.

Panigrahi, S., Parhi, P., Sarangi, K. and Nathsarma, K. 2009. A study on extraction of copper using LIX 84-I and LIX 622N. *Separation and Purification Technology*, 70, 58–62.

Park, Y. J. and Fray, D. J. 2009. Recovery of high purity precious metals from printed circuit boards. *Journal of Hazardous Materials*, 164, 1152–1158.

Peng, C., Jin, R., Li, G., Li, F. and Gu, Q. 2014. Recovery of nickel and water from wastewater with electrochemical combination process. *Separation and Purification Technology*, 136, 42–49.

Pillai, M. G., Regupathi, I., Kalavathy, M. H., Murugesan, T. and Miranda, L. R. 2009. Optimization and analysis of nickel adsorption on microwave irradiated rice husk using response surface methodology (RSM). *Journal of Chemical Technology and Biotechnology*, 84, 291–301.

Qin, B., Luo, H., Liu, G., Zhang, R., Chen, S., Hou, Y. and Luo, Y. 2012. Nickel ion removal from wastewater using the microbial electrolysis cell. *Bioresource Technology*, 121, 458–461.

Qiu, W. and Zheng, Y. 2009. Removal of lead, copper, nickel, cobalt, and zinc from water by a cancrinite-type zeolite synthesized from fly ash. *Chemical Engineering Journal*, 145, 483–488.

Rajapaksha, A. U., Vithanage, M., Weerasooriya, R. and Dissanayake, C. B. 2012. Surface complexation of nickel on iron and aluminum oxides: A comparative study with single and dual site clays. *Colloids and Surfaces A: Physicochemical and Engineering Aspects*, 405, 79–87.

Rajic, N., Stojakovic, D., Jovanovic, M., Logar, N. Z., Mazaj, M. and Kaucic, V. 2010. Removal of nickel (II) ions from aqueous solutions using the natural clinoptilolite and preparation of nano-NiO on the exhausted clinoptilolite. *Applied Surface Science*, 257, 1524–1532.

Raval, N. P., Shah, P. U. and Shah, N. K. 2016. Adsorptive removal of nickel (II) ions from aqueous environment: A review. *Journal of Environmental Management*, 179, 1–20.

Ren, J., Gao, Y., Zhao, Y. and Tao, L. 2015. Adsorption of Cu (II) from aqueous solution by sulfuric acid and thermally treated attapulgite. In: *Engineering Technology, Engineering Education and Engineering Management: Proceedings of the 2014 International Conference on Engineering Technology, Engineering Education and Engineering Management (ETEEEM 2014)*, Hong Kong, 15–16 November 2014, 2015. CRC Press, 431.

Ren, J., Song, X., Xu, D., Gou, D. and Tao, L. 2014. Adsorption of Ni (II) from aqueous solution by sulfuric acid and thermally treated attapulgite. *Advanced Materials Research*, 1073–1076, 90–94.

Rengaraj, S., Yeon, K.-H. and Moon, S.-H. 2001. Removal of chromium from water and wastewater by ion exchange resins. *Journal of Hazardous Materials*, 87, 273–287.

Rivas, B. L., Pereira, E. D., Palencia, M. and Sánchez, J. 2011. Water-soluble functional polymers in conjunction with membranes to remove pollutant ions from aqueous solutions. *Progress in Polymer Science*, 36, 294–322.

Robotin, B., Coman, V. and Ilea, P. 2011. Nickel recovery from electronic waste I. Nickel recovery from cathode ray tubes. *Studia Universitatis Babes-Bolyai, Chemia*, 56.

Rubio, J., Souza, M. and Smith, R. 2002. Overview of flotation as a wastewater treatment technique. *Minerals Engineering*, 15, 139–155.

Rydh, C. J. and Karlström, M. 2002. Life cycle inventory of recycling portable nickel–cadmium batteries. *Resources, Conservation and Recycling*, 34, 289–309.

Salmani, M. H., Davoodi, M., Ehrampoush, M. H., Ghaneian, M. T. and Fallahzadah, M. H. 2013. Removal of cadmium (II) from simulated wastewater by ion flotation technique. *Iranian Journal of Environmental Health Science and Engineering*, 10, 16.

Sanciolo, P., Harding, I. and Mainwaring, D. 1992. The removal of chromium, nickel, and zinc from electroplating wastewater by adsorbing colloid flotation with a sodium dodecylsulfate/dodecanoic acid mixture. *Separation Science and Technology*, 27, 375–388.

Schario, M. 2007. Troubleshooting decorative nickel plating solutions (Part I of III installments): Any experimentation involving nickel concentration must take into account several variables, namely the temperature, agitation, and the nickel-chloride mix. *Metal Finishing*, 105, 34–36.

Schaumlöffel, D. 2012. Nickel species: Analysis and toxic effects. *Journal of Trace Elements in Medicine and Biology*, 26, 1–6.

Sebba, F. 1962. *Ion Flotation*, Elsevier.

Shao, J., Qin, S., Davidson, J., Li, W., He, Y. and Zhou, H. S. 2013. Recovery of nickel from aqueous solutions by complexation-ultrafiltration process with sodium polyacrylate and polyethylenimine. *Journal of Hazardous Materials*, 244, 472–477.

Siemens, R., Jong, B. and Russell, J. 1986. Potential of spent catalysts as a source of critical metals. *Conservation & Recycling*, 9, 189–196.

Singh, B. 2009. Treatment of spent catalyst from the nitrogenous fertilizer industry—A review of the available methods of regeneration, recovery and disposal. *Journal of Hazardous Materials*, 167, 24–37.

Smith-Sivertsen, T., Lund, E., Thomassen, Y. and Norseth, T. 1997. Human nickel exposure in an area polluted by nickel refining: The Sør–Varanger study. *Archives of Environmental Health: An International Journal*, 52Z, 464–471.

Srivastava, N. and Majumder, C. 2008. Novel biofiltration methods for the treatment of heavy metals from industrial wastewater. *Journal of Hazardous Materials*, 151, 1–8.

Sudha, R., Srinivasan, K. and Premkumar, P. 2015. Removal of nickel (II) from aqueous solution using *Citrus limettioides* peel and seed carbon. *Ecotoxicology and Environmental Safety*, 117, 115–123.

Sullivan, J. L. and Gaines, L. 2012. Status of life cycle inventories for batteries. *Energy Conversion and Management*, 58, 134–148.

Szymczycha-Madeja, A. 2011. Kinetics of Mo, Ni, V and Al leaching from a spent hydrodesulphurization catalyst in a solution containing oxalic acid and hydrogen peroxide. *Journal of Hazardous Materials*, 186, 2157–2161.

Taha, S., Bouvet, P., Corre, G. and Dorange, G. 1996. Study and modelization of some heavy metals removal by ultrafiltration in the presence of soluble chitosan. *Advances in Chitin Science*, 1, 389.

Tao, W., Chen, G., Zeng, G., Yan, M., Chen, A., Guo, Z., Huang, Z., He, K., Hu, L. and Wang, L. 2016. Influence of silver nanoparticles on heavy metals of pore water in contaminated river sediments. *Chemosphere*, 162, 117–124.

Thomas, C. L. M. 1970. *Catalytic Processes and Proven Catalysts*. Academic Press, New York.

Tzanetakis, N., Taama, W., Scott, K., Jachuck, R., Slade, R. and Varcoe, J. 2003. Comparative performance of ion exchange membranes for electrodialysis of nickel and cobalt. *Separation and Purification Technology*, 30, 113–127.

Uddin, M. K. 2017. A review on the adsorption of heavy metals by clay minerals, with special focus on the past decade. *Chemical Engineering Journal*, 308, 438–462.

Valverde, I. M., Paulino, J. F. and Afonso, J. C. 2008. Hydrometallurgical route to recover molybdenum, nickel, cobalt and aluminum from spent hydrotreating catalysts in sulphuric acid medium. *Journal of Hazardous Materials*, 160, 310–317.

Varma, S., Sarode, D., Wakale, S., Bhanvase, B. and Deosarkar, M. 2013. Removal of nickel from waste water using graphene nanocomposite. *International Journal of Chemical and Physical Sciences*, 2, 132–139.

Wang, G., Li, A. and Li, M. 2010. Sorption of nickel ions from aqueous solutions using activated carbon derived from walnut shell waste. *Desalination and Water Treatment*, 16, 282–289.

Yang, Q., Qi, G., Low, H. and Song, B. 2011. Sustainable recovery of nickel from spent hydrogenation catalyst: Economics, emissions and wastes assessment. *Journal of Cleaner Production*, 19, 365–375.

Yang, S., Li, J., Lu, Y., Chen, Y. and Wang, X. 2009. Sorption of Ni (II) on GMZ bentonite: Effects of pH, ionic strength, foreign ions, humic acid and temperature. *Applied Radiation and Isotopes*, 67, 1600–1608.

Zahraei, F., Rahimi, K. and Yazdani, A. 2015. Preparation and characterization of graphene/nickel oxide nanorods composite. *International Journal of Nano Dimension*, 6, 371.

Zhang, P., Yokoyama, T., Itabashi, O., Wakui, Y., Suzuki, T. M. and Inoue, K. 1999. Recovery of metal values from spent nickel–metal hydride rechargeable batteries. *Journal of Power Sources*, 77, 116–122.

Zhou, Y., Ning, X.-A., Liao, X., Lin, M., Liu, J. and Wang, J. 2013. Characterization and environmental risk assessment of heavy metals found in fly ashes from waste filter bags obtained from a Chinese steel plant. *Ecotoxicology and Environmental Safety*, 95, 130–136.

15

Phytoextraction and Phytomining of Soil Nickel

Rufus L. Chaney

CONTENTS

15.1 Introduction

The concept of phytoextraction, the use of plants to remove metals from contaminated or mineralized soils, was first reported by Chaney (1983a). Practical phytoextraction and especially phytomining depend on the existence of plants, now called hyperaccumulators, which accumulate exceptionally high concentrations of metals in their shoots, typically 100-fold levels in crop plant species. As illustrated in Table 15.1, crop plants (e.g., maize [*Zea mays* L.]) cannot remove enough Ni to decontaminate soils even over centuries of removing aboveground crop residues. Actually, for highly contaminated or mineralized soils containing 2500 mg Ni kg^{-1} or higher, even phytoextraction using simple hyperaccumulators would be a very slow

TABLE 15.1

Potential of Crop Species Maize (Z. *mays*) and Ni Hypernickelophore
A. *murale* to Phytoextract Ni from Soils

Plant Species	Biomass Yield t/ha	Ni in Shoot Biomass			
		mh/kg	kg/ha	% of Soil	Ash-Ni %
Maize (control)	20	1	0.02	0.01	0.002
Maize (50% YD)	10	100	1	0.01	0.2
Alyssum in pasture	3	10,000	30	0.3	5–10
Wild *Alyssum* crop	10	15,000	150	1.5	20–30
Alyssum + Agric[a]	20	20,000	400	4.0	20–30
Alyssum cultivar[a,b]	20	30,000	600	6.0	25–30

Note: Assume soil contains 2500 mg Ni kg^{-1}, which equals 10,000 kg Ni ha^{-1} 30
cm deep.
[a] Appropriate agronomic practices: N, P, K, S, Ca, and B fertilizers; herbicides;
planted seeds.
[b] Improved cultivar bred to maximize annual shoot Ni content at annual
harvest.

process to deplete soil Ni. For some other elements, annual removals might
be useful for decontamination, for example, growing the southern France
genotypes of *Noccaea caerulescens* to remove soil Cd (Chaney and Baklanov
2017). However, using hyperaccumulators to remove Zn, Ni, and most other
metals from soils would not be rapid or cost-effective. Basic research on
phytoextraction was initially focused on Cd because of the health risks of
food-chain Cd, and on well-known Se hyperaccumulators, cases in which
the plants could remove a higher portion of soil total element per year.

Phytomining is a related concept in which hyperaccumulator plants are
used to both remove soil metals and produce crop biomass containing enough
metal to be economically profitable as a metal crop. Although phytomining
was discussed in the Chaney (1983a) report, it took some years to generate
interest and obtain funding to develop Ni phytomining technologies. The
alternative to phytoextraction, phytostabilization of phytotoxic levels of soil
Ni, was demonstrated in the 1950s (Crooke 1956) and is clearly capable of alleviating environmental risks from Ni-contaminated soils (Kukier and Chaney
2004; Siebielec et al. 2007). Phytostabilization can alleviate high phytoavailability and prevent Ni phytotoxicity and food-chain transfer of soil Ni in contaminated soils at low cost, much more easily than for other metals such as
Zn, Cd, and Pb. Because of the formation of very low solubility forms of Ni in
remediated alkaline soils, such sites could be readily phytostabilized, alleviating Ni environmental risks of the sites (Chaney et al. 2014).

Hyperaccumulators accumulate in harvestable shoot tissues about 100-
fold or more of an element than commonly accumulated in crop plants (see
van der Ent et al. 2012, 2015a). In the case of Ni, a hyperaccumulator is defined
as accumulating more than 1000 mg kg^{-1} in the dry matter of any aboveground tissue while completing its life cycle growing in its natural habitat

(see Reeves 1992; van der Ent et al. 2012). A subset of Ni hyperaccumulators that can accumulate more than 10,000 mg Ni kg^{-1} foliar DW (1%) when growing in their natural habitat was called "hypernickelophores" by the scientist who found such species in New Caledonia and stimulated interest in these species (Jaffré and Schmid 1974; see also Jaffré et al. 2013). As shown in Table 15.1, phytomining using hypernickelophore species may allow removal of soil Ni and support recycling of soil Ni biopurified by the plants, while crop plants cannot remove enough Ni to notice (Baker and Brooks 1989; Chaney and Mahoney 2014; Chaney et al. 1999, 2000, 2007b, 2010, 2014). Removal of Ni is not rapid, especially if accumulation from the subsoil is considered; decades of profitable phytomining could occur with most serpentine soils. Ni in ash of hypernickelophore plant shoots is in the matrix of plant nutrients (biopurified), not the matrix of Fe and Mn oxides and silicates of ultramafic soils. The matrix makes it expensive to extract Ni from ultramafic soils and ores, requiring heated sulfuric acid, and then expensive technology to separate the Ni from the other matter extracted.

It was evident when Chaney conceived the concept of phytoextraction that crop plants could not remove enough Ni annually to provide any remediation potential (maximum of 100 mg Ni kg^{-1} (=100 g t^{-1} dry biomass) at the point of significant phytotoxicity (Hunter and Vergnano 1952), and 10 t ha^{-1} = 1000 g or 1 kg of Ni ha^{-1} in the plant shoots for a high-yielding biomass crop such as maize (Z. *mays* L.). When the 15-cm surface layer of a Ni phytotoxic soil contains 2500 mg kg^{-1} (= 5000 kg Ni ha^{-1} 15 cm), even crops suffering phytotoxicity (~100 mg kg^{-1}) do not remove enough Ni to achieve useful removals (Table 15.1). However, by growing Ni hypernickelophore plant species with high biomass yield potential (e.g., *Alyssum murale* or *Phyllanthus* sp.) with appropriate fertilizers and weed control, one can phytoextract between 200 and 400 kg Ni ha^{-1} year^{-1}. Applying modern agricultural technology to genetic improvement and production practices, even higher annual phytomining might be attained (Table 15.1). With Ni metal selling at about $15 kg^{-1} on the London Metal Exchange during 2018, the value of Ni in a phytomining crop could reach $4800 ha^{-1} or higher, far more than that attainable with commercial agricultural crops (especially on the infertile serpentine soils where phytomining could be commercially conducted) (see also Robinson et al. (2003) for modeling of Ni phytomining values). Costs and markets for Ni-rich biomass will be discussed below.

Many *Alyssum* species have been found to hyperaccumulate Ni from serpentine soils of southern Europe (Reeves and Baker 2000; Robinson et al. 1997). Studies have also shown that the accumulation of Ni by *A. murale* is not due to the solubilization of forms of Ni that are unavailable to crop plants (non-labile) in these soils (Denys et al. 2002; Shallari et al. 2001). Rather, available data indicate that the hyperaccumulators are more effective in rapidly absorbing Ni from the soil "labile pool" of Ni. Dense fine roots give more surface area close to soil particles, which can accumulate soil Ni by diffusion of sorbed Ni to the root membrane Ni transporters. A large rooting volume in both topsoil and subsoil are important in the high levels of hyperaccumulation achieved by *Alyssum corsicum* and similar species (Chaney et al. 2019; Nkrumah et al. 2017; Paul et al. 2019).

I coordinated a large research team that conducted research to develop commercial Ni phytomining, and sought funding to develop and demonstrate a commercial technology. After confirmation that a patent would be issued on the technology we had developed (Chaney et al. 1998), we obtained funding to develop Ni phytomining technology under a Cooperative Research and Development Agreement (CRADA) between the United States Department of Agriculture (USDA) Agricultural Research Service and Viridian Resources LLC. The CRADA team that developed this technology included the USDA–Agricultural Research Service (R.L. Chaney and Y.-M. Li), the University of Maryland (J.S. Angle and E.P. Brewer), the Environmental Consultancy of the University of Sheffield (A.J.M. Baker and R.D. Reeves), and Oregon State University (V.V. Volk and R.J. Roseberg), and our graduate students and technical staff made many important contributions to the technology development.

In the research and development (R&D) program undertaken, we needed to (1) select for R&D hypernickelophore plant species with high promise for commercial phytomining (both high yield and high Ni concentration); (2) learn how to breed the species to develop improved cultivars; (3) collect diverse germplasm to support the breeding program; (4) identify agronomic practices needed to grow high yields of the selected species containing high levels of Ni (fertilizers and rates of application, planting methods, harvesting methods, plant density in the field, herbicides, etc.); (5) evaluate the ability of the collected germplasm to hyperaccumulate Ni from serpentine soils in the field and make genetic crosses to combine traits from outstanding genotypes in the collections; and (6) investigate all other agronomic, soil science, plant science, and metal recovery from the biomass (pyrometallurgy or hydrometallurgy) factors that could aid the success of phytomining. Our work was published in technical journals and review papers and in patents for most countries with extensive serpentine/ultramafic soils rich in Ni.

After the team selected *Alyssum* species and *Berkheya coddii* for possible development for use in the United States, Canada, and southern Europe based on their ecology, yield potential, and confirmed Ni hyperaccumulation ability, the CRADA research was conducted. Extensive germplasm collection of *Alyssum* species was made across southern Europe where these species occur naturally, led by Baker and Reeves (in Greece and Turkey) and Shallari (in Albania). Cold-tolerant *A. murale* from Bulgaria and Turkey were included in the germplasm collection so that improved cultivars adapted to the climates of U.S. and Canadian Ni-mineralized and -contaminated soils could be addressed with the developed phytomining technologies.

A. murale and *A. corsicum* were selected for commercial development after testing several species because, based on the judgment of the R&D team, they offered high probability of being able to be developed into a phytomining crop. Many of the Ni hyperaccumulator *Alyssum* species grow no taller than 15 cm above the soil, which would make harvest difficult and limit biomass yields (see also Dudley 1966). The shorter species also appeared to have considerably lower yield potential than the taller species. *A. murale* and *A. corsicum* are perennial and regrow after cutting at flowering to harvest biomass. These species accumulated up to 25,000 mg Ni kg^{-1} in aboveground biomass in our field test plots in Oregon without yield reduction from the Ni (compared to the 100 mg Ni kg^{-1}, which causes visible phytotoxicity in nearly all crop plants) (Table 15.1). Leaves contain about double the Ni level of stems plus petioles, so leaves of these plants can accumulate up to 40,000 mg Ni kg^{-1} DW without yield reduction.

Viridian LLC was the CRADA commercial cooperator, which allowed them to obtain exclusive licenses to all of the patented Ni and Co phytoextraction technologies and germplasm developed by the team during the CRADA (Angle et al. 2007; Chaney et al. 1998, 1999, 2004, 2007a). The team focused on producing *Alyssum* biomass on natural ultramafic soils in Oregon, USA, where such soils were extensive compared to other parts of the United States. Inco Ltd. contracted with Viridian LLC to conduct tests of phytoextraction of Port Colborne, Ontario, Canada, metal refinery contaminated soils, and then to conduct field tests of remediation of and phytoextraction of Ni from these soils and, after the CRADA terminated, at other INCO locations.

The cooperators collected diverse germplasm across southern Europe so that they could combine desirable properties by using standard crop breeding methods. Because *A. murale* and *A. corsicum* are genetically "self-incompatible" species, improved cultivars were bred by recurrent selection among the genotypes with properties judged to be best for phytomining (biomass yield, Ni concentration, and retention of leaves during early flowering), and were prepared for commercial use. Large variations of shoot Ni concentration and yield were observed when the diverse collected germplasm was grown at one location near Cave Junction, Oregon, on a fertilized Brockman variant silt loam soil (Typic Xerochrepts). Because plant breeding has to observe the whole life cycle of the germplasm to identify strains with valuable qualities, field testing was considered essential rather than the less expensive greenhouse pot testing. Figure 15.1 shows the field plots at early flowering stage in 2000. In this large field test, there were four replications. Each subplot included six lines being evaluated plus one *A. murale* and one *A. corsicum* genotype so that we could use covariance correction to correct for variability across the large field area.

In order to select genotypes with the desired properties for genetic crossing, we needed to obtain predictive tissue Ni analysis earlier in the growing season than flowering initiation. *Alyssum* genotypes were seeded in the field shortly before the rainy season in Josephine County, Oregon (late August); seeds germinated and plants thrived on the fertilized soil. Preplant herbicide was used to limit weedy grasses, and N, P, K, Ca, S, Mo, and B fertilizers were applied at appropriate rates based on previous greenhouse studies. The normal growth pattern of *Alyssum* Ni hyperaccumulators requires short cold days (winter) to vernalize the plants and induce flowering when the long

FIGURE 15.1
Genotype evaluation on serpentine soil testing growth and Ni accumulation by *Alyssum* species collected germplasm. Field test site on Brockman variant soil used to evaluate germplasm. Average test genotype contained 1.5% Ni (15 g kg⁻¹ dry weight) in shoot biomass at flowering in June.

warm days of spring arrive, so if possible, we preferred to collect and analyze the Ni concentration in pre-flowering shoots (stems plus leaves) before flowering started, so that genotypes that are not wanted in the genetic crossing to be conducted could be removed so their pollen would not contaminate the breeding work. In the field, we found that if secondary vernalized spiking stems were collected before flowers emerged, the Ni concentrations

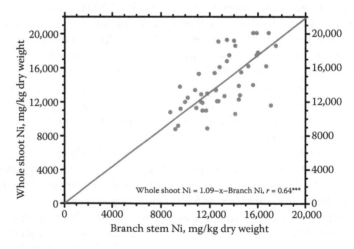

FIGURE 15.2
Strong relationship between Ni concentration in branch stems before flowering with Ni concentration in whole flowering shoots at harvest stage supports selection of parent plants for recurrent selection breeding of improved cultivars of *Alyssum* Ni hypernickelophore species.

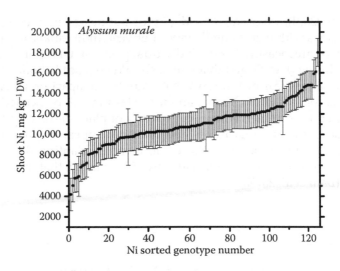

FIGURE 15.3
Genetic variation on shoot Ni concentration among diverse *A. murale* genotypes grown on an Oregon Brockman variant serpentine soil with 5500 mg Ni/kg. (From Li, Y.-M., R.L. Chaney, E.P. Brewer, J.S. Angle, and J.P. Nelkin. 2003a. Phytoextraction of nickel and cobalt by hyperaccumulator *Alyssum* species grown on nickel-contaminated soils. *Environ. Sci. Technol.* 37:1463–1468.)

were well correlated with the Ni concentrations in the whole aboveground biomass at harvest at the stage of mid-flowering (Figure 15.2). We observed wide genetic variation in Ni concentration (Figure 15.3), size/yield of the shoots, and retention of leaves into the flowering period, properties that were used to select the parent genotypes used to conduct the recurrent selection of improved cultivars. Crossing was conducted both by hand and by placing honeybee hives near the field plots. The bees thrived and seed set was excellent in the field. Ten plants of each of the best 10 genotypes of each *Alyssum* species were allowed to cross and their progeny were tested in the next generation of recurrent selection. *B. coddii* was tested in this work but found to accumulate only about 0.5% Ni compared to the average 1.5% Ni in the *Alyssum* entries.

15.2 Planting Density Evaluation

One of the agronomic variables that needed to be characterized experimentally was the planting density which gives the highest shoot Ni yield (concentration-X-biomass). Modern crop cultivars have been selected to grow high yields while close to neighboring plants to increase potential yields. At high density, plants compete for sun, nutrients, and water. As described in

Angle et al. (2001), we used a "Nelder" plot test design in which plants are placed in rows with increasing distance between individuals and rows, and yields and Ni are measured in the individual plants at harvest as replicates of spacing. This work indicated that plants should be placed about 30 cm apart in a filled pattern (5.6/m²) in rows that were 60 cm apart or so. Additional testing was conducted in Oregon, and more testing will be needed for cultivars selected for commercialization. A study of plant density by Bani et al. (2015b) used different approaches and does not appear to provide statistically validated justification of the recommended planting density (4/m²) (essentially two plants per meter with rows 50 cm apart) that they concluded was best (Bani et al. 2015b).

15.3 Agronomic Development of Ni Phytomining

Our agronomic testing identified important aspects of fertilizer requirements of *Alyssum* species growing on serpentine or smelter-contaminated soils. Natural serpentine soils are very infertile, and cause Ca, P, N, and K deficiency in crop plants and select the serpentine ecology of plants species adapted to these soils (Brady et al. 2005; Brooks 1987). Assuming that N and K would be applied at rates needed to support the biomass yield attained, greenhouse pot tests were conducted to test the phosphate and Ca fertilizer requirements for maximum growth of *Alyssum* species on a Brockman gravelly clay loam, and the effect of soil pH variation on Ni hyperaccumulation. Table 15.2 shows the yields and macroelement concentrations in the plant shoots after 120 days of growth post transplanting of seedlings into the prepared soils. Plants were grown in large pots (4 kg soil) to avoid effects of small pots on growth of *Alyssum* (see also below; Chaney et al. 2019). Several important results were obtained regarding macronutrients. First, amounts of N and K were applied to satisfy the crop requirement of these nutrients, which must be added regularly in fertilizers to support plant growth. Additional N fertilizer was added after 60 days of growth to maintain high N availability. The rates of P fertilizer tried to take into consideration the relatively high cost of P fertilizers, the strong binding of P in high Fe serpentine soils, and the known ability of *Alyssum* to obtain adequate P from low P serpentine soils (e.g., Gabbrielli et al. 1995). Even the first level of P addition (100 kg P ha^{-1}) gave full yield of shoot biomass and attained normal "adequate" levels of P in plant tissues while the zero P (Treatment 3) and control (Treatment 1) were clearly severely P-deficient. Yield was even lower when all the other fertilizers were applied (Treatment 3) than when no fertilizers were applied (Treatment 1).

The effect of treatments on plant Ca and Mg is also very interesting. *Alyssum* species are adapted to these very low Ca serpentine soils but

TABLE 15.2

Effect of Amending a Brockman Silt Loam from Near Cave Junction, Oregon, with $Ca(H_2PO_4)_2$ in kg ha^{-1}, pH Adjusting HNO_3 or $CaCO_3$, or Ca as $CaSO_4 \cdot 2H_2O$ in t ha^{-1} on Terminal Geometric Mean Soil pH, Geometric Mean Yield, and Macronutrient Composition of Shoots of *Alyssum* Species Grown for 120 days (GM Designates Geometric Mean)

Treatment		Final pH	GM-Yield	GM-P	Mg	Ca	K
			g/pot		g/kg		
1	None	6.56 a*	4.1 c	1.04 e	4.06 d	17.5 ab	9.1 d
Phosphate Treatments							
3	0 P	5.82 e	1.6 d	0.61 f	6.47a	17.5 ab	10.0 cd
2	100 P	6.24 b	24.5 a	2.16 cd	6.20 bc	17.1 ab	16.5 b
4	250 P	6.14 bcd	23.2 ab	3.00 b	6.46 bc	19.8 a	19.9 a
5	500 P	6.16 bc	26.5 a	3.59 a	6.40 bc	18.2 ab	19.8 a
pH Treatments							
6	Lo pH	5.42 g	27.4 a	2.03 d	4.92 cd	16.7 ab	18.4 ab
7	MLo pH	5.69 f	26.2 a	2.12 d	6.42 bc	18.5 ab	17.0 b
8	MHi pH	5.89 e	27.0 a	2.07 d	5.31 bcd	16.2 ab	18.4 ab
2	As is pH	6.24 b	24.5 a	2.16 cd	6.20 bc	17.1 ab	16.5 b
Ca:Mg Treatments							
9	0.0 Ca	6.10 cd	19.3 b	2.43 c	5.66 bc	14.8 b	12.4 c
2	1.0 Ca	6.24 b	24.5 a	2.16 cd	6.20 bc	17.1 ab	16.5 b
10	2.5 Ca	6.04 cd	25.2 a	2.10 d	6.74 b	18.4 ab	17.7 ab
11	5.0 Ca	6.03 d	24.2 a	1.94 d	6.26 bc	16.2 ab	17.2 ab

*Means followed by the same letter are not significantly different ($P < 0.05$ level) according to the Duncan–Waller K ratio t test.

accumulate "normal" (~2%) levels of Ca in shoots of dicot species. Mg is relatively high for crops (normal adequate Mg is 1.5 g kg^{-1} DW young leaves for crop plant species) but remained lower than the concentrations of Ca in the plants showing the remarkable ability of *Alyssum* to accumulate Ca and reject Mg from serpentine soils. One implication of these findings is that because so much Ca is removed in the shoot biomass in phytomining (20 g Ca kg^{-1} shoot DW × 20 t ha^{-1} = 400 kg ha^{-1}), the soil Ca fertility will need to be maintained over years of phytomining by applying Ca fertilizer. If the dissolved matter from biomass ash treated with acid to separate and concentrate the Ni can be returned to the field, Ca and some other nutrients could supply some of the required fertilizer nutrients to the soils if handling this liquid would be practical. With the decline in commercial availability of Ca-containing superphosphate P fertilizers, other Ca sources will be required; the high cost of superphosphate will minimize rates of this fertilizer, and monoammonium phosphate may be used instead, adding no Ca. Alternatively, Ca fertilizers such as gypsum ($CaSO_4 \cdot 2H_2O$) will need to be

Nickel in Soils and Plants

added periodically to maintain adequate Ca supply. Superphosphate fertilizer [Ca(H$_2$PO$_4$)$_2$] applied at 100 kg P ha^{-1} would supply less Ca than removed by one crop. Also, although high soil Mg:Ca ratios are believed to be part of the "serpentine factor," which limits vegetation adaptation for these soils, the *Alyssum* species rejected Mg compared to Ca, allowing the plant to maintain relatively normal Ca metabolism on these unusual soils.

Table 15.3 shows the soil pH, yields, and microelement concentrations for these treatments. In the beginning, we had expected the plants to accumulate higher concentrations of Ni in acidified soils than in limed soils (higher pH) because that is the response of crop plants discussed above (Kukier and Chaney 2004). Soil solution Ni concentrations are strongly increased when soil pH drops below about 6.2 (Figure 15.4), which greatly increases Ni accumulation by crop plants and the potential for Ni phytotoxicity. However, to our surprise, acidifying soils reduced *Alyssum* shoot Ni concentrations. Depending on the specific serpentine or contaminated soil under test, liming significantly

TABLE 15.3

Effect of Amending a Brockman Silt Loam from Near Cave Junction, Oregon, with Ca(H$_2$PO$_4$)$_2$ in kg ha^{-1}, pH Adjusting HNO$_3$ or CaCO$_3$, or Ca as CaSO$_4 \cdot$ 2H$_2$O in t ha^{-1} Geometric Mean Soil pH, GM Yield, and Element Concentrations in Shoots of Two Species of *Alyssum* after 120 days Growth (GM is Geometric Mean Because of Lognormal Distribution of These Data)

Treatment		Final pH	GM-Yield	GM-Ni	GM-Co	GM-Mn	GM-Zn	GM-Fe	Cu
			g DM/pot	g/kg DM		mg/kg DW			
1	None	6.56*	4.1 c	14.7 a	34.3	56	63	154	3.0
Phosphate Treatments									
3	0 P	5.82	1.6 d	6.25 cd	19.4	62	118	273	2.8
2	100 P	6.24	24.5 a	6.27 cd	19.9	61	59	112	3.6
4	250 P	6.14	23.2 ab	6.81 bc	22.6	65	60	104	4.2
5	500 P	6.16	26.5 a	5.69 d	18.1	67	55	92	4.0
pH Treatments									
6	Lo pH	5.42	27.4 a	6.15 cd	224	462	63	144	4.4
7	MLo pH	5.69	26.2 a	6.80 bc	50.4	132	68	117	4.6
8	MHi pH	5.89	27.0 a	5.99 cd	28.8	73	58	96	3.6
2	As is pH	6.24	24.5 a	6.27 cd	19.9	61	59	112	3.6
Ca:Mg Treatments									
9	0.0 Ca	6.10	19.3 b	7.86 b	21.1	56	49	87	3.1
2	1.0 Ca	6.24	24.5 a	6.27 cd	19.9	61	59	112	3.6
10	2.5 Ca	6.04	25.2 a	6.05 cd	18.4	58	59	87	3.8
11	5.0 Ca	6.03	24.2 a	5.63 d	24.4	78	63	93	3.6

Note: Rate of P in kg P ha^{-1}.

*Means followed by the same letter are not significantly different ($P <$ 0.05 level) according to the Duncan–Waller K ratio t test.

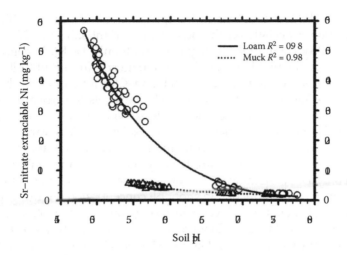

FIGURE 15.4
Effect of harvest soil pH on Ni extractable with 0.01 M Sr(NO₃)₂ from Welland loam and Quarry muck soils contaminated by emissions of a Ni refinery at Port Colborne, Ontario. (From Kukier, U. and R.L. Chaney. *In situ* remediation of Ni-phytotoxicity for different plant species. *J. Plant Nutr.* 27:465–495.)

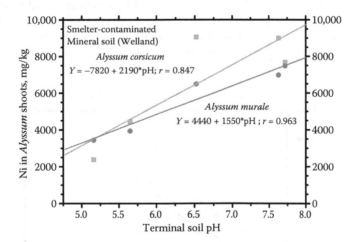

FIGURE 15.5
Effect of soil pH on accumulation of Ni in shoots of two *Alyssum* species grown on Welland soil from Port Colborne, Ontario, Canada. (See Li, Y.-M., R.L. Chaney, E. Brewer, R.J. Roseberg, J.S. Angle, A.J.M. Baker, R.D. Reeves and J. Nelkin. 2003b. Development of a technology for commercial phytoextraction of nickel: Economic and technical considerations. *Plant Soil* 249:107–115.)

increased shoot Ni concentration and shoot Ni content. Figure 15.5 shows the strong increase in shoot Ni of both *Alyssum* species grown on the Port Colborne Welland mineral soil as pH was decreased by HNO₃ or increased by limestone application (Li et al. 2003b). Shoot Ni increased all the way to pH ≥7 where the soil was made calcareous, while with serpentine soils rich in Fe, liming above about pH 6.4 was counterproductive (Kukier et al. 2004) to Ni phytoextraction especially on the Brockman variant soil with 22% Fe compared to the Brockman gravelly clay loam with only 4% Fe. In contrast with serpentine soils, the highly fertile Welland loam soil from Port Colborne had substantial levels of organic matter, high P and K fertility, and only the normal 1% Fe of Canadian soils in that region. The Fe levels in serpentine soils are much higher than the Port Colborne 3 soils, so Ni is not so strongly adsorbed and occluded in normal soils as in serpentine soils. The "Vineyard" Brockman variant soil used in our studies contained 22% Fe, which appears to have made higher soil pH less successful for phytomining than the Port Colborne soils with only 1% Fe. Acidification of other serpentine soils from Oregon and Maryland often cause reduction in shoot Ni concentration (Figure 15.6).

Other microelements are at levels commonly found in crop plants grown on such soils. Shoot Fe, Mn, and Zn were at "adequate" levels, far below phytotoxic levels, while Cu was present at quite low levels in plant shoots. The Port Colborne soils were at least somewhat contaminated by Ni, Cu, and Co, but plant levels were not unusual and *Alyssum* shoot Cu was near the edge of deficiency for crop plants. As expected for Mn, Zn, and Co, lower soil pH in the acidified treatments caused significant increases in shoot levels of these elements in great contrast with Ni. Shoot Cu and Fe levels are tightly regulated by plant metabolism and were hardly affected by the applied treatments.

Another research program worked to develop Ni phytomining in Albania (e.g., Bani et al. 2014). They confirmed that planting a crop gave better yields

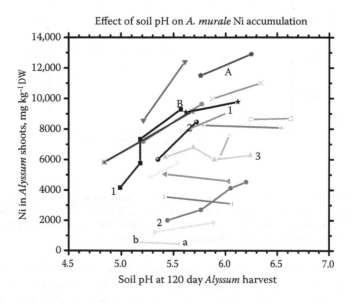

FIGURE 15.6
(See color insert.) Effect of variation in pH of serpentine or Ni refinery contaminated soils (1, 2) on Ni concentration in shoots of *A. murale*.

than relying on volunteer plants and that fertilization increased Ni yields due to higher biomass yield. They confirmed that if grassy weeds were controlled with herbicides, Ni yields were improved. As discussed below, they have also tested production of Ni products from the biomass. Their tests often involved less than a full crop year of growth and may underestimate potential Ni yield potential.

Other research programs are working to develop Ni phytomining for tropical serpentine soils in Malaysia, Indonesia, New Caledonia, the Philippines, and similar areas (van der Ent et al. 2015a). Many new Ni hyperaccumulator species have been identified recently in these nations (e.g., Losfeld et al. 2014b, 2015; Nkrumah et al. 2016; Perrier et al. 2004; Quimado et al. 2015; van der Ent et al. 2015a), offering climate-adapted species. Some even appear to have high annual biomass yield and hypernickelophore properties and regrow after cutting as perennials (*Phyllanthus rufuschaneyi*), or accumulate appreciable amounts of Ni in the tree bole over time (*Rinorea bengalensis*), allowing intermittent harvest and lower production costs than annual harvested species. Significant field and large pot growth tests are being conducted to identify agronomic management practices required for tropical Ni phytomining (Nkrumah et al. 2017). In most hypernickelophore species, foliar Ni concentration is much higher than stem and petiole Ni concentration, so leafy crops appear to offer the best opportunity for phytomining. For example, *Streptanthus polygaloides* is mostly only a stem at the time of flowering, so the whole shoot Ni is much lower than the leaf Ni concentration usually reported.

15.4 Soil Co Phytoextraction

Soil Co could also be phytoextracted using *A. murale*, but soil or solution Ni inhibits Co phytoextraction, and annual Co removal was greatest at low pH (Malik et al. 2000) in contrast to the higher Ni removal at higher soil pH (Kukier et al. 2004; Li et al. 2003a). Interestingly, studies by Tappero et al. (2007) showed that although *A. murale* accumulated and translocated Co from roots to shoots similarly to Ni, Co was not effectively accumulated into leaf cell vacuoles, but accumulated at the leaf tips outside of cells. It will be interesting to learn how transmembrane transporters that process Ni to achieve the hypernickelophore property carry both Ni and Co at the root epidermal cells and at the xylem parenchyma transfer cells that pump Ni into the xylem. Whether the difference between Ni and Co metabolism occurs in the leaf cell membrane or in the vacuolar membrane Ni transporters is not yet known. Malik et al. (2000) discuss some other Co accumulating plant species that might be used for Co phytoextraction and which might be applied to ^{60}Co contamination remediation.

As with crop plants, soil properties may strongly influence plant growth. Phytomining plants must be fertilized and managed for optimum economic production of Ni and Co. Soil drainage may require adaptation of the farming technology. In the case of the poorly drained soils contaminated by Ni near the Port Colborne refinery, soils were very wet in spring. After snow melt, rainfall caused water to stand in the field. In early trials with *A. murale* at Port Colborne, some genotypes tested were too stressed by the long anaerobic soil period and their roots died with fungal diseases before they would have normally been harvested. In a different approach tested in the 2001–2002 crop season, improved surface drainage was installed and *A. murale* was planted on ridge-tilled plots so that a considerable part of the root system could be in aerobic soils at all times. With this management, plants survived the wet spring season and gave excellent yields and metal accumulation.

15.5 Ni Phytomining Development

Ni is one element for which the value of the plant ash as a metal ore should offset the costs of soil remediation and provide profit in environmental remediation of areas near point source stack emissions and similar high contamination of soils (Chaney and Mahoney 2014; van der Ent et al. 2015a). Further, if phytoextraction is conducted on arable sites with anthropogenic Ni enrichment, the fraction of total soil Ni that could be removed may be higher than for ancient serpentine soils where part of the Ni is occluded in the Fe oxides. To maintain high Ni phytoavailability during the phytoextraction period, growers would adjust soil pH and add appropriate fertilizers to obtain maximal yields. At the end of the remediation period, soil pH would be restored to the range for production of common farm crops, and the soils would have restored or improved fertility for normal crop production.

The CRADA team have reported development of a complete Ni phytoextraction technology (Angle et al. 2001; Chaney et al. 1999, 2007b, 2010; Li et al. 2003a,b). The cooperators selected Ni hyperaccumulator species (*A. murale* and *A. corsicum*) because of their great potential for developing a phytoextraction crop (strong hypernickelophore, tall to facilitate harvest, etc.), collected wild germplasm across southern Europe (Reeves and Baker 2000), and tested the genotypes under uniform soil conditions in the field in southwest Oregon on serpentine soil and at Port Colborne, Ontario, on Ni refinery-contaminated soil. Both *A. murale* and *A. corsicum* grew to about 1 m at flowering and could be mechanically harvested (Figure 15.7), allowed to air dry in the field for several days, and allowed to be "baled" using mechanical equipment. *A. murale* survived Canadian winters appreciably better than *A. corsicum*; genotypes from locations with hard winters were collected to support breeding improved cultivars of *A. murale*. Fertilizer requirements are not unusual because the plants were adapted to low-fertility serpentine soils (see above). Increased N fertilization increased biomass yield and increased

FIGURE 15.7
Mechanical harvesting of dried *Alyssum* biomass at the Vineyard site near Cave Junction, Oregon. Left: Mowing flowering *Alyssum*; Right: Baled dried *Alyssum* hay. (Photos from Richard Roseberg, Oregon State University.)

biomass Ni concentration. Because of the growth pattern of the *Alyssum* species with a winter vernalization causing rapid growth and flowering in the spring, N fertilizer application is split with 1/3 to 1/2 of the annual N applied with planting or transplanting, and the other 1/2 to 2/3 applied when growth begins again in the spring. N, Ca, K, B, and Mo fertilizers may be applied to the soil surface for effective fertilization, but P fertilizer is better incorporated in the rooting zone to aid root access to the P (larger volume of P-enriched soil), which will be sorbed to soil particles. Gypsum Ca is also likely better incorporated than surface applied in practice because gypsum dissolves only slowly. Table 15.2 clearly shows the strong response of *Alyssum* to P fertilizer on the serpentine farm Brockman gravelly clay loam soil.

15.6 Why Did Nickel Hyperaccumulators Evolve?

Many plant species evolved to tolerate the very low Ca and P fertility of serpentine soils (see Brady et al. 2005; Proctor and Woodell 1975). Fertilization with Ca and P removes the "serpentine factor" of plant ecology (Koide and Mooney 1987). As noted above, few serpentinophytes had to be tolerant of high levels of soluble Ni in soils (e.g., Proctor and Baker 1994); with the moderately high pH and high soil levels of Fe and Mn oxides, phytoavailable levels of Ni remain low. As noted by Chaney et al. (2009), shoots of serpentine adapted plants (serpentinophytes) growing on the Soldiers Delight Serpentine area near Baltimore, Maryland, contain low levels of leaf Ni (1–10 mg Ni kg^{-1} DW) despite the soil containing 2000 mg Ni^{-1}. Thus, among plants adapted to serpentine infertility, some species evolved Ni hyperaccumulation as a further benefit from adaptation to these soils.

The Ni hyperaccumulation trait is useless on soils without Ni mineralization or contamination.

Most evidence indicates that these plants evolved to hyperaccumulate Ni in order to protect themselves from chewing insects and disease organisms (Boyd 1998, 2007, 2009, 2012). On the other hand, increased Ni supply increased the susceptibility of *S. polygaloides* to turnip mosaic virus (Davis et al. 2001a). Davis et al. (2001b) compared the role of organic compounds in plants (which can limit insect feeding) and Ni accumulation on insect defense. Gonçalves et al. (2007) found that organic compounds contributed to the defense of several Ni hyperaccumulator species.

Some have expressed concern about risks to insects from metals accumulated in hyperaccumulator plant leaves (Peterson et al. 2003), but the evidence suggests that higher foliar Ni evolved in these species to give them an advantage when other nearby plants can tolerate serpentine infertility, but accumulating Ni from the soil provides defense benefit as well. The study by Coleman et al. (2005) indicates that defense against chewing insects could be obtained by concentrations of foliar Ni appreciably below the "hyperaccumulator" threshold levels, suggesting that this could support evolution of the trait. Research remains active in sorting out the role of hyperaccumulation in defense against insects and pathogens, and most research teams have seen strong defense against both during their research.

15.7 Soil Microbial Inoculation to Promote Phytomining

Another interesting finding regarding maximization of Ni accumulation by hypernickelophores from Ni mineralized ultramafic soils has been reported. Abu-Shanab et al. (2003) reported that when selected microbes collected from the rhizosphere of *A. murale* growing on a high-Ni soil were used to inoculate seedling roots and the plants were transplanted into both sterile and non-sterile serpentine soils, inoculated *A. murale* plants accumulated about 30% more Ni than without inoculation. Cabello-Conejo et al. (2013, 2014a,b) selected rhizobacteria that were tolerant of serpentine soils, produced growth-stimulating compounds, and/or increased soil Ni solubility and tested their effect on Ni phytoextraction by *Alyssum pintodasilvae* grown on serpentine soil and biosolids Ni-contaminated soil. Rhizobacterial species had different effects on the two soils, generally increasing the quantity of Ni in shoots (either yield or concentration was improved), which suggests additional research might increase annual Ni phytoextraction.

Mycorrhizae are soil fungi that form symbiotic relationships with roots of many plant species and increase the surface area of the root system to absorb nutrients that slowly diffuse from soil surfaces to membranes for uptake. Mycorrhizae provide phosphate and some other nutrients to most plants when soil phytoavailable levels are low. Because *Alyssum* species belong to the *Brassica* family, they do not form symbiotic relationships with mycorrhizae. However, other species of hypernickelophores in other plant families have been found to have mycorrhizae-infected roots in serpentine soils where the plants occur naturally. This was first reported for some New Caledonian species by Amir et al. (2007). The role of mycorrhizae in Ni hyperaccumulation has also been studied in *B. coddii*, a species that supports mycorrhizal infection (Orłowska et al. 2011). Testing has not revealed any specific role of mycorrhizae in Ni hyperaccumulation by *B. coddii*, but the much higher quantities of shoot Ni reported in field-collected *B. coddii* in South Africa than in test pots (Brooks et al. 2001; Li, Chaney, and Chen, unpublished work) or fields with non-native soils (about two- to fivefold lower Ni) suggest that mycorrhizae may be involved. The supply of phosphate by mycorrhizae may strongly affect yields of this species on the naturally low phytoavailable phosphate serpentine soils of South Africa and mycorrhizae will likely be important in phytomining management with non-*Brassica* species. Inoculation of rooting media requires a living root mass with the organisms, so selection of highly effective strains of mycorrhizae may be a key agronomic practice for tropical phytomining in addition to selection of the plant species and breeding improved cultivars.

Soil microbes that might stimulate growth of Ni-hyperaccumulating plants were different from microbial strains reported by Burd et al. (2000), which could reduce the susceptibility to Ni phytotoxicity of several plant species grown in acidic Ni-contaminated soils. Several reports on soil microbes that improved plant growth under Ni phytotoxic conditions have been published, but considering that a very small increase in pH would give an equal or a higher yield increase on Ni phytotoxic soils, it is hard to argue that field inoculation with metal-tolerant microbes would be of value in soil remediation or phytoextraction. Further management practices to increase annual Ni phytomining potential may be identified in the future if basic research is applied in this area of soil microbiology.

Questions about the effect of normal shoot biomass returning to the soil surface below hypernickelophore plants in the field were addressed by research. Several field tests indicated that surface soil Ni concentration is increased under hypernickelophore species (e.g., Ma et al. 2015). The question of whether this deposited Ni could achieve allelopathic control of less Ni-tolerant plant species was addressed by the research of Zhang et al. (2007). They mixed 20 t ha^{-1} dry *Alyssum* biomass rich in Ni into both an ultramafic Brockman soil and a low-Ni Maryland farm soil, the Christiana fine sandy loam, compared with equal Ni added from Ni salt, and tested germination and growth of common species and weeds. In the Christiana soil, both the Ni salt and *Alyssum* biomass caused Ni phytotoxicity in many of the test species. However, in the higher-pH, high-Fe Brockman soil, no adverse effects of the added Ni were observed. Serpentine soils are only Ni phytotoxic when soil pH drops to lower levels increasing soil Ni solubility (e.g., Crooke 1956).

15.8 Plant Growth Regulators and Phytomining

Another possible approach to improve phytomining yield with *Alyssum* hyperaccumulators is the application of plant growth regulators. Recently, Cassina et al. (2010) found that application of a commercial cytokinin product increased the yield of *A. murale* growing on a serpentine soil, while not reducing the Ni concentration in the plants. Thus, application of the plant growth regulator raised Ni yield by as much as 75% above the control plants. Cabello-Conejo et al. (2013, 2014a,b) later reported finding no significant effect of growth hormones on yield or Ni concentration in a longer growth period test on serpentine soil. In a subsequent paper, Cabello-Conejo et al. (2014b) reported that a different combination of growth-promoting compounds did improve yields but not Ni concentration, suggesting that additional research might identify growth-promoting compounds that could increase annual Ni yield more than the cost of using the treatment.

15.9 Potential Need for Sterile Phytomining Genotypes

One issue raised with growing *Alyssum* Ni hyperaccumulators in Oregon was the danger of spreading the non-native species into the general ecosystem. We had shown that if plants were harvested by mid-flowering, there were no viable seeds formed. However, after the CRADA ended, local Viridian staff failed to harvest all flowering *Alyssum* plants in time, and seed did escape to nearby public lands (Strawn 2013). Oregon regulators have since declared *Alyssum* a noxious weed in Oregon.

One way to prevent this problem is to develop sterile genotypes using radiation or other mutation technologies to select individuals that do not form viable seeds. The development of sterile genotypes would start with cultivars that had high yield and Ni accumulation ability. Reproduction of sterile valuable plants would use vegetative propagation or cell culture methods.

15.10 Hyperaccumulator Biomass Processing

Ni hypernickelphore plants achieve remarkable biopurification of soil Ni during their life cycle. Serpentine soils contain much higher levels of Fe than Ni; on the Vineyard Brockman variant soil with 4 g Ni kg^{-1} and 220 g Fe kg^{-1}, the plants contained 15 g Ni kg^{-1} with only 0.15 g Fe kg^{-1}, so the ratio of Ni:Fe rose from that of the soil (0.018) to that of the hypernickelophore plant shoots (100), or an increase of 550. This biopurification of soil/ore Ni is important to the value of Ni in phytomining biomass because much energy is expended separating Ni from Fe in traditional metallurgy.

Several approaches have been tested for recovery of Ni from the phytomining plant biomass: incineration, pyrolysis, and direct acid extraction. Recovery of the biopurified Ni can use different metallurgical approaches. In the CRADA research, dried *Alyssum* biomass "hay" (Figure 15.7) was burned and the ash was processed. Burning could be conducted at a biomass generator that could provide energy value equal to at least the cost of growing the crop. The ash is a high-grade Ni ore. Because of the high purity of hypernickelophore biomass Ni compared to the ores and soils where they can be grown, many processing methods can be used to recover the Ni for markets. Both pyrometallurgical methods and hydrometallurgical methods might be used.

Inco Ltd. and Viridian LLC conducted a test of recovery of Ni from Ni-phytomining biomass ash at the Sudbury, Ontario, "Copper Cliffs" smelter of Inco. A large amount of *Alyssum* ash (500 kg) was processed in an electric arc furnace (pyrometallurgical) quite successfully according to Inco. Because Ni has value for industrial use, and plant ash is richer in Ni than alternative ores commonly used in Ni production, the plant ash should become an article of commerce. After all, ores from New Caledonia with only 2.5% Ni were shipped around the world. Plant ash contains 10%–20% Ni before removal of carbonates, and this can give washed acidified ore residue with even higher Ni concentration if carbonate is removed by treatment with acid or high temperature.

Others tested pyrolysis of biomass (metal salt amended biomass, or bioengineered hairy root *Alyssum* biomass grown with soluble salt Ni), which can remove most of the carbon (Boominathan et al. 2004; Koppolu et al. 2003). It seems possible that unique Ni catalysts for the organic chemical synthesis industry might be produced from such pyrolysis char, but few studies on high-value catalyst production have been reported (Lerch et al. 2010). Several low-value catalysts have been prepared from Ni-rich biomass, but used the Ni^{2+} ions as the catalyst (Losfeld et al. 2014a). Although the market for Ni in commercial metal salts such as $(NH_4)_2Ni(SO_4)_2 \cdot 6H_2O$ may exist, and the production from biomass has been demonstrated (Barbaroux et al. 2012; Zhang et al. 2014, 2016), the cost-effectiveness has not been established.

Standard hydrometallurgical methods should be successful with biomass ash (e.g., van der Ent et al. 2015a; Zhang et al. 2014, 2016; Vaughan et al. 2017). The water-soluble materials in the ash could be removed, and even the mild acid-soluble and carbonate materials could be removed, which might recover many of the nutrients in the plant ash. The Ni could then be processed similarly to high-pressure acid leaching used with laterite ores by industry. The Ni could be dissolved and re-precipitated as a purification step, or organic solvents could be used to extract chelates of Ni, which are then stripped back into aqueous solution to prepare pure Ni products.

Another potential use of Ni-rich *Alyssum* biomass is as an "organic" Ni fertilizer. Production of "organic" crops limits the kinds of fertilizers one may apply. Deficiency of Ni in field-grown pecan [*Carya illinoinensis* (Wangenh.) K. Koch] trees growing in Georgia, USA, was demonstrated by Wood et al. (2006b). We tested using water extracts of *Alyssum* Ni-rich biomass compared with NiSO$_4$ as a foliar Ni fertilizer and found that the extract of biomass was an excellent Ni fertilizer (Wood et al. 2006a). Alternatively, ground *Alyssum* biomass could be applied to soil to provide "organic" Ni fertilizer to prevent Ni deficiency rather than spraying leaves after symptoms appear.

15.11 Ni Physiology and Chemistry of Hypernickelophores

One of the most unexpected aspects found during development of this technology was the effect of soil pH on Ni hyperaccumulation by *Alyssum* species. As shown in Figure 15.4, soil solution Ni falls with increasing soil pH. Accumulation of Ni by crop plants always falls as pH is increased (Kukier and Chaney 2004). However, *Alyssum* species had increased plant Ni as soil pH was raised (Kukier et al. 2004; Li et al. 2003a). This pattern was observed with most serpentine soils studied (Figure 15.5), and with all of the Port Colborne contaminated soils examined (Figure 15.6). Further, it was confirmed by nutrient solution tests in which solution pH was buffered at four levels from 5.5 to 7.5 (Figure 15.8) and Ni^{2+} activity was maintained by using FeHBED as the Fe source (based on Geochem-PC calculations) (Peters et al. 2000).

Another agronomic aspect of Ni phytomining is the potential for mixed culture versus pure culture of Ni hypernickelophore species. Especially in southern Europe and northwestern United States, Mediterranean climate limits water availability for crop growth where serpentine soils occur. The presence of weeds interferes with the supply of soil water and nutrients for the phytomining crop, so weed control has been shown to be important in both the United States and Albanian field tests (Bani et al. 2015a; Li et al 2003b). Others have evaluated the use of native serpentine legumes to supply fixed N, which would support *Alyssum* production (Durand et al. 2016); although the legumes could provide the required N fertilizer, they would use soil water and nutrients and inhibit *Alyssum* yields compared to use of herbicides that control both grasses and legumes in *Brassica* crops. Co-crops or weeds would reduce the concentration of Ni in ash of harvested biomass of mixed species, increasing costs of Ni recovery from the biomass. In other tests, the potential for co-culture with a grass was examined by Broadhurst

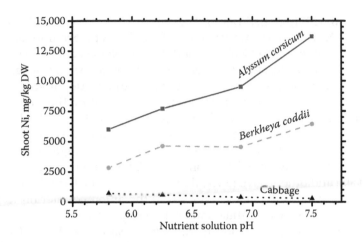

FIGURE 15.8
Effect of nutrient solution pH on plant uptake of Ni from pH-buffered serpentine nutrient solution using FeHBED, which prevents interaction of added Ni with Fe chelation. (From PhD thesis of C.A. Peters.) Both log and linear scales for shoot Ni are shown for perspective.

and Chaney (2016). Grasses secrete phytosiderophores (nonspecific chelating agents that chelate Ni, Zn, Cu, etc. in addition to Fe^{3+}) to dissolve soil Fe, which is the form absorbed by grasses. Growing *Alyssum* as monoculture vs. mixed root co-cultures showed no benefit for co-culture, only reduced yield of the hypernickelophore.

A number of groups are conducting research to better understand how hyperaccumulator plants achieve this remarkable accumulation from soils compared to other plant species. In general, no ligand has been found to be secreted by the roots to solubilize Ni. Rather, it appears that the transport protein that accumulates Ni into epidermal root cells is very selective and very fast for uptake of Ni^{2+} and Co^{2+}. Chelation that lowers Ni^{2+} activity in solution reduces Ni uptake proportionally; chelator-induced phytoextraction is counterproductive for Ni. Although addition of EDTA to soils promoted "induced phytoextraction" of Pb by *Brassica juncea* (Blaylock et al. 1997), addition of EDTA to soil actually reduced Ni accumulation by *A. murale* (Li, Chen et al., unpublished work) while Robinson et al. (1999a,b) found that added EDTA reduced Ni accumulation in *B. coddii* [addition of EDTA was never cost-effective and caused leaching of metal to groundwater such that it is not allowed in the United States or the European Union (Chaney et al. 2010)]. Another transporter is believed to pump Ni from xylem parenchyma cells into xylem fluid. Leaf cells have both the transport protein to accumulate Ni into the cells and another to pump cytoplasmic Ni into vacuoles. It appears that guard cells of *Thlaspi montanum* var. *siskiyouense* (Heath et al. 1998) may actively excrete Ni to protect their function. Epidermal cells on top and bottom of leaves accumulated Ni before mesophyll cells, and both these types could accumulate high concentrations of Ni in vacuoles (Broadhurst et al. 2004a,b).

Persans et al. (2001) cloned one of these proteins (*TgMTP1*) believed to be involved in Ni transport into vacuoles of the Ni hyperaccumulator *Thlaspi goesingense*. Krämer et al. (2000) reported that the Ni in vacuoles was chelated with organic acids. Although Kramer et al. initially reported that leaf Ni was within the trichome rays on the leaf surface, an additional study showed that the trichome-associated Ni was in the vacuole within the base of the trichome and not present in the trichome ray (Broadhurst et al. 2004b; Tappero et al. 2007). Although others have suggested that shoot Ni is associated with organic acids, especially malate, research shows little change in shoot malate in relation to shoot Ni levels from control to Ni phytotoxic levels. Broadhurst et al. (2004a) and others have noted that sulfur is associated with high levels of Ni in epidermal cells; this suggests that the anion counterion to Ni^{2+} in epidermal cell vacuoles is sulfate rather than organic acids.

In a study of Ni translocation in *Alyssum lesbiacum*, Krämer et al. (1996) reported that when this species was translocating Ni, the histidine level in the xylem exudate was substantially increased. Careful examination of the data shows that Ni was in molar excess of histidine, and chemical modeling indicated that glutamine, citrate, and malate also chelated Ni. A subsequent study by Kerkeb and Krämer (2003) showed that exposure to Ni over time resulted in the decline of the histidine concentration in xylem exudate compared to the Ni concentration, so that an even lower portion of the xylem Ni could be chelated with histidine. In recent studies by Centofanti et al. (2013), several *Alyssum* species (*A. murale*, *A. corsicum*, *A. montanum*) grown with steady-state Ni supply to achieve 1.5% Ni in shoots were examined for evidence of a role of histidine in Ni translocation. Although histidine was somewhat increased by Ni supply, it seldom exceeded 50 µM when exudate Ni was 3000 µM, while citrate, malate, other amino acids, and even nicotianamine remained at low concentrations (<300 µM), far lower than the Ni present.

Others have used several techniques to try to identify the chemical forms of Ni present in plant tissues and in xylem and phloem fluids. One approach was X-ray spectroscopy (extended X-ray absorption fine structure and X-ray absorption near edge structure), which can identify specific chemical forms of elements in samples. Application of these techniques to living plant samples is complex and interpretation may still be confused at this stage of the development of the techniques. Montargès-Pelletier et al. (2008) examined Ni speciation in *Leptoplax emarginata* and found that citrate and malate were the dominant ligands with Ni, while no Ni–histidine complex was observed. Sagner et al. (1998) examined the forms of Ni in the phloem sap of *Sebertia acuminata*, a Ni hypernickelophore tree from New Caledonia. They showed that the only chelating agent in the phloem sap was citrate and used NMR to verify the chelation with citrate. Only about 40% of the Ni could be chelated by the citrate present, and the rest was assumed to be free ionic Ni^{2+}. Subsequently, Schaumloffel et al. (2003) used new column methods to separate the forms of Ni and measured the organic compounds in the Ni peaks using electrospray MS-MS. This work showed that 99.4% of the Ni present in phloem latex of *S. acuminata* was Ni-citrate, while 0.3% was chelated with nicotianamine, a natural non-protein amino acid in all plants. It was believed that the more stable Ni(citrate)$_2$ form of the Ni chelate was present. They also looked in extracts of *Thlaspi caerulescens* tissues especially for nicotianamine chelation of Ni because it forms a strong chelate with Ni, and found a small part of the Ni was chelated with nicotianamine (Ouerdane et al. 2006). The work by Centofanti et al. (2013) clearly showed that xylem exudate of steady-state Ni supply grown *Alyssum* genotypes translocating high amounts of Ni had increased levels of several chelating natural organic and amino acids present in cells, but under full hyperaccumulation conditions, none of the organic or amino acids, or nicotianamine, could quantitatively chelate all the Ni present. The chemical forms of Ni in the plant tissues and in the vacuoles of epidermal cells that accumulate such high levels of Ni have not been identified. Considering the low pH of vacuoles, it seems likely that Ni would hardly be chelated even if substantial levels of malate were present as suggested by early authors.

Other patterns of Ni localization in hypernickelophore tissues have also been reported. Although *Alyssum* and *Thlaspi* species have shown accumulation of Ni in vacuoles of epidermal cells, other species (e.g., *Phyllanthus balgooyi*) show accumulation of Ni in lactifers so that their phloem sap is green (Mesjasz-Przybylowicz et al. 2016).

15.12 Which Ni Chemical Species in Soils Are the Source of Hypernickelophore Absorbed Ni?

Several groups have worked to identify the chemical forms of Ni that are accessed by hyperaccumulator plants. The Morel et al. team looked at the chemical speciation of Ni in serpentine soils and the chemical extractability of different forms of soil Ni (Bani et al. 2014; Massoura et al. 2004, 2006; Pinel et al. 2003). As noted above, they found that these plants accumulated Ni from the same "labile" pools that were the source of Ni accumulated by crop plants (Shallari et al. 2001).

In a different approach, Centofanti et al. (2012) prepared different Ni compounds and added them to serpentine nutrient solutions and grew *A. corsicum* to evaluate the availability of Ni in the compounds. Some compounds dissolved rapidly enough that considerable Ni was dissolved, while others did not dissolve but still released Ni to the plant roots fast enough to support hyperaccumulation. Only one compound, NiO, was very unbioavailable due to the very slow kinetics of NiO dissolution (see also Fellet et al. 2009).

Several recent studies examined the evidence for convection versus diffusion control of Ni uptake by hypernickelophores, and the role of soil volume and access to serpentine subsoil on Ni phytomining. Chaney et al. (2019)

noted that *A. murale* grown on Brockman variant gravelly loam soil (4600 mg Ni kg^{-1}) accumulated >15 g Ni kg^{-1} dry shoots from a soil with a soil saturation extract Ni of 0.047 µg mL^{-1}. Plants commonly evapotranspire about 250 mL of water per 1 g of shoot dry matter produced. The 250 mL would contain only 11.8 µg Ni, but the plant shoots contained 15,000 µg Ni g^{-1}; thus, convection could supply only 0.08% of the Ni actually accumulated. Hence, diffusion to root surfaces of Ni in or on soil particles is the major source of Ni accumulated by *Alyssum* species hypernickelophores.

The implications of the importance of diffusion are many. One is that the volume of soil available to roots will strongly affect the amount of Ni accumulated; this was confirmed by Chaney et al. (2019) who found that soil or pot volume should be at least 2.24 kg of serpentine soil for the result to reasonably estimate potential Ni hyperaccumulation and that higher soil volumes gave even higher Ni yield. Chaney et al. (2019), Nkrumah et al. (2017), and Paul et al. (2019) also tested the role of subsoil access and found that root access to real serpentine subsoil increased the amount of Ni in *Alyssum* shoots significantly. This observation has implications for phytomining of Ni from industrially contaminated soils where the Ni is only mixed into the tillage depth compared with serpentine soil where Ni concentration increases with depth. Our Port Colborne field tests showed poorer Ni hyperaccumulation in field-grown *Alyssum* species than in plants grown in large pots of the tillage depth of the same contaminated soil. These soils contained some NiO, a kinetically inert Ni species that has persisted long after aerosol contamination of the soils (McNear et al. 2007), while most of the Ni was the Ni-LDH and organic matter or Fe oxide adsorbed Ni.

The actual chemical species of soil Ni that support diffusion of Ni to the root surface remain unclear. The Centofanti et al. (2012) study noted above showed that Ni species that have low enough solubility to prevent phytotoxicity (such as the Ni-Al layered double hydroxide) are readily phytoavailable to *Alyssum* roots, likely because of the dense lateral roots and root hairs of *Alyssum* hypernickelophores. With diffusion control of most Ni uptake, longer root length can occur in larger soil volumes. Perhaps the most important lesson of these studies is that growing hypernickelophores in small pots can strongly underestimate their potential for hyperaccumulation.

15.13 Potential Risks of Ni Hyperaccumulator Biomass to Wildlife

One concern expressed about growing hyperaccumulator plants commercially is the potential risk to animals, which could consume the plants in the field. Our research and observations contradict this possibility. In the case of Ni-rich *A. murale*, goats, sheep, and cattle have been observed to avoid eating these plants by choice in large areas of serpentine soils used as pastures in Albania, Turkey, Greece, and so on. During our field studies in Oregon, deer and rabbits grazed on weeds or crop plants and grasses around the *Alyssum* plots, but did not eat the *Alyssum*. It appears that *Alyssum* shoots are unpalatable, perhaps due to the wiry texture of leaves densely covered with trichomes (e.g., Broadhurst et al. 2004a), which substantially reduces the risk of harm to wildlife. Further, their seeds have little feed value because they are so small. As commercial phytomining further develops, testing should be conducted to understand why animals do not eat the plants, and whether they might be harmed if wildlife had nothing else to eat.

In order to test the potential toxicity of Ni to livestock, several classes of livestock and common laboratory test animals have been fed Ni salt amended diets. Ruminants were considerably more sensitive to Ni salts added to their diets than were monogastric animals, so risk assessment is commonly based on toxicity to ruminants (cattle, sheep). However, toxicity to cattle is complicated. The initial tests added $NiCl_2 \cdot 6H_2O$ to the diets and the livestock started to lose weight at 100 mg Ni kg^{-1} diet (O'Dell et al. 1970a). Testing showed that this resulted from the animals not consuming the diet (O'Dell et al. 1970b), rather than from toxicity of absorbed Ni within the animal. Hence, they tested other forms of Ni, including $NiCO_3$, which caused no change in gain rate or milk production of cows fed as high as 250 mg Ni kg^{-1} diet (O'Dell et al. 1971).

These results were the basis for concluding that Ni soil does not violate the "soil–plant barrier" that protects animals from soil metals (Chaney 1983b). If livestock and wildlife can consume chronic lifetime diets with higher Ni concentration than is accumulated in plant shoots when common plants suffer Ni phytotoxicity (<100 mg kg^{-1}), then animals are not at risk from soil Ni accumulation in plants. Some have questioned the validity of the conclusion that the soil–plant barrier protects animals from excessive soil Ni based on a species that accumulated higher than 100 mg Ni kg^{-1} DW when the plant was grown in nutrient solutions (Kopittke et al. 2008, 2010). The report that high levels (250 mg Ni kg^{-1}) of $NiCO_3$ could be added to feeds and not interfere with feed intake or affect life processes of cattle indicates that the limit of 50 or 100 mg Ni kg^{-1} diet recommended by the National Research Council (NRC) committee (NRC 1980) is very conservative. As noted above, monogastric animals tolerate much higher levels of Ni and the levels of Ni in feed grains are far below levels needed to cause toxicity (e.g., Alexander et al. 1979); this was also supported by the study of feeding biosolid-grown vegetables by Boyd et al. (1982) and Swiss chard grown on soils amended with several types of biosolids (Chaney et al. 1978). It is likely that if Ni were incorporated into the crops during growth, Ni would have considerably less potential to cause toxicity because it would enter the animal in chelated forms rather than soluble salts. Unfortunately, no feeding tests of crops with high levels of intrinsic Ni have been reported other than Alexander et al. (1979) where Ni level was not high enough to test the question relevant to wildlife risks from hypernickelophores.

15.14 Conclusions

Great progress has been made in the development of commercial Ni phytomining technologies using hypernickelophore species adapted to local climates. Serpentine soils are extensive on earth (e.g., Nkrumah et al. 2017) and are very poor soils for crop production. However, phytomining could be conducted by local farmers using improved cultivars of hypernickelophores bred for commercial use. All agronomic management practices needed for *Alyssum* species production have been demonstrated for Mediterranean climate areas, and for *Phyllanthus* for Malaysian and possibly other tropical serpentine soils. Even at the present value of Ni metal, Ni phytomining would be profitable. Although phytomining failed in Oregon, the French–Albanian team is working with local farmers to crop serpentine soils and market the biomass-derived Ni. Only time will tell if phytomining becomes fully commercialized. Present evidence suggests it should succeed.

References

Abu-Shanab, R.A., J.S. Angle, T.A. Delorme, R.L. Chaney, P. van Berkum, H. Moawad, K. Ghanem and H.A. Ghozlan. 2003. Rhizobacterial effects on nickel extraction from soil and uptake by *Alyssum murale*. *New Phytol.* 158:219–224.

Alexander, J., R. Koshut, R. Keefer, R. Singh, D.J. Horvath and R.L. Chaney. 1979. Movement of nickel from sewage sludge into soil, soybeans, and voles. *Trace Subst. Environ. Health* 12:377–388.

Amir, H., N. Perrier, F. Rigault and T. Jaffré. 2007. Relationships between Ni-hyperaccumulation and mycorrhizal status of different endemic plant species from New Caledonian ultramafic soils. *Plant Soil* 293:23–35.

Angle, J.S., R.L. Chaney, R.A. Abou-Shanab and P. Van Berkum, 2007. Bacterial effects on metal accumulation by plants. U.S. Patent 7,214,516.

Angle, J.S., R.L. Chaney, A.J.M. Baker, Y. Li, R. Reeves, V. Volk, R. Roseberg, E. Brewer, S. Burke and J. P. Nelkin. 2001. Developing commercial phytoextraction technologies: Practical considerations. *S. Afr. J. Sci.* 97:619–623.

Baker, A.J.M. and R.R. Brooks. 1989. Terrestrial higher plants which hyperaccumulate metal elements—A review of their distribution, ecology, and phytochemistry. *Biorecovery* 1:81–126.

Bani, A., G. Echevarria, E. Montargès-Pelletier, F. Gjoka, S. Sulçe and J.L. Morel. 2014. Pedogenesis and nickel biogeochemistry in a typical Albanian ultramafic toposequence. *Environ. Monit. Assess.* 186:4431–4442.

Bani, A., G. Echevarria, S. Sulçe and J.L. Morel. 2015a. Improving the agronomy of *Alyssum murale* for extensive phytomining: A five-year field study. *Int. J. Phytoremed.* 17:117–127.

Bani, A., G. Echevarria, X. Zhang, B. Laubie, J.L. Morel and M.-O. Simonnot. 2015b. The effect of plant density in nickel phytomining field experiments with *Alyssum murale* in Albania. *Aust. J. Bot.* 63:72–77.

Barbaroux, R., E. Plasari, G. Mercier, M.O. Simonnot, J.L. Morel and J.F. Blais. 2012. New process for nickel ammonium disulfate production from ash of the hyperaccumulating plant *Alyssum murale*. *Sci. Total Environ.* 423:111–119.

Blaylock, M.J., D.E. Salt, S. Dushenkov, O. Zakharova, C. Gussman, Y. Kapulnik, B.D. Ensley and I. Raskin. 1997. Enhanced accumulation of Pb in Indian mustard by soil-applied chelating agents. *Environ. Sci. Technol.* 31:860–865.

Boominathan, R., N.M. Saha-Chaudhury, V. Sahajwalla and P.M. Doran. 2004. Production of nickel bio-ore from hyperaccumulator plant biomass: Applications in phytomining. *Biotechnol. Bioeng.* 86:243–250.

Boyd, J.N., G.S. Stoewsand, J.G. Babish, J.N. Telford and D.J. Lisk. 1982. Safety evaluation of vegetables cultured on municipal sewage sludge-amended soil. *Arch. Environ. Contam. Toxicol.* 11:399–405.

Boyd, R.S. 1998. Hyperaccumulation as a plant defensive strategy. pp. 181–201. *In* R.R. Brooks (Ed.) *Plants That Hyperaccumulate Heavy Metals*. CAB Intern., Wallingford, UK.

Boyd, R.S. 2007. The defense hypothesis of elemental hyperaccumulation: Status, challenges and new directions. *Plant Soil* 293:153–176.

Boyd, R.S. 2009. High-nickel insects and nickel hyperaccumulator plants: A review. *Insect Sci.* 16:19–31.

Boyd, R.S. 2012. Plant defense using toxic inorganic ions: Conceptual models of the defensive enhancement and joint effects hypotheses. *Plant Sci.* 195:88–95.

Brady, K.U., A.R. Kruckeberg and H.D. Bradshaw, Jr. 2005. Evolutionary ecology of plant adaptation to serpentine soils. *Annu. Rev. Ecol. Evolut. Syst.* 36:243–266.

Broadhurst, C.L. and R.L. Chaney. 2016. Growth and metal accumulation of an *Alyssum murale* nickel hyperaccumulator ecotype co-cropped with *alyssum montanum* and perennial ryegrass in serpentine soil. *Front. Plant Sci.* 7:451 (doi.org/10.3389/fpls.2016.00451).

Broadhurst, C.L., R.L. Chaney, J.S. Angle, E.F. Erbe and T.K. Maugel. 2004a. Nickel localization and response to increasing Ni soil levels in leaves of the Ni hyperaccumulator *Alyssum murale* 'Kotodesh'. *Plant Soil* 265:225–242.

Broadhurst, C.L., R.L. Chaney, J.S. Angle, T.K. Maugel, E.F. Erbe and C.A. Murphy. 2004b. Simultaneous hyperaccumulation of nickel, manganese and calcium in *Alyssum* leaf trichomes. *Environ. Sci. Technol.* 38:5797–5802.

Brooks, R.R. 1987. *Serpentine and Its Vegetation: A Multidisciplinary Approach*. Dioscorides Press, Portland, OR.

Brooks, R.R., B.H. Robinson, A.W. Howes and A. Chiarucci. 2001. An evaluation of *Berkheya coddii* Roessler and *Alyssum bertolonii* Desv. for phytoremediation and phytomining of nickel. *S. Afr. J. Sci.* 97:558–560.

Burd, G.I., G.D. Dixon and B.R. Glick. 2000. Plant growth-promoting bacteria that decrease heavy metal toxicity in plants. *Can. J. Microbiol.* 46:237–245.

Cabello-Conejo, M.I., C. Becerra-Castro, A. Prieto-Fernández, C. Monterroso, A. Saavedra-Ferro, M. Mench and P.S. Kidd. 2014a. Rhizobacterial inoculants can improve nickel phytoextraction by the hyperaccumulator *Alyssum pintodasilvae*. *Plant Soil* 379:35–50.

Cabello-Conejo, M.I., T. Centofanti, P.S. Kidd, A. Prieto-Fernandez and R.L. Chaney. 2013. Evaluation of plant growth regulators to increase Ni phytoextraction by *Alyssum* species. *Int. J. Phytorem.* 15:365–375.

Cabello-Conejo, M.I., A. Prieto-Fernandez and P.S. Kidd. 2014b. Exogenous treatments with phytohormones can improve growth and nickel yield of hyperaccumulating plants. *Sci. Total Environ.* 494:1–8.

Cassina, L., E. Tassi, E. Morelli, L. Giorgetti, D. Remorini, R. Massai, R.L. Chaney and M. Barbafieri. 2010. Exogenous cytokinin treatments of a Ni hyper-accumulator, *Alyssum murale*, grown in a serpentine soil: Implications for phytoextraction. *Int. J. Phytoremed.* 13:(S1):90–101.

Centofanti, T., M.G. Siebecker, R.L. Chaney, A.P. Davis and D.L. Sparks. 2012. Hyperaccumulation of nickel by *Alyssum* species is related to solubility of Ni mineral species. *Plant Soil* 359:71–83.

Centofanti, T., Z. Sayers, M.I. Cabello-Conejo, P. Kidd, N.K. Nishizawa, Y. Kakei, A.P. Davis, R.C. Sicher and R.L. Chaney. 2013. Xylem exudate composition and root-to-shoot nickel translocation in *Alyssum* species. *Plant Soil* 373:59–75.

Chaney, R.L. 1983a. Plant uptake of inorganic waste constituents. pp. 50–76. *In* J.F. Parr, P.B. Marsh and J.M. Kla (Eds.) *Land Treatment of Hazardous Wastes.* Noyes Data Corp., Park Ridge, NJ.

Chaney, R.L. 1983b. Potential effects of waste constituents on the food chain. pp. 152–240. *In* J.F. Parr, P.B. Marsh and J.M. Kla (Eds.) Land Treatment of Hazardous Wastes. Noyes Data Corp., Park Ridge, NJ.

Chaney, R.L. and I.A. Baklanov. 2017. Phytoremediation and Phytomining: Status and Promise. *Adv. Botan. Res.* 83:189–221.

Chaney, R.L., J.S. Angle, A.J.M. Baker and Y.-M. Li. 1998. Method for phytomining of nickel, cobalt and other metals from soil. U.S. Patent 5,711,784 (issued Jan. 27, 1998).

Chaney, R.L., J.S. Angle, A.J.M. Baker and Y.-M. Li. 1999. Method for phytomining of nickel, cobalt and other metals from soil. U.S. Patent No. 5,944,872 (issued Aug. 31, 1999).

Chaney, R.L., J.S. Angle and Y.-M. Li. 2004. Method for phytomining of nickel, cobalt and other metals from soil. U.S. Patent No. 6,786,948. 22 pp.

Chaney, R.L., J.S. Angle, Y.-M. Li and A.J.M. Baker. 2007a. Recovering metals from soil. U.S. Patent 7,268,273 (issued September 11, 2007).

Chaney, R.L., J.S. Angle, C.L. Broadhurst, C.A. Peters, R.V. Tappero and D.L. Sparks. 2007b. Improved understanding of hyperaccumulation yields commercial phytoextraction and phytomining technologies. *J. Environ. Qual.* 36:1429–1443.

Chaney, R.L., I.A. Baklanov, T.C. Ryan and A.P. Davis. 2019. Effect of soil volume on Ni hyperaccumulation from serpentine soil by *Alyssum corsicum. Int. J. Phytoremed.* In press.

Chaney, R.L., C.L. Broadhurst and T. Centofanti. 2010. Phytoremediation of soil trace elements. Chapter 17. pp. 311–352. *In* P. Hooda (Ed.) *Trace Elements in Soils.* Blackwell Publ., Oxford, UK.

Chaney, R.L., G. Fellet, R. Torres, T. Centofanti, C.E. Green and L. Marchiol. 2009. Using chelator-buffered nutrient solutions to limit Ni phytoavailability to the Ni-hyperaccumulator *Alyssum murale. Northeast. Natural.* 16(Special Issue 5): 215–222.

Chaney, R.L., Y.-M. Li, J.S. Angle, A.J.M. Baker, R.D. Reeves, S.L. Brown, F.A. Homer, M. Malik and M. Chin. 2000. Improving metal hyperaccumulator wild plants to develop commercial phytoextraction systems: Approaches and progress. pp. 131–160. *In* N Terry and G.S. Banuelos (Eds.) *Phytoremediation of Contaminated Soil and Water.* CRC Press, Boca Raton, FL.

Chaney, R.L. and M. Mahoney. 2014. Phytostabilization and phytomining: Principles and successes. Paper 104.Proc. Life of Mines Conf. (July 15–17, 2014; Brisbane Australia). Aust. Inst. Mining Metallurgy, Brisbane, Australia.

Chaney, R.L., G.S. Stoewsand, A.K. Furr, C.A. Bache and D.J. Lisk. 1978. Elemental content of tissues of Guinea pigs fed Swiss chard grown on municipal sewage sludge-amended soil. *J. Agr. Food Chem.* 26:944–997.

Chaney, R.L., R.D. Reeves, I.A. Baklanov, T. Centofanti, C.L. Broadhurst, A.J.M. Baker, J.S. Angle, A. van der Ent and R.J. Roseberg. 2014. Phytoremediation and phytomining: Using plants to remediate contaminated or mineralized environments. pp. 365–391. *In* R. Rajakaruna, R.S. Boyd and T. Harris (Eds.) *Plant Ecology and Evolution in Harsh Environments.* Nova Science Publishers, NY.

Coleman, C.M., R.S. Boyd and M.D. Eubanks. 2005. Extending the elemental defense hypothesis: Dietary metal concentrations below hyperaccumulator levels could harm herbivores. *J. Chem. Ecol.* 31:1669–1681.

Crooke, W.M. 1956. Effect of soil reaction on uptake of nickel from a serpentine soil. *Soil Sci.* 81:269–276.

Davis, M.A., J.F. Murphy and R.S. Boyd. 2001a. Nickel increases susceptibility of a nickel hyperaccumulator to turnip mosaic virus. *J. Environ. Qual.* 30:85–90.

Davis, M.A., S.G. Pritchard, R.S. Boyd and S.A. Prior. 2001b. Developmental and induced responses of nickel-based and organic defences of the nickel-hyperaccumulating shrub, *Psychotria douarrei. New Phytol.* 150:49–58.

Denys, S., G. Echevarria, E. Leclerc-Cessac, S. Massoura and J.L. Morel. 2002. Assessment of plant uptake of radioactive nickel from soils. *J. Environ. Radioact.* 62:195–205.

Dudley, T.R. 1966. Ornamental madworts (*Alyssum*) and the correct name of the goldentuft alyssum. *Arnoldia* 26:33–45.

Durand, A., S. Piutti, M. Rue, J.L. Morel, G. Echevarria and E. Benizri. 2016. Improving nickel phytoextraction by co-cropping hyperaccumulator plants inoculated by plant growth promoting rhizobacteria. *Plant Soil* 399:179–192.

Fellet, G., T. Centofanti, R.L. Chaney and C.E. Green. 2009. NiO(s) (bunsenite) is not available to *Alyssum* species. *Plant Soil* 319:219–223.

Gabbrielli, R., T. Pandolfini and B. Pucci. 1995. Physiological role of root surface phosphatases in adaptation strategies of *Alyssum bertolonii* Desv. to serpentine edaphic conditions. *Phyton* 35:187–197.

Gonçalves, M.T., S.C. Gonçalves, A. Portugal, S. Silva, J.P. Sousa and H. Freitas. 2007. Effects of nickel hyperaccumulation in *Alyssum pintodasilvae* on model arthropods representatives of two trophic levels. *Plant Soil* 293:177–188.

Heath, S.M., D. Southworth and J.A. D'Allura. 1998. Localization of nickel in epidermal subsidiary cells of leaves of *Thlaspi montanum* var. *siskiyouense* (Brassicaceae) using energy-dispersive X-ray microanalysis. *Int. J. Plant Sci.* 158:184–188.

Hunter, J.G. and O. Vergnano. 1952. Nickel toxicity in plants. *Ann. Appl. Biol.* 39:279–284.

Jaffré, T., Y. Pillon, S. Thomine and S. Merlot. 2013. The metal hyperaccumulators from New Caledonia can broaden our understanding of nickel accumulation in plants. *Front. Plant Sci.* 4:279. doi: 10.3389/fpls.2013.00279.

Jaffré, T. and M. Schmid. 1974. Accumulation du nickel par une Rubiacée de Nouvelle Calédonia: *Psychotria douarrei* (G. Beauvisage) Däniker (in French). *Compt. Rendus Acad. Sci. Paris* 278:D1727–1730.

Kerkeb, L. and U. Krämer. 2003. The role of free histidine in xylem loading of nickel in *Alyssum lesbiacum* and *Brassica juncea*. *Plant Physiol.* 131:716–724.

Koide, R.T. and H.A. Mooney. 1987. Revegetation of serpentine substrates: Response to phosphate application. *Environ. Manag.* 11:563–567.

Kopittke, P.M., C.J. Asher, F.P.C. Blamey and N.W. Menzies. 2008. Tolerance of two perennial grasses to toxic levels of Ni^{2+}. *Environ. Chem.* 5:426–434.

Kopittke, P.M., F.P.C. Blamey, R.A. Kopittke, C.J. Asher and N.W. Menzies. 2010. Tolerance of seven perennial grasses to high nickel in sand culture. *Environ. Chem.* 7:279–286.

Koppolu, L., F.A. Agblevor and L.D. Clements. 2003. Pyrolysis as a technique for separating heavy metals from hyperaccumulators. Part II: Lab-scale pyrolysis of synthetic hyperaccumulator biomass. *Biomass Bioenergy* 25:651–663.

Krämer, U., J.D. Cotter-Howells, J.M. Charnock, A.J.M. Baker and J.A.C. Smith. 1996. Free histidine as a metal chelator in plants that accumulate nickel. *Nature* 379:635–638.

Krämer, U., I.J. Pickering, R.C. Prince, I. Raskin and D.E. Salt. 2000. Subcellular localization and speciation of nickel in hyperaccumulator and non-accumulator *Thlaspi* species. *Plant Physiol.* 122:1343–1353.

Kukier, U. and R.L. Chaney. 2004. In situ remediation of nickel phytotoxicity for different plant species. *J. Plant Nutr.* 27:465–495.

Kukier, U., C.A. Peters, R.L. Chaney, J.S. Angle and R.J. Roseberg. 2004. The effect of pH on metal accumulation in two *Alyssum* species. *J. Environ. Qual.* 32:2090–2102.

Lerch, M., T. Ressler, F. Krumeich, J.-P. Cosson, E. Hnawia and A. Grohmann. 2010. Carbon-supported nickel nanoparticles from a wood sample of the tree *Sebertia acuminata* Pierre ex. Baillon. *Aust. J. Chem.* 63:830–835.

Li, Y.-M., R.L. Chaney, E.P. Brewer, J.S. Angle and J.P. Nelkin. 2003a. Phytoextraction of nickel and cobalt by hyperaccumulator *Alyssum* species grown on nickel-contaminated soils. *Environ. Sci. Technol.* 37:1463–1468.

Li, Y.-M., R.L. Chaney, E. Brewer, R.J. Roseberg, J.S. Angle, A.J.M. Baker, R.D. Reeves and J. Nelkin. 2003b. Development of a technology for commercial phytoextraction of nickel: Economic and technical considerations. *Plant Soil* 249:107–115.

Losfeld, G., V. Escande, P.V. De La Blache and C. Grison. 2014a. Chemical exploitation of metal contaminated biomass produced in phytoextraction. *Int. J. Sustain. Develop. Plan.* 9:400–416.

Losfeld, G., L. L'Huillier, B. Fogliani, T. Jaffré and C. Grison. 2015. Mining in New Caledonia: Environmental stakes and restoration opportunities. *Environ. Sci. Pollut. Res.* 22:5592–5607.

Losfeld, G., R. Mathieu, L. L'Huillier, B. Fogliani, T. Jaffré and C. Grison. 2014b. Phytoextraction from mine spoils: Insights from New Caledonia. *Environ. Sci. Pollut. Res. Int.* 22:5608–5619.

Ma, R., J. Vallance, G. Echevarria, P. Rey and E. Benizri. 2015. Phytoextraction of nickel and rhizosphere microbial communities under mono- or multispecies hyperaccumulator plant cover in a serpentine soil. *Aust. J. Bot.* 60:92–102.

Malik, M., R.L. Chaney, E.P. Brewer and J.S. Angle. 2000. Phytoextraction of soil cobalt using hyperaccumulator plants. *Int. J. Phytoremed.* 2:319–329.

Massoura, S.T., G. Echevarria, T. Becquer, J. Ghanbaja, E. Leclerc-Cessac and J.-L. Morel. 2006. Control of nickel availability by nickel bearing minerals in natural and anthropogenic soils. *Geoderma* 136:28–37.

Massoura, S.T., G. Echevarria, E. Leclerc-Cessac and J.-L. Morel. 2004. Response of excluder, indicator, and hyperaccumulator plants to nickel availability in soils. *Aust. J. Soil Res.* 42:933–938.

McNear, D.H., Jr., R.L. Chaney and D.L. Sparks. 2007. The effects of soil type and chemical treatment on nickel speciation in refinery enriched soils: A multi-technique investigation. *Geochim. Cosmochim. Acta* 71:2190–2208.

Mesjasz-Przybylowicz, J., W. Przybylowicz, A. Barnabas and A. van der Ent. 2016. Extreme nickel hyperaccumulation in the vascular tracts of the tree *Phyllanthus balgooyi* from Borneo. *New Phytol.* 209:1513–1526.

Montargès-Pelletier, E., V. Chardot, G. Echevarria, L.J. Michot, A. Bauer and J.-L. Morel. 2008. Identification of nickel chelators in three hyperaccumulating plants: An X-ray spectroscopic study. *Phytochemistry* 69:1695–1709.

Nkrumah, P.N., A.J.M. Baker, R.L. Chaney, P.D. Erskine, G. Echevarria, J.L. Morel and A. van der Ent. 2016. Current status and challenges in developing Ni phytomining: An agronomic perspective. *Plant Soil* 406:55–69.

Nkrumah, P.N., R.L. Chaney and J.L. Morel. 2017. Agronomy of 'metal crops' used in agromining. pp. 19–38. *In* A. van der Ent et al. (Eds.), *Agromining: Farming for Metals. Mineral Resource Reviews.* Springer. London. DOI 10.1007 /978-3-319-61899-9_2.

NRC (National Research Council). 1980. Nickel. pp. 345–363. *In Mineral Tolerance of Domestic Animals.* National Academy of Sciences, Washington, DC.

O'Dell, G.D., W.J. Miller, W.A. King, S.L. Moore and D.M. Blackmon. 1970a. Nickel toxicity in the young bovine. *J. Nutr.* 100:1447–1454.

O'Dell, G.D., W.J. Miller, S.L. Moore and W.A. King. 1970b. Effect of nickel as the chloride and the carbonate on palatability of cattle feed. *J. Dairy Sci.* 53:1266–1269.

O'Dell, G.D., W.J. Miller, S.L. Moore, W.A. King, J.C. Ellers and H. Jurecek. 1971. Effect of dietary nickel level on excretion and nickel content of tissues of male calves. *J. Anim. Sci.* 32:769–773.

Orłowska, E., W. Przybyłowicz, D. Orlowski, K. Turnau and J. Mesjasz-Przybyłowicz. 2011. The effect of mycorrhiza on the growth and elemental composition of Ni-hyperaccumulating plant *Berkheya coddii* Roessler. *Environ. Pollut.* 159:3730–3738.

Ouerdane, L., S. Mari, P. Czernic, M. Lebrun and R. Lobinski. 2006. Speciation of non-covalent nickel species in plant tissue extracts by electrospray Q-TOFMS/MS after their isolation by 2D size exclusion-hydrophilic interaction LC (SEC-HILIC) monitored by ICP-MS. *J. Anal. Atom. Spectrom.* 21:676–683.

Paul, A.L.D., I.A. Baklanov and R.L. Chaney. 2019. Influence of subsoil and soil volume on the accumulation of nickel by *Alyssum corsicum* grown on a serpentine soil. *Int. J. Phytoremed.* In press.

Perrier, N., F. Colin, T. Jaffré, J.-P. Ambrosi, J. Rose and J.-Y. Bottero. 2004. Nickel speciation in *Sebertia acuminata*, a plant growing on a lateritic soil of New Caledonia (in French and English). *C.R. Geosciences* 336:567–577.

Persans, M.W., K. Nieman and D.E. Salt. 2001. Functional activity and role of cation-efflux family members in Ni hyperaccumulation in *Thlaspi goesingense*. *Proc. Natl. Acad. Sci. USA* 98:9995–10000.

Peters, C.A., R.L. Chaney, J.S. Angle and R.J. Roseberg. 2000. Effect of the pH of pH-buffered nutrient solutions on Ni accumulation by hyperaccumulator species. *Agron. Abstr.* 2000:50.

Peterson, L.R., V. Trivett, A.J.M. Baker, C. Aguiar and A.J. Pollard. 2003. Spread of metals through an invertebrate food chain as influenced by a plant that hyper-accumulates nickel. *Chemoecology* 13:103–108.

Pinel, F., E. Leclerc-Cessac and S. Staunton. 2003. Relative contributions of soil chem-istry, plant physiology and rhizosphere induced changes in speciation on Ni accumulation in plant shoots. *Plant Soil* 255:619–629.

Proctor, J. and A.J.M. Baker. 1994. The importance of nickel for plant growth in ultra-mafic (serpentine) soils. pp. 417–432. *In* S.M. Ross (Ed.) *Trace Metals in Soil–Plant Systems.* John Wiley & Sons, London.

Proctor, J. and S.R.J. Woodell. 1975. The ecology of serpentine soils. *Adv. Ecol. Res.* 9:255–366.

Quimado, M.O., E.S. Fernando, L.C. Trinidad and A. Doronil. 2015. Nickel-hyperaccumulating species of *Phyllanthus* (Phyllanthaceae) from the Philippines. *Aust. J. Bot.* 63:103–110.

Reeves, R.D. 1992. The hyperaccumulation of nickel by serpentine plants. pp. 253–277. *In* A.J.M. Baker, J. Proctor and R.D. Reeves (Eds.) *The Vegetation of Ultramafic (Serpentine) Soils.* Intercept, Andover, Hampshire, UK.

Reeves, R.D. and A.J.B. Baker. 2000. Metal accumulating plants. pp. 193–229. *In* I. Raskin and B. Ensley (Eds.). *Phytoremediation of Toxic Metals: Using Plants to Clean Up the Environment.* Wiley, New York.

Robinson, B.H., R.R. Brooks and B.E. Clothier. 1999a. Soil amendments affecting nickel and cobalt uptake by *Berkheya coddii*: Potential use for phytomining and phytoremediation. *Ann. Bot.* 84:689–694.

Robinson, B.H., R.R. Brooks, P.E.H. Gregg and J.H. Kirkman. 1999b. The nickel phy-toextraction potential of some ultramafic soils as determined by sequential extraction. *Geoderma* 87:293–304.

Robinson, B.H., A. Chiarucci, R.R. Brooks, D. Petit, J.H. Kirkman, P.E.H. Gregg and V. De Dominicis. 1997. The nickel hyperaccumulator plant *Alyssum bertolonii* as a potential agent for phytoremediation and phytomining of nickel. *J. Geochem. Explor.* 59:75–86.

Robinson, B., J.-E. Ferñandez, P. Madeĵon, T. Marañon, J.M. Murillo, S. Green and B. Clothier. 2003. Phytoextraction: An assessment of biogeochemical and eco-nomic viability. *Plant Soil* 249:117–125.

Sagner, S., R. Kneer, G. Wanner, J.-P. Cosson, B. Deus-Neuman and M.H. Zenk. 1998. Hyperaccumulation, complexation and distribution of nickel in *Serbertia acumi-nata*. *Phytochemistry* 47:339–347.

Schaumloffel, D., L. Ouerdane, B. Bouyssiere and R. Lobinski. 2003. Speciation anal-ysis of nickel in the latex of a hyperaccumulating tree *Sebertia acuminata* by HPLC and CZE with ICP MS and electrospray MS-MS detection. *J. Anal. Atomic Spect.* 18:120–127.

Shallari, S., G. Echevarria, C. Schwartz and J.L. Morel. 2001. Availability of nickel in soils for the hyperaccumulator *Alyssum murale* Waldst. & Kit. *S. Afr. J. Sci.* 97:568–570.

Siebielec, G., R.L. Chaney and U. Kukier. 2007. Liming to remediate Ni contaminated soils with diverse properties and a wide range of Ni concentration. *Plant Soil* 299:117–130.

Strawn, K.E. 2013. Unearthing the habitat of a hyperaccumulator: Case study of the invasive plant yellowtuft (*Alyssum*; Brassicaceae) in Southwest Oregon, USA. *Manage. Biol. Invas.* 4:249–259.

Tappero, R.V., E. Peltier, M. Gräfe, K. Heidel, M. Ginder-Vogel, K.J.T. Livi, M.L. Rivers, M.A. Marcus, R.L. Chaney and D.L. Sparks. 2007. Hyperaccumulator *Alyssum murale* relies on a different metal storage mechanism for cobalt than for nickel. *New Phytol.* 175:641–654.

van der Ent, A., A.J.M. Baker, R.D. Reeves, R.L. Chaney, C.W.N. Anderson, J.A. Meech, P.D. Erksine, M.-O. Simonnot, J. Vaughan, J.-L. Morel, G. Echevarria, B. Fogliani, R.-L. Qiu and D.R. Mulligan. 2015a. Agromining: Farming for metals in the future? *Environ. Sci. Technol.* 49:4773–4780.

van der Ent, A., A.J.M. Baker, R.D. Reeves, A.J. Pollard and H. Schat. 2012. Hyperaccumulators of metal and metalloid trace elements: Facts and fiction. *Plant Soil* 362:319–334.

van der Ent, A., P. Erskine and S. Sumail. 2015b. Ecology of nickel hyperaccumulator plants from ultramafic soils in Sabah (Malaysia). *Chemoecology* 25:243–259.

Vaughan, J., J. Riggio, J. Chen, H. Peng, H.H. Harris and A. van der Ent. 2017. Characterisation and hydrometallurgical processing of nickel from tropical agromined bio-ore. *Hydrometallurgy* 169:346–355.

Wood, B.W., R.L. Chaney and M. Crawford. 2006a. Correcting micronutrient deficiency using metal hyperaccumulators: *Alyssum* biomass as a natural product for nickel deficiency correction. *Hort. Sci.* 41:1231–1234.

Wood, B.W., C.C. Reilly and A.P. Nyczepir. 2006b. Field deficiency of nickel in trees: Symptoms and causes. *Acta Hortic.* 721:83–97.

Zhang, L., J.S. Angle and R.L. Chaney. 2007. Do high-nickel leaves shed by the Ni-hyperaccumulator *Alyssum murale* inhibit seed germination of competing plants? *New Phytol.* 173:509–516.

Zhang, X., V. Houzelot, A. Bani, J.L. Morel, G. Echevarria and M.-O. Simonnot. 2014. Selection and combustion of Ni-hyperaccumulators for the phytomining process. *Int. J. Phytoremed.* 16:1058–1072.

Zhang, X., B. Laubie, V. Houzelot, E. Plasari, G. Echevarria and M.-O. Simonnot. 2016. Increasing purity of ammonium nickel sulfate hexahydrate and production sustainability in a nickel phytomining process. *Chem. Eng. Res. Design* 106:26–32.

Index

Page numbers followed by f and t indicate figures and tables, respectively.

9 780367 571207